Methods in Molecular Biology

Series Editor
John M. Walker
School of Life Sciences
University of Hertfordshire
Hatfield, Hertfordshire, AL10 9AB, UK

For further volumes:
http://www.springer.com/series/7651

Bacterial Artificial Chromosomes

Second Edition

Edited by

Kumaran Narayanan

Monash University Malaysia, Bandar Sunway, Malaysia

Icahn School of Medicine at Mount Sinai, New York, USA

Editor
Kumaran Narayanan
Monash University Malaysia
Bandar Sunway, Malaysia

Icahn School of Medicine at Mount Sinai
New York, USA

ISSN 1064-3745 ISSN 1940-6029 (electronic)
ISBN 978-1-4939-1651-1 ISBN 978-1-4939-1652-8 (eBook)
DOI 10.1007/978-1-4939-1652-8
Springer New York Heidelberg Dordrecht London

Library of Congress Control Number: 2014948259

Printed on acid-free paper

Humana Press is a brand of Springer
Springer is part of Springer Science+Business Media (www.springer.com)

Preface

The Bacterial Artificial Chromosome (BAC) system was developed in the early 1990s—at about the same time the Human Genome Project (HGP) was officially launched—as a vector for cloning high molecular weight genomic DNA. BAC clones were large enough for recovering full-length genes that included exons, introns, and distal regulatory elements. Moreover their low-copy replication conferred stability to complex human DNA, including repetitive sequences that would ordinarily rearrange in most plasmids. These qualities quickly made BACs indispensable for sequencing the human and other genomes.

At the same time these very features acted like double-edged swords—their large size and low-copy nature made it challenging to work with BACs for functional studies. Routine plasmid techniques such as DNA isolation and transfer into *E. coli* and mammalian cells became extremely tricky when attempted with BACs because of their low yield and degradation issues. And since restriction enzyme-based modifications were limited when applied to large DNA, mutagenesis of BAC clones became laborious. However, with time new protocols were developed that overcame these obstacles, gaining BACs widespread use in many new applications, including for deciphering gene structures and function, understanding disease phenotypes, and developing treatment approaches in cell and animal models.

Bacterial Artificial Chromosomes (2nd edition) presents a blend of some of the most important methods developed for working with BACs. It begins with fundamental techniques used for BAC construction and characterization (Chapters 1–3) and follows up with more advanced procedures for introducing modifications (Chapters 4–6) and for achieving gene expression from these vectors (Chapters 7–10). The book ends with chapters describing applications of BACs in model organisms (Chapters 11 and 12) and in medical genetics and drug discovery (Chapters 13–18).

These chapters follow the proven format of the Methods in Molecular Biology Series: a brief introduction that provides an overview of the method, a materials section, followed by a series of detailed protocols divided into clear sections and subsections that are supported with timely Notes to guide the reader. The Notes section is the hallmark of the Series, supplying the reader with valuable tips and tricks that aim to help them execute the method successfully. This section may include information that are traditionally not found in the methods section of research papers and therefore not widely known. Or they may detail specific modifications that are developed and applied by an expert lab or experienced researcher with extensive familiarity with the technique.

This book would not have been possible without the support of many people. I am indebted to John Walker for inviting me to this project and for all his advice and help with the overall editing of the chapters. My deep gratitude goes to all the authors who graciously contributed their time and effort to this project and for following through their original drafts with many rounds of editing to bring their chapters to the final form that appears in the pages of this book. At Humana Press, Patrick Marton and David Casey have been a great source of help with editing and for bringing the book to print. Finally, I would like to thank my wife Stella Salleh for selflessly letting me spend many hours in solitude working on my writing. Her patience, understanding, and support inspired me to finish this book.

Kumaran Narayanan

Contents

Part IV Applications of BACs in Model Organisms, Medical Genetics, and Drug Discovery

Contributors

SHAWKAT ALI • *Agriculture & Agri-Food Canada, Pacific Agri-Food Research Centre, Summerland, BC, Canada; Department of Botany, University of British Columbia, Vancouver, BC, Canada; Division of Biological and Environmental Sciences and Engineering, Center for Desert Agriculture, 4700 King Abdullah University of Science and Technology, Thuwal, Saudi Arabia*

JOSE M. ALONSO • *Department of Plant and Microbial Biology, North Carolina State University, Raleigh, NC, USA*

GUUS BAKKEREN • *Agriculture & Agri-Food Canada, Pacific Agri-Food Research Centre, Summerland, BC, Canada*

TAHSIN STEFAN BARAKAT • *Department of Reproduction and Development, Erasmus MC, University Medical Center, Rotterdam, The Netherlands*

SAMARTH BHATT • *Jena University Hospital, Institute of Human Genetics, Friedrich Schiller University, Jena, Germany*

PRADEEP K. CHATTERJEE • *Department of Chemistry, Julius L. Chambers Biomedical/ Biotechnology Research Institute, North Carolina Central University, Durham, NC, USA*

QINGWEN CHEN • *School of Science, Monash University Malaysia, Bandar Sunway, Malaysia*

JRGANG CHENG • *The Neuroscience Center, University of North Carolina at Chapel Hill, Chapel Hill, NC, USA*

XIAOBO FAN • *Jena University Hospital, Institute of Human Genetics, Friedrich Schiller University, Jena, Germany*

F. ALEX FELTUS • *Clemson University Genomics Institute, Clemson, SC, USA*

JELENA FILIPOVIC • *Department of Physical Chemistry, Vinča Institute for Nuclear Sciences, Belgrade, Serbia*

CORNEL FRAEFEL • *Institute for Virology, Vetsuisse Faculty, University of Zurich, Zurich, Switzerland*

FRANCESCA ROMANA GRATI • *Research and Development, Cytogenetics and Molecular Biology, TOMA Advanced Biomedical Assays, Busto Arsizio (VA), Italy*

JOOST GRIBNAU • *Department of Reproduction and Development, Erasmus MC, University Medical Center, Rotterdam, The Netherlands*

SUSAN GROSS • *Department of Obstetrics & Gynecology and Women's Health, Albert Einstein College of Medicine, Bronx, NY, USA*

NIINA HAIMINEN • *IBM T.J. Watson Research Center, Yorktown Heights, NY, USA*

AHMED B. HAMID • *Jena University Hospital, Institute of Human Genetics, Friedrich Schiller University, Jena, Germany*

SCOTT HOLMES • *The Neuroscience Center, University of North Carolina at Chapel Hill, Chapel Hill, NC, USA*

SARA HOWDEN • *Cell and Gene Therapy Research Group, Murdoch Childrens Research Institute, Royal Children's Hospital, Parkville, Melbourne, VIC, Australia; Department of Paediatrics, University of Melbourne, Melbourne, VIC, Australia*

JEN-KANG HSU • *The Neuroscience Center, University of North Carolina at Chapel Hill, Chapel Hill, NC, USA*

GORDANA JOKSIC • *Department of Physical Chemistry, Vinča Institute for Nuclear Sciences, Belgrade, Serbia*
IVANA JOKSIC • *Department of Physical Chemistry, Vinča Institute for Nuclear Sciences, Belgrade, Serbia*
BETTY R. KAO • *Cell and Gene Therapy Research Group, Murdoch Childrens Research Institute, Royal Children's Hospital, Parkville, Melbourne, VIC, Australia; Department of Paediatrics, University of Melbourne, Melbourne, VIC, Australia*
TATYANA KARAMYSHEVA • *Laboratory of Morphology and Function of Cell structure, Institute of Cytology and Genetics, Russian Academy of Sciences, Siberian Branch, Novosibirsk, Russian Federation*
ARTEM V. KONONENKO • *Laboratory of Molecular Pharmacology, National Cancer Institute (NCI, NIH), Bethesda, MD, USA*
NADEZDA KOSYAKOVA • *Jena University Hospital, Institute of Human Genetics, Friedrich Schiller University, Jena, Germany*
ATHANASSIOS KOTSINAS • *Department of Histology and Embryology, School of Medicine University of Athens, Athens, Greece*
GEORGE KOTZAMANIS • *Department of Histology and Embryology, School of Medicine, University of Athens, Athens, Greece*
NATALAY KOUPRINA • *Laboratory of Molecular Pharmacology, National Cancer Institute (NCI, NIH), Bethesda, MD, USA*
PETRA KRAUS • *Department of Biology, Clarkson University, Potsdam, NY, USA*
VLADIMIR LARIONOV • *Laboratory of Molecular Pharmacology, National Cancer Institute (NCI, NIH), Bethesda, MD, USA*
NICHOLAS C.O. LEE • *Laboratory of Molecular Pharmacology, National Cancer Institute (NCI, NIH), Bethesda, MD, USA*
ANDREJA LESKOVAC • *Department of Physical Chemistry, Vinča Institute for Nuclear Sciences, Belgrade, Serbia*
THOMAS LIEHR • *Jena University Hospital, Institute of Human Genetics, Friedrich Schiller University, Jena, Germany*
SHARON LÖHMER • *Jena University Hospital, Institute of Human Genetics, Friedrich Schiller University, Jena, Germany*
THOMAS LUFKIN • *Department of Biology, Clarkson University, Potsdam, NY, USA*
SUZANNE LYMAN • *The Neuroscience Center, University of North Carolina at Chapel Hill, Chapel Hill, NC, USA*
TIMOTHY J. MAHONY • *Queensland Alliance for Agriculture and Food Innovation, Centre for Animal Science, The University of Queensland, St Lucia, QLD, Australia*
BRADLEY MCCOLL • *Cell and Gene Therapy Research Group, Murdoch Childrens Research Institute, Royal Children's Hospital, Parkville, Melbourne, VIC, Australia; Department of Paediatrics, University of Melbourne, Melbourne, VIC, Australia*
KRISTIN MRASEK • *Jena University Hospital, Institute of Human Genetics, Friedrich Schiller University, Jena, Germany*
KUMARAN NARAYANAN • *Monash University Malaysia, Bandar Sunway, Malaysia; Icahn School of Medicine at Mount Sinai, New York, NY, USA*
MONEEB A.K. OTHMAN • *Jena University Hospital, Institute of Human Genetics, Friedrich Schiller University, Jena, Germany*
LAXMI PARIDA • *IBM T.J. Watson Research Center, Yorktown Heights, NY, USA*
LUCIANA G. QUINTANA • *Jena University Hospital, Institute of Human Genetics, Friedrich Schiller University, Jena, Germany*

GOPAKUMAR RADHAKRISHNAN • *Jena University Hospital, Institute of Human Genetics, Friedrich Schiller University, Jena, Germany*

KARL E. ROBINSON • *Queensland Alliance for Agriculture and Food Innovation, Centre for Animal Science, The University of Queensland, St Lucia, QLD, Australia*

ALEXANDER SAMOSHKIN • *Laboratory of Molecular Pharmacology, National Cancer Institute (NCI, NIH), Bethesda, MD, USA*

CHRISTOPHER A. SASKI • *Clemson University Genomics Institute, Clemson, SC, USA*

ANNA N. STEPANOVA • *Department of Plant and Microbial Biology, North Carolina State University, Raleigh, NC, USA*

LUISE THEUSS • *Jena University Hospital, Institute of Human Genetics, Friedrich Schiller University, Jena, Germany*

KURT TOBLER • *Institute for Virology, Vetsuisse Faculty, University of Zurich, Zurich, Switzerland*

JIM VADOLAS • *Cell and Gene Therapy Research Group, Murdoch Childrens Research Institute, Royal Children's Hospital, Parkville, Melbourne, VIC, Australia; Department of Paediatrics, University of Melbourne, Melbourne, VIC, Australia*

FRANÇOIS VIALARD • *Department of Cytogenetics, Fetopathology, Obstetrics and Gynaeacology, CHI Poissy, St Germain, France*

ANJA WEISE • *Jena University Hospital, Institute of Human Genetics, Friedrich Schiller University, Jena, Germany*

KATHLEEN WILHELM • *Jena University Hospital, Institute of Human Genetics, Friedrich Schiller University, Jena, Germany*

CECILIA L. WINATA • *Human Genetics, Genome Institute of Singapore, Singapore, Singapore*

Part I

BAC Construction and Characterization

Chapter 1

From Selective Full-Length Genes Isolation by TAR Cloning in Yeast to Their Expression from HAC Vectors in Human Cells

Natalay Kouprina, Nicholas C.O. Lee, Artem V. Kononenko, Alexander Samoshkin, and Vladimir Larionov

Abstract

Transformation-associated recombination (TAR) cloning allows selective isolation of full-length genes and genomic loci as large circular Yeast Artificial Chromosomes (YACs) in yeast. The method has a broad application for structural and functional genomics, long-range haplotyping, characterization of chromosomal rearrangements, and evolutionary studies. In this paper, we describe a basic protocol for gene isolation by TAR as well as a method to convert TAR isolates into Bacterial Artificial Chromosomes (BACs) using a retrofitting vector. The retrofitting vector contains a 3′ HPRT-loxP cassette to allow subsequent gene loading into a unique loxP site of the HAC-based (Human Artificial Chromosome) gene delivery vector. The benefit of combining the TAR gene cloning technology with the HAC gene delivery system for gene expression studies is discussed.

Key words YAC, BAC, TAR cloning, Gene isolation, Human artificial chromosome (HAC)

1 Introduction

TAR cloning allows entire genes and large chromosomal regions to be selectively and accurately isolated from total genomic DNA by in vivo recombination in yeast [1–5]. Figure 1 shows a step-by step scheme for selectively isolating a gene as a circular YAC by a TAR cloning vector and the following steps up to its expression from a HAC-based vector in gene-deficient human cells. Figure 2 shows the schematic diagram of both the TAR vector (pVC604) used for gene cloning and a scheme of the retrofitting vector (pJBRV1) used for conversion of a TAR-isolated YAC into a BAC. The TAR cloning vector contains gene-specific targeting sequences (hooks) at both ends (the hooks can be as short as 60 bp) [6], a *CEN* sequence, and a yeast selectable marker. Recombination between the TAR cloning vector and homologous sequences in the co-transformed

Kumaran Narayanan (ed.), *Bacterial Artificial Chromosomes*, Methods in Molecular Biology, vol. 1227, DOI 10.1007/978-1-4939-1652-8_1, © Springer Science+Business Media New York 2015

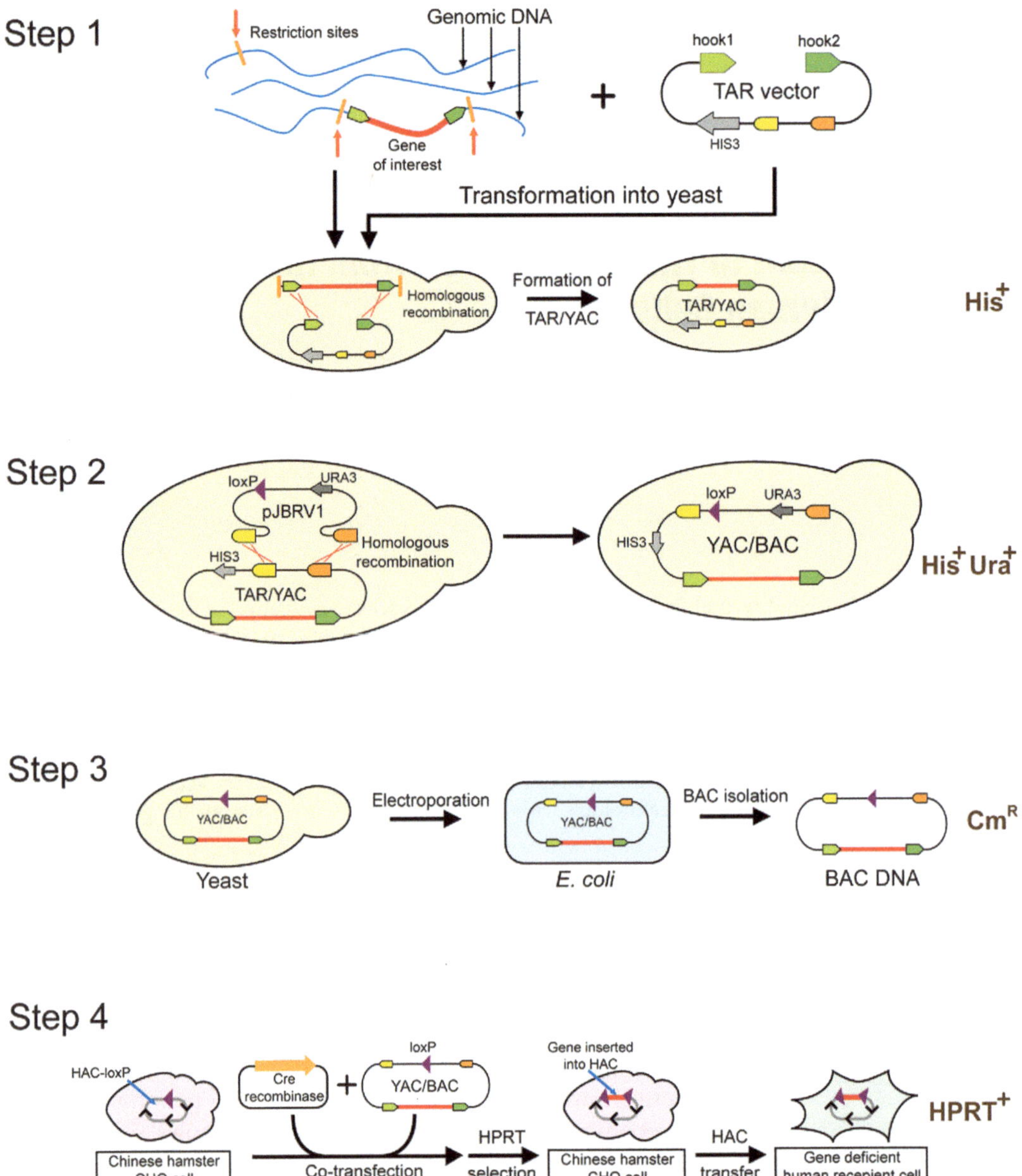

Fig. 1 The four consecutive experimental steps from selective gene isolation to its expression in gene-deficient human cells. *Step 1*: The diagram shows TAR cloning of a gene of interest from total genomic DNA with a linearized TAR vector containing two unique targeting sequences (hooks) homologous to a gene of interest. After transformation into yeast *Saccharomyces cerevisiae* cells (in *yellow*), recombination between targeting sequences in the vector and the targeted sequences of the genomic DNA fragment leads to the rescue of the fragment (gene) as a circular TAR/YAC molecule. The *orange arrows* indicate random positions of an endonuclease recognition site in throughout the genomic DNA and near the targeted sequences of the desirable genomic fragment. Restriction of the genomic DNA before yeast transformation by such an endonuclease may increase the yield of gene-positive TAR/YAC clones up to 20–30 %. For TAR cloning experiments, the pVC604 vector DNA is linearized by a unique endonuclease located between the hooks to expose targeting sequences (hooks). *Step 2*: Schematic representation of retrofitting a circular TAR/YAC containing a gene of interest into a

mammalian (or human) DNA can result in a circular YAC molecule that is able to replicate, segregate, and be selected for in yeast. A retrofitting vector contains a 3′-*HPRT*-loxP cassette for insertion of a TAR/YAC isolate into the HAC vector by Cre-lox-mediated recombination. Typically, in TAR cloning, the desired gene is produced at frequency of ~1 %. Yield of positive clones may be increased up to 20–30 % if before TAR cloning experiments genomic DNA is treated by restriction endonuclease(s) that cuts DNA near the targeted sequences [7]. The size of the isolated genomic fragment may be as large as 250 kb (summarized in [4]).

TAR cloning, as described above, requires the cloned DNA fragment to carry at least one *ARS*-like sequence that can function as an origin of replication site in yeast. On average, the mammalian genome contains one *ARS*-like sequence (WWWWTTTAYRT TTWGTT) every 40 kb, and thus this strategy is applicable for most euchromatic regions [8, 9]. However, in chromosomal regions with multiple repetitive elements, however, such as the centromere and telomere, the *ARS* frequency may be reduced, precluding isolation by the above method. Consequently, a modified TAR cloning methodology was developed to allow isolation of regions regardless of the presence of *ARS* elements [10, 11]. This modified system greatly expands the utility of TAR cloning to include characterization of telomeric and centromeric regions. Moreover, it allows selective cloning of fragments of bacterial genomes up to several hundred kilobases that are poor in *ARS*-like sequences.

When only limited sequence information is available, such as a 3′-flanking EST, TAR cloning with vectors that have two specific targeting sequences is impossible. To circumvent this limitation, we

Fig. 1 (continued) YAC/BAC molecule using the pJBRV1 vector that allows subsequent propagation of a TAR clone in a BAC form. The *Bam*HI linearized pJBRV1 vector is transformed into yeast cells carrying a circular TAR-isolated YAC containing a gene of interest. Targeting sequences correspond to two regions flanking the *ColE1* origin of replication in pVC604-TAR cloning vector. Recombination between the *Bam*HI-linearized pJBRV1 vector and a TAR/YAC during yeast transformation leads to replacement of the *ColE1* origin of replication in the pVC0604-TAR cloning vector by a cassette containing the *F′* factor origin of replication (BAC), the chloramphenicol acetyltransferase (Cm^R) gene, the *URA3* yeast selectable marker and 3′ HPRT-loxP-eGFP cassette for further loading of YAC/BAC into the human artificial chromosome (HAC-based) gene delivery vector. YAC retrofitting is highly efficient: more than 95 % of Ura^+His^+ transformants obtained with pJBRV1 contain retrofitted YACs. *Step 3*: Transferring YAC/BAC molecules from yeast cell to bacterial cells (in *light blue*). The YAC/BAC molecules are moved to *E. coli* by electroporation for further BAC DNA isolation by a standard procedure. *Step 4*: Loading of a TAR-isolated gene along with a YAC/BAC vector into the unique loxP site of the HAC gene delivery vector in hamster CHO (hprt−/−) cells (in *pink*). A desired gene, TAR-cloned into the YAC/BAC vector, is loaded into the HAC by Cre-loxP mediated recombination followed by reconstitution of the *HPRT* gene. After that, the HAC carrying a gene of interest can be transferred from CHO cells to desired gene-deficient recipient cells (in *green*) via microcell-mediated chromosome transfer technique (MMCT) for complementation analysis

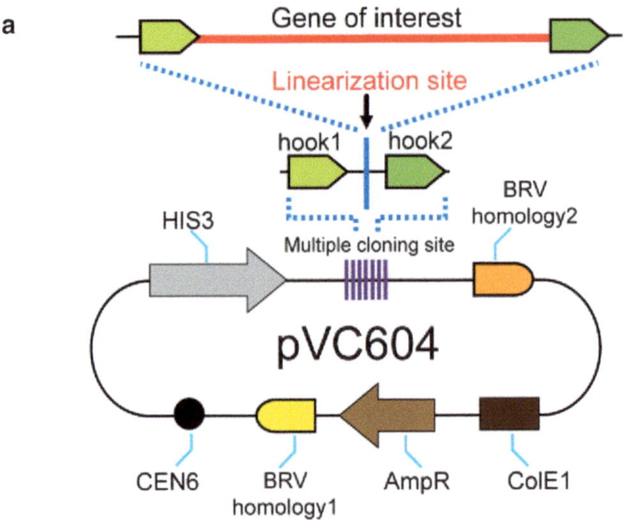
a
Gene of interest
Linearization site
hook1
hook2
HIS3
BRV
homology2
Multiple cloning site
pVC604
CEN6
BRV
homology1
AmpR
ColE1

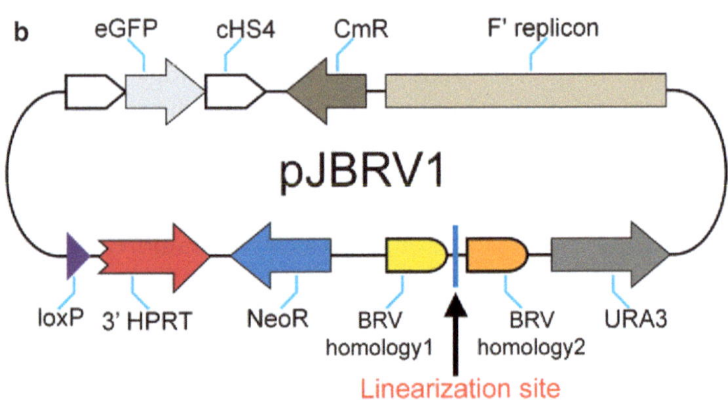
b
eGFP
cHS4
CmR
F' replicon
pJBRV1
loxP
3' HPRT
NeoR
BRV
homology1
BRV
homology2
URA3
Linearization site

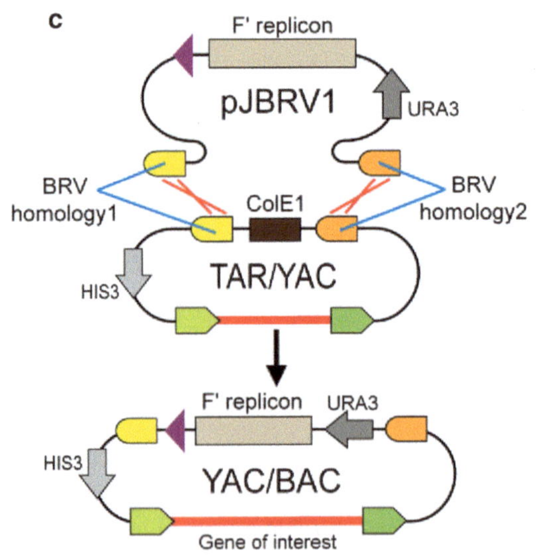
c
F' replicon
pJBRV1
URA3
BRV
homology1
ColE1
BRV
homology2
HIS3
TAR/YAC
F' replicon
URA3
HIS3
YAC/BAC
Gene of interest

developed another way to isolate a single copy gene directly from genomic DNA [12]. This TAR approach uses a vector that has one unique sequence hook and one repeated sequence hook (*Alu* or *B1* repeats for human or mouse DNA, respectively). The repeated element makes it possible to isolate a set of nested overlapping fragments that extend from the specific hook to different upstream or downstream repeat sequence positions. Because one of the ends is fixed, this approach is called radial TAR cloning. We emphasize the radial nature of this cloning scheme because by simply changing the arrangement of the unique targeting hook, it is possible to clone large regions that extend in both directions along the chromosome from the specific targeting sequence. The size of YACs obtained by radial cloning varies from 40 to 250 kb [12–17]. Yield of gene positive clones for radial TAR cloning is approximately 1 %, the same as that for cloning with two unique targeting hooks.

At the beginning of the Human Genome Project, TAR cloning was used to construct chromosome-specific YAC libraries from human–rodent monochromosomal cell lines [18, 19] and for selective isolation of human DNA from radiation hybrids [2, 20]. In the post-genomic era, TAR cloning has many other applications [4]. Among them is the isolation of full-length genes for functional studies [3, 4, 12, 13, 21–23] and the isolation of centromeric regions that are poorly cloned by a routine ligation method [24]. TAR cloning provides a tool for the selective isolation of the same specific chromosomal segment or gene from a representative

Fig. 2 A scheme of the basic TAR cloning vector and its retrofitting counterpart. (**a**) A scheme of the TAR cloning vector pVC604. This vector contains *CEN6* that corresponds to the yeast chromosome VI centromere, the *HIS3* marker for selection in yeast, the *Amp*^{R} gene for selection in *E. coli*, two sequences (BRV homology 1 and BRV homology 2) homologous to the pJBRV1 vector used for replacement of the *ColE1* origin of replication in pVC604 by the *F′* factor origin of replication from the pJBRV1 vector for further propagation as a single copy BAC molecule. In addition, pVC604 contains multiple cloning sites (an extensive polylinker consisting of 14 restriction endonuclease with 6- and 8 bp recognition sites) where two unique targeting sequences (hook 1 and hook 2) homologous to the 5′ and 3′ ends of a gene (or a genomic fragment) of interest may be inserted. Hooks are cloned into the vector in such a way that after linearization of the vector, orientation of the hooks should correspond to their orientation in genomic DNA. Before TAR cloning, the TAR vector is linearized. (**b**) A scheme of the retrofitting vector pJBRV1. This vector contains a 3′ end of the *HPRT* gene, the *eGFP* transgene flanked by cHS4 insulators, a lox P sequence and two targeting sequences, BRV homology1 and BRV homology 2, (approximately 300 bp each) separated by a unique *Bam*HI site. In addition, the vector contains a mammalian selectable marker, the *Neo*^{R} gene, and a yeast selectable marker, the *URA3* gene, and the *Cm*^{R} gene for selection in *E. coli*. (**c**) A scheme of the retrofitting of a TAR/YAC by homologous recombination in yeast. As a result of recombination, the *ColE1* origin of replication is replaced by *F′* factor origin of replication

sample of individuals for mutational analysis [25]. Because TAR cloning produces multiple gene isolates, it allows isolation of both parental alleles of a gene that can then be used for haplotype analysis [26]. In many cases, when a gene of interest is represented in multiple copies, physical separation of alleles is the only way to determine a haplotype phase. This technique also provides a tool for the isolation of rearranged chromosomal regions, such as translocations from patients and model organisms, without the need to construct a new genomic library of random fragments [27]. The tolerance of TAR cloning to DNA divergence expands the potential applications of the technique for comparative genomics [28]. Given that 15 % divergence is characteristic for mammalian gene homologues, most homologous regions in different mammals can be selectively cloned by in vivo recombination in yeast using targeting hooks developed from human exon sequences. This has been demonstrated for several nonhuman primates and mouse genes which were isolated using human-specific targeting sequences [29–31]. Another application of TAR cloning is the isolation of chromosomal regions that are unclonable in bacterial vectors. More specifically, recombinational cloning system in yeast could accelerate work on closing the remaining gaps within the human genome and the genome of other species [32, 33].

For gene functional studies, the human artificial chromosomes (HACs) represent a novel system for gene delivery and expression with a potential to overcome many of the problems caused by the commonly used viral-based gene transfer systems [34, 35]. Firstly, the presence of a functional centromere in the HACs enables their long-term stable maintenance as a single copy episome without integration into the host chromosomes. This not only minimizes such complications as silencing of the therapeutic gene, it allows a single copy of the gene to be delivered into the cell. Secondly, there is no upper size limit to DNA cloned in HACs. Therefore, not only single genes up to 2–3 Mb with all regulatory elements, but groups of genes encoding complex pathways can be carried on a single HAC that would mimic the normal pattern of the natural gene(s) expression. Thirdly, the HACs can be transferred from one cell to another. Finally, because of the lack of viral sequences, HAC vectors minimize adverse host immunogenic responses and the risk of cellular transformation. A new generation HAC, alphoid^{tetO}-HAC, was engineered in human cells using a synthetic alphoid DNA array containing tetracycline operator (tetO) sequences embedded into the alphoid DNA [36]. A powerful advantage of this HAC is that its centromere can be inactivated by expression of tet-repressor (tetR) fusion proteins. Such inactivation results in HAC loss during cell divisions. This feature of the HAC provides a unique possibility to compare the phenotypes of the recipient gene-deficient human cells with and without a functional copy of a gene that provides a control for phenotypic changes attributed to expression of a HAC-encoded gene. To adopt this HAC for gene delivery and expression studies, a single gene acceptor site was inserted into the HAC ([23] and references therein). The capacity of the HAC to deliver genomic copies of several human full-length genes and complement genetic deficiencies in cell lines derived from the patients with gene deficiencies was demonstrated [23]. Thus, the combination of gene isolation by the TAR cloning technology and the HAC-based gene delivery vector that matches all features required for gene function studies represents a powerful tool for functional genomic studies and potentially for gene therapy.

In this paper, we present updated protocols of TAR cloning with its further application. The chapter includes four protocols. The first protocol describes preparation of highly competent yeast spheroplasts and transformation of the spheroplasts by genomic DNA along with a TAR vector (Subheadings 3.1 and 3.2) (Fig. 1, Step 1). The second protocol describes identification of gene positive clones among primary yeast transformants (Subheadings 3.3 and 3.4). The third protocol provides a method for retrofitting of TAR-isolated YACs containing a gene of interest into BACs and transferring these YAC/BACs into *E. coli* cells for further BAC DNA isolation (Subheadings 3.5 and 3.6) (Fig. 1, Steps 2 and 3). (Alternatively, YAC molecules up to 600 kb in size can be high-quality purified from yeast cells directly [37]). The fourth protocol describes insertion of a gene containing BAC into the HAC gene delivery vector (Subheading 3.7) (Fig. 1, Step 4 and Fig. 3). Thus, a complete cycle from selective gene isolation in yeast to gene insertion into the HAC vector for further functional analysis is described.

2 Materials

2.1 Strains and Vectors

1. A highly transformable *Saccharomyces cerevisiae* strain VL6-48 (*MAT alpha, his3-Δ 200, trp1-Δ 1, ura3-Δ 1, lys2, ade2-101, met14*) that has *HIS3*, *TRP1*, and *URA3* deleted is used as a host for TAR cloning experiments. This strain is available on request from the Laboratory of Molecular Pharmacology, National Cancer Institute (NCI, NIH). This strain is also available from the American Type Culture Collection (ATCC Number MYA-3666™).
2. The basic TAR cloning vector pVC604 contains a yeast selectable marker (*HIS3*) and an yeast centromeric sequence (*CEN6*). Before use, the TAR vector is "activated" by insertion of the targeting sequences (hooks) specific for a gene of interest (5′ and 3′ ends of the gene) into the polylinker. Such vector should be constructed before TAR cloning experiments. Hooks should be unique sequences; no repeated sequences

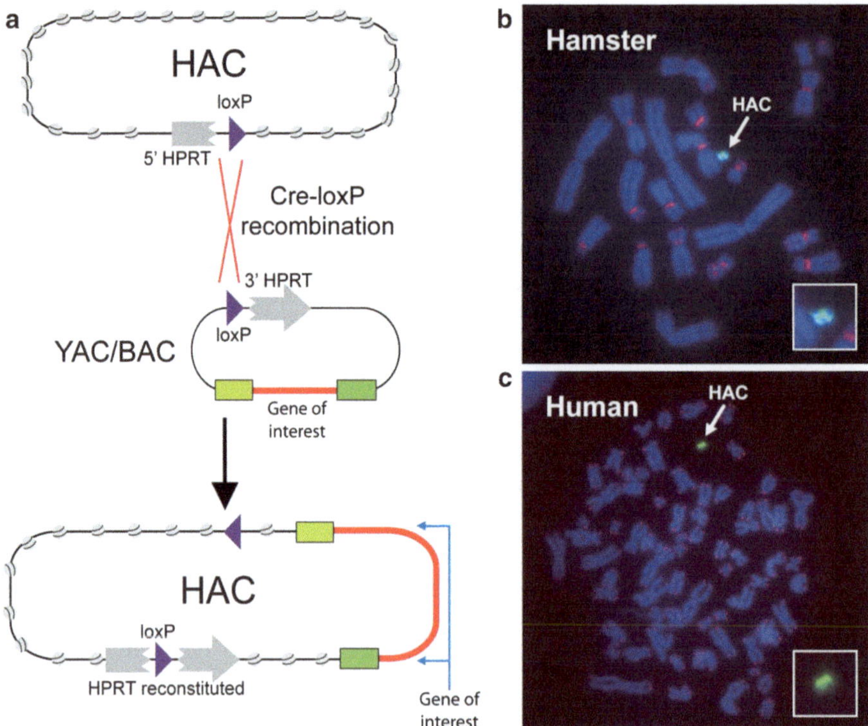

Fig. 3 Loading of a gene of interest into the HAC vector. (**a**) A scheme of insertion of a gene isolated by TAR cloning into a unique gene acceptor loxP site of the HAC in hamster CHO cells. Loading of a gene of interest by Cre-loxP mediated recombination is accompanied by reconstitution of the *HPRT* gene. (**b**) FISH analysis of the HAC carrying a gene of interest in CHO cells using a specific probe for BAC vector or for gene sequences (in *light blue*). DAPI staining of the nucleus is in *dark blue*. (**c**) FISH analysis of the HAC carrying a gene of interest in gene-deficient human cells using a specific probe for BAC vector or for gene sequences (in *green*). DAPI staining of the nucleus is in *dark blue*. The probe for human telomere sequences is in *red*. It is worth noting that this probe hybridizes with centromere sequences of some hamster chromosomes (*see* **b**)

should be present in the hooks (*see* **Notes 1–4**). The uniqueness of hooks can be checked by blasting against a genome sequence. The minimal size of the targeting sequences is 60 bp. Longer targeting sequences (up to 1 kb) can also be used. However increase of the hook size does not increase the yield of positive clones. Before TAR cloning, the vector should be linearized by cutting between the targeting hooks with a restriction enzyme. Concentration of the linearized TAR vector should be 0.5–1 μg/μl (the linearized vector can be kept at −20 °C up to 2–3 months). Vector DNA may be isolated by DNA Maxi kit (Qiagen). The basic pVC604 vector and its description are available from the American Type Culture Collection (ATCC Number MBA-212™) and also available on request from the Laboratory of Molecular Pharmacology, National Cancer Institute (NIH).

3. A yeast-bacteria-mammalian cell shuttle vector, pJBRV1, for retrofitting circular YAC/TAR isolates into a BAC form contains a 3′ HPRT-loxP-eGFP cassette, allowing gene loading into a unique loxP site within the HAC-based gene delivery vector. An F′ factor origin of replication allows YAC propagation as a single copy BAC molecule. pJBRV1 is available on request from the Laboratory of Molecular Pharmacology, National Cancer Institute (NIH).
4. Diagnostic PCR primers. Diagnostic PCR primers to screen yeast transformants for gene positive clones may be chosen from any unique internal gene sequences (e.g., exonic regions).
5. The iCre plasmid. The iCre plasmid is derived from the pCpGfree-vitroHLacZ plasmid (InvivoGen). It contains a hygromycin resistance gene that allows selection of the plasmid in both bacteria and vertebrate cells. The optimized *Cre* gene is under the EF1 promoter and flanked by Matrix Attachment Region (MAR) elements. The MAR elements together with the absence of CpG sites in promoter or vector sequences greatly improve the transcription efficiency of the *Cre* gene. The iCre plasmid also contains a conditional *R6K* bacterial origin, which requires a pir + expressing *E. coli* strain to propagate. This plasmid is available on request from the Laboratory of Molecular Pharmacology, National Cancer Institute (NIH).
6. Primers to verify insertion of a gene of interest into the HAC by reconstitution of the *HPRT* gene: HPRT-R 5′-AGCCTTCTGTACACATTTCTTCTC-3′ and HPRT-F 5′-GCTCTACTAAGCAGATGGCCACAGAACTAG-3′.
7. The PNA-tetO probe FITC-OO-ACCACTCCCTATCAG to detect the HAC by FISH analysis. This probe may be purchased from PANAgene Inc. Follow the PNA-FISH protocol as directed by the company.

2.2 Preparation of Highly Competent Yeast Spheroplasts and Transformation of the Spheroplasts by Genomic DNA Along with a TAR Vector

1. 1 M Sorbitol.
2. SPE solution: 1 M Sorbitol, 10 mM Na_2 EDTA, 0.01 M Na phosphate, pH 7.5.
3. SOS solution: 1 M Sorbitol, 6.5 mM $CaCl_2$, 0.25 % yeast extract, 0.5 % Bacto Peptone.
4. STC solution: 1 M Sorbitol, 10 mM $CaCl_2$, 10 mM Tris–HCl, pH 7.5.
5. Zymolyase solution: 10 mg/ml of zymolyase 20T in 20 % glycerol (keep as frozen aliquots at −20 °C).
6. PEG 8000 solution: 20 % (w/v) polyethylene glycol 8000 (Sigma), 10 mM $CaCl_2$, 10 mM Tris–HCl, pH 7.5.

7. YPD medium: 2 % D-glucose, 2 % Bacto Peptone, 1 % yeast extract.
8. TOP agar-His: 1 M Sorbitol, 2 % D-glucose, 0.17 % Yeast Nitrogen Base, 0.5 % $(NH_4)_2SO_4$, 3 % agar containing the following supplements: 0.006 % adenine sulfate, 0.006 % uracil, 0.005 % L-arginine.HCl, 0.008 % L-aspartic acid, 0.01 % L-GLUTAMIC acid, 0.005 % L-isoleucine, 0.01 % L-leucine, 0.012 % L-lysine.HCl, 0.002 % L-methionine, 0.005 % L-phenylalanine, 0.0375 % L-serine, 0.01 % L-threonine, 0.005 % L-tryptophan, 0.005 % L-tyrosine, 0.015 % L-valine. Alternatively, the medium can be purchased from Teknova, Inc. (www.teknova.com).
9. SORB-His plates: 1 M Sorbitol, 2 % D-glucose, 0.17 % Yeast Nitrogen Base 0.5 % $(NH_4)_2SO_4$, 2 % agar supplemented as described above. Alternatively, synthetic complete medium can be purchased from Teknova, Inc. (www.teknova.com).
10. SD-His plates: 2 % D-glucose, 0.17 % Yeast Nitrogen Base 0.5 % $(NH_4)_2SO_4$, 2 % agar supplemented as described above. Alternatively, synthetic complete medium can be purchased from Teknova, Inc. (www.teknova.com).
11. A linearized TAR vector with the appropriate hooks; concentration 0.5–1 μg/μl (keep at –20 °C).

2.3 Preparation of Genomic DNA for TAR Cloning

Blood and Cell Culture DNA Maxi kit (Qiagen), may be used for preparation of genomic DNA in solution (*see* **Note 5**). DNA isolated by this method allows TAR cloning of fragments with size not bigger than 150 kb. For isolation of bigger size fragments (up to 250–300 kb), genomic DNA should be prepared in agarose blocks as previously described [38].

2.4 Isolation of DNA from a Pool of Yeast Transformants

1. 1 M Sorbitol.
2. SP solution: 1.2 M Sorbitol, 0.1 M Na phosphate, pH 7.5.
3. SPE solution: 1 M Sorbitol, 10 mM Na_2 EDTA, 0.01 M Na phosphate, pH 7.5.
4. Zymolyase solution: 10 mg/ml of zymolyase 20T in 20 % glycerol (keep as frozen aliquots at –20 °C).
5. 5 M KAc (potassium acetate).
6. Diethylpyrocarbonate (DEPC).
7. 14.3 M beta-mercaptoethanol (ME).
8. 100 % isopropanol (keep at room temperature).

2.5 Identification of Gene-Positive Clones in the Pools

1. Zymolyase solution: 10 mg/ml of zymolyase 20T in 20 % glycerol (keep as frozen aliquots at –20 °C).
2. 2 % SDS.
3. 5 M KAc (potassium acetate).

4. 14.3 M beta-mercaptoethanol (ME).
5. 100 % isopropanol (keep at room temperature).

2.6 Retrofitting of Circular YACs into BACs

1. pJBRV1 vector linearized at a *Bam*HI site; concentration 0.2–1.0 μg/μl (keep at −20 °C).
2. LiAc solution: 100 mM lithium acetate, 10 mM Tris–HCl, 0.1 mM EDTA, pH 7.5.
3. PEG 4000 solution: 40 % (w/v) polyethylene glycol 4000 (Fluka) aqueous solution.
4. Carrier salmon DNA: 10 mg/ml sonicated salmon sperm DNA (Stratagene) denaturated by boiling for 10 min every time before the experiment.
5. SD-Ura plates: 2 % D-glucose, 0.17 % Yeast Nitrogen Base, 0.5 % $(NH_4)_2SO_4$, 0.006 % adenine sulfate, 0.005 % L-arginine. HCl, 0.008 % L-aspartic acid, 0.01 % L-glutamic acid, 0.004 % L-histidine.HCl, 0.005 % L-isoleucine, 0.01 % L-leucine, 0.012 % L-lysine.HCl, 0.002 % L-methionine, 0.005 % L-phenylalanine, 0.0375 % L-serine, 0.01 % L-threonine, 0.005 % L-tryptophan, 0.005 % L-tyrosine, 0.015 % L-valine, 2 % agar. Alternatively, synthetic complete medium can be purchased from Teknova, Inc. (www.teknova.com).
6. SD-His synthetic liquid medium: 2 % D-glucose, 0.17 % Yeast Nitrogen Base, 0.5 % $(NH_4)_2SO_4$, 0.006 % adenine sulfate, 0.006 % uracil, 0.005 % L-arginine.HCl, 0.008 % L-aspartic acid, 0.01 % L-glutamic acid, 0.005 % L-isoleucine, 0.01 % L-leucine, 0.012 % L-lysine.HCl, 0.002 % L-methionine, 0.005 % L-PHENYLALANINE, 0.0375 % L-serine, 0.01 % L-threonine, 0.005 % L-tryptophan, 0.005 % L-tyrosine, 0.015 % L-valine. Alternatively, synthetic complete medium can be purchased from Teknova, Inc. (www.teknova.com).

2.7 Transferring of a TAR/YAC/BAC Gene Positive Clone from Yeast to E. coli Cells

1. DH10B *E. coli* competent cells.
2. LET solution: 0.5 M EDTA, 0.01 M Tris–HCl, pH 7.5.
3. SOC solution: 2 % Bacto Tryptone, 0.5 % Bacto Yeast Extract, 10 mM NaCl, 2.5 mM KCl, 10 mM $MgCl_2$, 10 mM $MgSO_4$, 20 mM D-glucose.
4. LMP agarose (low gelling/melting temperature agarose): 1 % of agarose gel prepared in 0.125 M EDTA, pH 7.5.
5. EDTA mix: 0.05 M EDTA, 0.01 M Tris–HCl, pH 7.5.
6. NDS cell lysis buffer: 0.39 M EDTA, 0.01 M Tris–HCl, pH 7.5, 1 % *N*-lauroyl sarcosine, 2 mg/ml proteinase K (keep at −20 °C).
7. LB Cm plates supplemented with 12.5 μg/ml chloramphenicol.
8. β-agarase (keep at −20 °C).

2.8 Insertion of the TAR/BAC Clone Containing a Gene of Interest into a Single Gene Loading Site of the HAC in Hamster CHO Cells

1. Lipofectamine® 2000 Transfection Reagent.
2. Opti-MEM® Reduced Serum Medium.
3. Hypoxanthine-Aminopterin-Thymidine (HAT) Supplement (50×).
4. Ham's F-12 Nutrient Mix, GlutaMAX™.
5. Fetal Bovine Serum—Optima.
6. Phosphate Buffered Saline (PBS), pH 7.4.
7. Penicillin-Streptomycin (100×), Liquid.
8. Blasticidin S HCl, 10 mg/ml in 20 mM HEPES (pH 7.2–7.5).
9. Cloning Cylinder, 8 mm × 8 mm diameter.
10. 0.25 % Trypsin-EDTA (1×), Phenol Red.
11. HPRT-deficient hamster CHO cells (JCRB0218) carrying alphoid^{tetO}-HAC with a single gene loading site (a 5′ HPRT-loxP cassette) [24]. This cell line is available on request from the Laboratory of Molecular Pharmacology, National Cancer Institute (NIH).

3 Methods

3.1 Preparation of Competent Yeast Spheroplasts

1. One day before the TAR cloning experiments, inoculate 50 ml aliquots of YPD medium in three 250-ml Erlenmeyer flasks. Inoculate the flasks with three different size single-cell colonies of the host yeast strain VL6-48 freshly grown on YEPD plate with agar, and grow the cultures overnight (for 14–16 h) at 30 °C with vigorous shaking (200 rpm) to assure good aeration.
2. The following morning, measure optical density of the cultures with 20 min intervals. Choose the flask with an OD_{660} of 2.0–4.0 [For actual measurement, dilute the culture 1/10 in water; the density should be between 0.2 and 0.4 (*see* **Note 6**)]. The culture with such an optical density is ready for the preparation of highly competent spheroplasts. This optical density corresponds to approximately 2×10^7 cells per ml.
3. Transfer the yeast culture from this flask into a 50-ml Falcon conical tube and pellet the cells by centrifugation 5 min at 1,000 × *g* at 5 °C. Remove and discard the supernatant.
4. Resuspend a cell pellet in 30 ml of sterile water by vortexing and centrifuge 5 min at 3,000 × *g* at 5 °C. Remove and discard the supernatant.
5. Resuspend a cell pellet in 20 ml of 1 M Sorbitol by vortexing and centrifuge 5 min at 3,000 × *g* at 5 °C. Remove and discard the supernatant. (Yeast cells in 1 M Sorbitol may be kept overnight.)

6. Resuspend a cell pellet in 20 ml of SPE solution. Add into a tube 20 μl of zymolyase solution, 40 μl of ME, mix well, and incubate at 30 °C for ~20 min with a slow shaking (*see* **Note 7**).
7. Check the level of spheroplasting by comparison of optical densities of the cell suspension in 1 M Sorbitol versus 2 % SDS. To measure the OD_{660} difference, take two 200 μl aliquots of the zymolyase treated cell suspension and dilute tenfold by 1 M Sorbitol and 2 % SDS. The spheroplasts are determined to be ready when the difference between the Sorbitol and SDS OD_{660} readings is three to fivefold (e.g., OD = 0.844 vs. OD = 0.187). Both underexposure and overexposure to zymolyase greatly affects transformation efficiency. From this point on, extreme care must be taken to avoid lysing the delicate spheroplasts: very slow, gentle resuspensions are necessary.
8. Centrifuge spheroplasts for 10 min at 570 × *g* at 5 °C. Decant the supernatant, add 50 ml of 1.0 M Sorbitol, then rock very gently to resuspend the pellet. Pellet the spheroplasts again by centrifugation for 10 min at 300–600 × *g* at 5 °C.
9. Repeat the wash with 50 ml of 1 M Sorbitol one more time and gently resuspend the final pellet in 2.0 ml of STC solution. The spheroplasts are ready for transformation and are stable at room temperature for at least 1 h.

3.2 Transformation of Spheroplasts by Genomic DNA Along with a TAR Vector

1. Take 200 μl from 2.0 ml of spheroplast suspension in STC solution and mix gently with 1–2 μg of genomic DNA and 1.0 μg of the linearized TAR vector in 2.0 ml Eppendorf tube. Incubate for 10 min at room temperature. (2.0 ml of spheroplasts in STC solution is enough to make ten 200 μl samples).
2. Add 800 μl of PEG 8000 solution into each Eppendorf tube, gently mix by inverting, and incubate for 10 min at room temperature.
3. Pellet the spheroplasts by centrifugation in the Eppendorf microfuge for 5 min at 300–500 × *g* at 5 °C. Remove the supernatant and gently resuspend the spheroplasts in each tube with 800 μl of SOS solution using Pipetman.
4. Incubate the spheroplasts for 40 min at 30 °C without shaking.
5. Transfer the spheroplasts from each tube into a 15-ml Falcon conical tube containing 7.0 ml of melted TOP agar-His (equilibrated at 50 °C) using Pipetman, gently mix, and quickly pour agar onto a SORB-His plate with selective medium containing 1 M Sorbitol.
6. Keep the plates at 30 °C for 5–7 days until all the transformants become visible. For transformation conditions described above (i.e., with 1 μg of a vector, 2 μg of genomic DNA, and

~1×10^8 spheroplasts), the yield of transformants varies from 10 to 200 colonies per plate dependently from the hooks used. Typically the higher yield of transformants is observed with the hooks containing nonunique sequences.

3.3 Identification of Gene-Positive Pools

1. Transfer ~300 primary transformants by toothpicks on SD-His plates lacking histidine. Streak 30 transformants per each master plate.
2. Incubate the plates with pools of transformants at 30 °C overnight and replica plate on new plates with SD-His selective medium. Use the master plates for detection of gene-positive pools. (Incubate replica plates at 30 °C overnight and later use the plate(s) with the positive pool(s) for detection of gene-positive individual colonies by a second round of PCR).
3. Wash the yeast cells out from the replica plates with 5 ml water into 15-ml Falcon conical tubes, and pellet the cells by centrifugation 5 min at 1,000×*g* at 5 °C. Remove and discard the supernatant.
4. Resuspend each cell pellet in 1 ml of 1 M Sorbitol by vortex, transfer the suspension to 2.0 ml Eppendorf microfuge tube, and spin at 20,800×*g* for 30 s. Remove and discard the supernatant.
5. Resuspend cells in 0.5 ml of SPE solution containing ME (1/1,000 dilution), add into each tube 20 μl of zymolyase solution and incubate for 2 h at 30 °C.
6. Harvest the spheroplasts by centrifugation for 5 min at 2,700×*g* on the Eppendorf microfuge and resuspend the pellets in 0.5 ml of 50 mM EDTA solution containing 0.2 % SDS with Pipetman.
7. Add 1 μl of DEPC at room temperature and vortex well.
8. Lyse the spheroplasts completely by incubation at 70 °C for 15 min.
9. Add 50 μl 5 M KAc to lysate, mix well and let the tubes sit on ice for 30 min.
10. Pellet the precipitate by centrifugation for 15 min at maximum Eppendorf minifuge speed (20,800×*g*).
11. Transfer the supernatant to fresh microfuge tubes, fill the tubes with room temperature ethanol, mix and pellet the DNA by centrifugation for 5 min. Remove the supernatant as much as possible and dry the tubes by inverting on blotting paper.
12. Resuspend each damp DNA pellet in 0.4 ml water. (Samples can be incubated at 4 °C overnight to completely dissolve.)
13. Add 0.5 ml of isopropanol. Mix well and immediately pellet the DNA precipitate by centrifugation for 5 min at maximum Eppendorf minifuge speed at room temperature.

14. Remove the supernatant as much as possible and dry the tubes well.
15. Dissolve the final pellet of DNA in 0.3 ml of water.
16. Use 1 μl of the DNA solution in a 50 μl PCR reaction with appropriate diagnostic primers to identify gene-positive pools (*see* **Note 8**). For this PCR, the standard protocol suggested by the Taq polymerase manufacturer should work.

3.4 Identification of Individual Gene-Positive Clones in Pools

Individual transformants from each positive pool are screened by a second round of PCR with the same diagnostic primers to identify gene positive clones.

1. Touch the streak of each His+ transformant from a replica plate with "a positive pool" with a sterile disposable pipette tip and then rinse the tip thoroughly in 100 μl mixture of 80 μl water plus 20 μl zymolyase solution plus 1 μl of ME.
2. Incubate the resulting suspension for 1 h at 30 °C.
3. Add 10 μl of 2 % SDS. Incubate for 15 min at 70 °C.
4. Add 10 μl of 5 M KAc and let the tubes sit on ice for 15 min.
5. Spin at 20,800 × *g* for 2 min at room temperature.
6. Transfer the supernatant to a new Eppendorf tube and add equal volume of isopropanol. Precipitate at maximum Eppendorf minifuge speed for 5 min.
7. Dissolve the pellets in 30 μl water.
8. Use 1 μl of the DNA solution in a 50 μl PCR reaction with appropriate diagnostic primers to identify gene-positive clones.

3.5 Retrofitting of TAR YACs into BACs with the pJBRV1 Vector for Gene Loading into the HAC

This protocol describes an efficient and accurate procedure for retrofitting circular TAR/YACs containing a gene of interest into BACs using the pJBRV1 vector (Fig. 2b, c). The vector contains a 3′ HPRT-loxP-eGFP cassette, allowing gene loading into a unique loxP site of the resulting HAC in CHO cells. An F′ factor origin of replication allows YAC propagation as a BAC molecule. The retrofitted YAC/BACs can be moved to *E. coli* by electroporation for a standard BAC DNA isolation.

1. Inoculate 5 ml of SD-His synthetic liquid medium without histidine with one individual colony containing a TAR/YAC and grow overnight at 30 °C with vigorous shaking to assure a good aeration.
2. Transfer the yeast culture into 50-ml YPD and grow for additional 4–5 h at 30 °C with vigorous shaking.
3. Pellet the culture by centrifugation for 5 min at 1,000 × *g* at 5 °C in the 50-ml Falcon conical tube. Remove and discard the supernatant.

4. Resuspend the cell pellet in 10 ml of sterile water by vortex, transfer into an Eppendorf tube and pellet the cells by centrifugation for 1 min at maximum speed. Remove and discard the supernatant.
5. Resuspend the cells in 10 ml of LiAc solution. Incubate at 30 °C for 1 h with a slow shaking. Alternatively, cells can be stored at 5 °C during 2–3 days with no effect on transformation efficiency.
6. Collect the cells by centrifugation.
7. Decant the supernatant and resuspend the cells in 100 μl of LiAc solution using Pipetman.
8. Add 1 μg of a *Bam*HI linearized pJBRV1 vector DNA and 5 μl of carrier salmon DNA to the cells and mix well.
9. Add 0.45 ml of fresh PEG 4000 solution, mix by vortexing and incubate for 1 h at 30 °C.
10. Heat-shock the cells in a 42 °C heating block for 15 min.
11. Top off the tube with sterile, distilled water and mix by inversion.
12. Collect the cells by centrifugation at a high speed for 1 min.
13. Decant supernatant and resuspend the cells in 1 ml of water using a sterile toothpick.
14. Collect the cells by centrifugation for 1 min.
15. Decant the supernatant and resuspend the cells in 400 μl of water and spread the suspension of 50, 100 and 200 μl on 100-mm SD-Ura plates.
16. Incubate the plates at 30 °C. Colonies of Ura^+ transformants should be visible in 2–3 days. With 1 μg of vector, the yield of Ura^+ transformants varies from 50 to 200 colonies. More than 90 % of the transformants derive from recombination between a pJBRV1 vector and a circular YAC.

3.6 Transferring a YAC/BAC Containing a Gene of Interest from Yeast into E. coli Cells

This protocol below describes the preparation of yeast DNA in agarose plugs to keep the YAC DNA intact if the size of the cloned DNA molecule is bigger than 50 kb (*see* **Notes 9** and **10**). If the size of TAR-cloned YAC DNA is less than 50 kb, the DNA isolated from the yeast gene-positive clones may be directly transformed into *E. coli* (to isolate yeast DNA, follow the protocol in Subheading 3.3). It is worth noting that a circular YAC DNA molecule can be high-quality purified directly from yeast cells [37] (*see* **Note 11**). In this case, Subheadings 3.5 and 3.6 are not required. Also, it is important to check the integrity of YAC/BAC clones before functional analysis (*see* **Note 12**).

1. Inoculate two to three His^+Ura^+ transformants separately in 5 ml of YPD medium in 50-ml Falcon conical tubes and grow overnight at 30 °C with a vigorous shaking (200 rpm).

2. Pellet the cells by centrifugation. Remove and discard the supernatant.
3. Resuspend the cells in 100 μl EDTA mix (vortex well) and transfer into 1.5 ml Eppendorf tubes. Add 100 μl of zymolyase solution, vortex the cells for 4 s, and incubate the suspension for 30 min at 37 °C.
4. Melt an appropriate quantity of LMP agarose and place it in a 50 °C water bath to cool.
5. Transfer the melted agarose and resuspended cells to a 42 °C heating block and equilibrate for 15 min.
6. Add to the cell suspension an equal volume of the melted agarose and mix well by vortexing. Keep the cell/agarose suspension at 42 °C. It is important that the final concentration of agarose be equal to 0.5 %. With a higher concentration of agarose it is impossible to completely melt the plugs for electroporation.
7. Take 50 μl aliquots of the cell/agarose suspension and gently place each into Ultra Micro tips. Keep the tips horizontal for 10 min at 5 °C until the agarose is completely solidified.
8. Transfer the agarose plugs into Eppendorf tubes. To do this, take up LET solution in a 6 cc syringe without a needle, place the tip of the Ultra Micro tip into the syringe lure and gently apply pressure. The plug should slide out into the tube. Make three to four 50 μl agarose plugs and incubate them for 1 h at 37 °C.
9. Remove LET and add enough NDS solution to cover the plugs. Incubate the plugs for 1 h at 55 °C.
10. Remove the NDS solution carefully and wash the plugs three times with EDTA mix (20 min each time at room temperature). Dialyzed plugs may be stored at 5 °C in EDTA mix for several months.
11. Incubate the plugs overnight at room temperature in water before melting and use for electroporation.
12. To electroporate YAC/BACs into *E. coli*, the plugs are melted at 68 °C for 15 min, cooled to 42 °C for 10 min, treated with 1.5 units of β-agarase for 1 h at 42 °C, and chilled on ice for 10 min.
13. Dilute the treated plug twofold with sterile water.
14. Use 1 μl of the mixture to electroporate 20 μl of the *E. coli* DH10B competent cells using a Bio-Rad Gene Pulser with the settings 2.5 kV, 200 Ω and 25 μF.
15. After electroporation, add 1 ml of SOC into a cuvette, mix well by Pipetman, and transfer into a microfuge tube.
16. Incubate the cells for 1 h at 37 °C.

17. Spread 30, 100, and 300 µl of the cell suspension onto LB-Cm plates.
18. Incubate the plates at 37 °C overnight (*see* **Note 9**).

3.7 Insertion of the YAC/BAC Clone Containing a Gene of Interest into a Single Gene Loading Site of the HAC in Hamster CHO Cells

To insert a genomic copy of a gene into the HAC, the appropriate BAC construct with a gene is co-transfected with a Cre-recombinase expression vector into a hprt-minus hamster CHO cells carrying the HAC and HPRT-plus colonies are selected on HAT medium (Fig. 1, Step 4 and Fig. 3).

1. Purify BAC DNA containing a genomic copy of the gene by Large-Construct Kit (Qiagen) and iCre-plasmid by Spin Miniprep Kit (Qiagen), respectively.
2. One day before transfection, plate CHO cells in six-well plate of 2 ml of growth medium (F12 + 10 % FBS + 1 % PenStrep + 8 µg/ml blasticidin) so that the cells will be about 70–80 % confluent at the time of transfection.
3. Next day: dilute DNAs (BAC ~50–100 kb in size 20–30 µg and iCre 1 µg) (*see* **Note 13**) in 100 µl of Opti-MEM medium without serum (final volume 100 µl). Mix gently by tapping.
4. Mix Lipofectamine 2000 gently before use, then dilute 10 µl amount in 90 µl of Opti-MEM medium. Incubate for 5 min at room temperature.
5. Combine the diluted DNA (**step 3**) with diluted Lipofectamine 2000 (**step 4**). Total volume is 200 µl. Mix gently and incubate for 20–30 min at room temperature.
6. At the same time wash the CHO cells in six-well plate one time by PBS, next rinse with 2 ml of Opti-MEM medium, aspirate and add 2 ml Opti-MEM medium again (three washes in total).
7. Incubate the cells at 37 °C for 20–30 min.
8. After incubation aspirate the cells and add 800 µl Opti-MEM medium into the DNA-Lipofectamine containing tube, pipette several times and apply to cells (aspirate the cells before applying).
9. Incubate the cells for about 12 h at 37 °C.
10. Wash the cells with growth media one to two times and incubate the CHO cells for 12–16 h at 37 °C in 2 ml of growth media.
11. Split the cells into the 10-cm plate and apply F12-HAT media (F12 + 10 % FBS + 1 % PenStrep + 8 µg/ml blasticidin + 1× HAT).
12. Pick up the colonies in 2–3 weeks after HAT selection using cloning cylinders and 0.25 % Trypsin. Transfer the colony to a six-well plate and later to 10-cm plate for further colony expansion.

13. For each colony, collect cells from a confluent 10-cm plate to make frozen stocks and add more culture medium to the plate to grow more cells.
14. To proceed for verification of the gene integration into the HAC (*see* **Note 14**), isolate genomic DNA using any standard genomic DNA purification protocol and use 1 μl of the DNA solution in a 50 μl PCR reaction with the diagnostic primers (*see* Subheading 2 for the primers) to confirm the *HPRT* gene reconstitution.

4 Notes

1. An important step is the selection of specific hook(s) for a TAR vector. Hooks should be unique sequences, with no repeated sequences present. For human and rodent genomes, the uniqueness of hooks can be checked by blasting against genome sequences. We demonstrated that the size of a hook can be as small as 60 bp [6]. A further increase of the length of a targeting sequence had no effect on selectivity of gene isolation. Hooks should be also free of yeast *ARS*-like sequences. Potential *ARS*-like sequences in hooks can be identified based on the presence of a 17 bp *ARS* core consensus, WWWWTTTAYRTTTWGTT, where W = A or T; Y = T or C; R = A or G [39]. The final conclusion about the absence of the yeast origin of replication in a hook(s) can be obtained only by yeast transformation assay. No, or only a few, His^+ transformants should appear when the TAR cloning vector (with its hooks) is transformed into LiAc treated yeast cells deficient in *HIS3*.
2. One of the potential mistakes during TAR cloning is the incorrect orientation of the hooks in the TAR vector that are used to isolate target gene by recombination in yeast. Hooks should be cloned into the vector in such a way that after linearization of the vector, the orientation of the hooks should correspond to that illustrated in Fig. 1, Step 1. For TAR cloning experiments, the vector DNA should not be contaminated by bacterial chromosomal DNA, and the completeness of the vector linearization by endonuclease digestion should be carefully checked by electrophoresis. Nonlinearized vector molecules cannot participate in homologous targeting of input chromosomal DNA. In addition, they can induce circularization of linear vector molecules through a gap repair mechanism when both molecules present in the same cell.
3. If a gene (region) is unclonable by a TAR vector with two specific hooks, the radial TAR cloning approach may be exploited. A TAR vector with one unique hook and a common repeat as a second hook can target a region that is up to 250 kb far away

from a specific sequence, increasing the probability of *ARS*-like sequence capture. In contrast to TAR cloning with two specific hooks, where chimeras are never observed, radial TAR cloning may produce a few chimeras [32]. These cloning artifacts may be due to the presence in genomic DNA of multiple targets for the common repeat hook. Because TAR cloning produces multiple gene isolates, chimeric clones can be easily identified by the sequencing of the ends of the inserts. Then they can be eliminated from a further analysis.

4. It is notable that isolation of gene homologues may be carried out using the targeting hooks developed from human genome sequence. Examples of TAR cloning of primate gene homologous are presented in several publications [29–31].

5. The quality of genomic DNA is also critical for TAR cloning. Preparation of genomic DNA in agarose plugs is one option as it has been previously described [7, 38]. However, agarose fragments inhibit transformation efficiency [7] and therefore at present we prefer to use DNAs isolated in aqueous solution (Qiagen DNA Maxi kit). Size of genomic DNA should be checked by pulse gel electrophoresis before its use. Recently we have discovered that human and mouse DNA purchased from Promega can be also used for TAR cloning when a targeted gene is smaller than ~100 kb. Several human genes were successfully cloned by TAR in our laboratory using Promega genomic DNA (60 kb *ASPM*, 50 kb *NBS1*, 50 kb *TERT*, 140 kb *ATM*). Size of genomic Promega DNA checked by CHEF was between 50 and 100 kb.

6. Optical density of the culture to make competent spheroplasts may vary depending on (1) medium for growth, (2) conditions of growing, and (3) a type of spectrophotometer used. Thus optimum optical density should be determined empirically in an OD_{660} interval of 2.0–5.0.

7. The treatment time varies depending on the zymolyase stock. When new stocks of enzyme are made, even if from the same lot number of powder, they should be re-titrated to confirm or adjust the amount of enzyme necessary for optimal transformation efficiency since reduction of activity of the enzyme during storage at 4 °C can occur.

8. Typically, one to five among 100–300 primary His^+ transformant colonies contains a gene of interest. To increase the yield of gene-positive colonies up to 20–30 %, before TAR cloning the genomic DNA may be treated by the restriction endonucleases. The restriction endonucleases used should cut near the target sequences but not within it. Hence this strategy requires sequence information of the target to be known. 8 bp cutter restriction endonucleases are preferred as

polymorphism within the target sequence is less likely to create a 8 bp restriction site compared to a 6 bp restriction site; however, a 6 bp restriction endonuclease could also be tried. To identify positive colonies, primary transformants are combined into pools and examined for the presence of the gene by PCR using a pair of primers specific for its internal sequence. We recommend using pools containing not more than 30 transformants. Usually analysis of three to four pools results in identification of a gene-positive pool. The yield might be lower if the targeting hook sequences are not unique.

9. Because some YAC/BACs may be truncated during electroporation, it is necessary to compare size of inserts in yeast and in *E. coli* cells. To estimate the size of circular YACs or YAC/BACs in yeast and BACs in *E. coli*, they should be linearized by endonuclease digestion (unique *Not*I site is present in pVC604 vector) separated by pulse-field gel electrophoresis and visualized by EtBr staining for BACs or blot-hybridized with a TAR vector-specific probe for YACs [33].
10. TAR/YAC/BAC isolates with sizes up to 250 kb can be efficiently and faithfully transferred from yeast cells into *E. coli* cells by electroporation. However, approximately 5 % of human DNA fragments cloned in YAC/BAC vectors exhibit abnormally low transformation efficiency during electroporation into *E. coli* cells [32, 33]. Clones with such inserts should be analyzed in yeast.
11. Large circular YAC molecules up to 600 kb in size can be high-quality purified in microgram quantities from yeast [37]. The purified YAC DNA is suitable for restriction enzyme digestion, DNA sequencing and functional studies. For example, YACs carrying full-size genes can be purified and used for transfection into mammalian cells or for direct insertion into the HAC for further functional analyses.
12. Although TAR cloning produces a low level of artifacts (such as deletions or chimeras), we recommend to physically characterize two to three independently isolated clones in a YAC or a BAC form before their use for functional/structural studies. The size of DNA inserts in the clones and their restriction profiles should be identical. Conditions for physical characterization of TAR isolates are previously described [3, 22].
13. Given the large size of the BACs (50–150 kb), more DNA (20–30 μg) is required in the transformation compared to smaller plasmids (~10–15 kb), where 2–3 μg is sufficient because efficiency of transformation drops as the size of the BAC molecule increases and we aim to use a comparable number of BAC molecules in transformation experiments.

14. Once colonies have been obtained, they have to be checked by PCR to verify *HPRT* reconstitution (*see* Subheading 2.1, **item 6**). Autonomous form of the HAC is verified by FISH using the HAC-specific PNA-tetO probe following the company protocol (*see* Subheading 2.1, **item 7**). From CHO cells the HAC with a gene can be transferred to a recipient gene-deficient cell line for complementation analysis by microcell-mediated chromosome transfer (MMCT) ([23] and references therein).

Acknowledgement

This work was supported by the Intramural Research Program of the NIH, National Cancer Institute, Center for Cancer Research, USA.

References

1. Larionov V, Kouprina N, Graves J, Chen XN, Korenberg J, Resnick MA (1996) Specific cloning of human DNA as YACs by transformation-associated recombination. Proc Natl Acad Sci U S A 93:491–496
2. Larionov V, Kouprina N, Graves J, Resnick MA (1996) Highly selective isolation of human DNAs from rodent-human hybrid cells as circular YACs by TAR cloning. Proc Natl Acad Sci U S A 93:13925–13930
3. Larionov V, Kouprina N, Solomon G, Barrett JC, Resnick MA (1997) Direct isolation of human BRCA2 gene by transformation-associated recombination in yeast. Proc Natl Acad Sci U S A 94:7384–7387
4. Kouprina N, Larionov V (2006) TAR cloning: insights into gene function, structural variants, long-range haplotypes and genome evolution. Nat Genet Rev 8:805–812
5. Kouprina N, Larionov V (2008) Selective isolation of genomic loci from complex genomes by transformation-associated recombination (TAR cloning) in yeast *Saccharomyces cerevisiae*. Nat Protoc 3:371–377
6. Noskov V, Koriabine M, Solomon G, Randolph M, Barrett JC, Leem SH, Stubbs L, Kouprina N, Larionov V (2001) Defining the minimal length of sequence homology required for selective gene isolation by TAR cloning. Nucleic Acids Res 29:E62
7. Leem S-H, Noskov V, Park J-E, Kim S-I, Larionov V, Kouprina N (2003) Optimum conditions for selective isolation of genes from complex genomes by transformation-associated recombination cloning. Nucleic Acids Res 31:e29
8. Stinchomb DT, Thomas M, Kelly I, Selker E, Davis RW (1980) Eukaryotic DNA segments capable of autonomous replication in yeast. Proc Natl Acad Sci U S A 77:4559–4563
9. Noskov V, Kouprina N, Leem SH, Koriabine M, Barrett JC, Larionov V (2002) A genetic system for direct selection of gene-positive clones during recombinational cloning in yeast. Nucleic Acids Res 30:e8
10. Raymond CK, Sims EH, Kas A, Spencer DH, Kutyavin TV, Ivey RG, Zhou Y, Kaul R, Clendenning JB, Olson MV (2002) Genetic variation at the O-antigen biosynthetic locus in *Pseudomonas aeruginosa*. J Bacteriol 184: 3614–3622
11. Noskov V, Kouprina N, Leem SH, Ouspenski I, Barrett JC, Larionov V (2003) A general transformation-associated recombination cloning system to selectively isolate any eukaryotic or prokaryotic genomic region. BMC Genomics 4:16
12. Kouprina N, Annab L, Graves J, Afshari C, Barrett JC, Resnick MA, Larionov V (1998) Functional copies of a human gene can be directly isolated by TAR cloning with a small 3′ end target sequence. Proc Natl Acad Sci U S A 95:4469–4474
13. Kouprina N, Graves J, Resnick MA, Larionov V (1997) Specific isolation of human rDNA genes by TAR cloning. Gene 197:269–276
14. Cancilla M, Tainton K, Barry A, Larionov V, Kouprina N, Resnick M, Du Sart D, Choo A

(1998) Direct cloning of human 10q25 neocentromere DNA transformation-associated recombination (TAR) in yeast. Genomics 47: 399–404
15. Humble M, Kouprina N, Noskov V, Graves J, Garner E, Tennant R, Resnick MA, Larionov V, Cannon RE (2000) Radial TAR cloning from the TgAC mouse. Genomics 70: 292–299
16. Kim J, Noskov V, Lu X, Bergmann A, Ren X, Warth T, Richardson P, Kouprina N, Stubbs L (2000) Discovery of a novel, paternally expressed ubiquitin-specific processing protease gene through comparative analysis of an imprinted region of mouse chromosome 7 and human chromosome 19q13.4. Genome Res 10:1138–1147
17. Kouprina N, Larionov V (2003) Exploiting the yeast *Saccharomyces cerevisiae* for the study of the organization of complex genomes. FEMS Microbiol Rev 27:629–649
18. Kouprina N, Campbell M, Graves J, Campbell E, Meincke L, Tesmer J, Grady D, Doggett N, Moyzis R, Deaven L, Larionov V (1998) Construction of human chromosome 16- and 5- specific YAC/BAC libraries by *in vivo* recombination in yeast (TAR cloning). Genomics 53:21–2818
19. Zeng C, Kouprina N, Zhu B, Cairo A, Hoek M, Cross G, Larionov V, de Jong P (2001) New BAC/YAC libraries allowing to selectively re-isolate a desired genomic region by in vivo recombination in yeast. Genomics 77:27–34
20. Nihei N, Kouprina N, Larionov V, Oshima J, Martin GM, Ichikawa T, Barrett JC (2002) Functional evidence for a metastasis suppressor gene for rat prostate cancer within a 60 kb region on human chromosome 8p21-p12. Cancer Res 62:367–370
21. Annab L, Kouprina N, Solomon G, Cable L, Hill D, Barrett JC, Larionov V, Afshari C (2000) Isolation of functional copy of the human *BRCA1* gene by TAR cloning in yeast. Gene 250:201–208
22. Leem SH, Londono-Vallejo JA, Kim JH, Bui H, Tubacher E, Solomon G, Park JE, Horikawa I, Kouprina N, Barrett JC, Larionov V (2002) Cloning of the human telomerase gene: complete genomic sequence and analysis of tandem repeats polymorphisms in intronic regions. Oncogene 21:769–777
23. Kim JH, Kononenko A, Erliandri I, Kim TA, Nakano M, Iida Y, Barrett JC, Oshimura M, Masumoto H, Earnshaw WC, Larionov V, Kouprina N (2011) Human artificial chromosome (HAC) vector with a conditional centromere for correction of genetic deficiencies in human cells. Proc Natl Acad Sci U S A 108:20048–20053
24. Kouprina N, Ebersole T, Koriabine M, Pak E, Rogozin IB, Katoh M, Oshimura M, Ogi K, Peredelchuk M, Solomon G, Brown W, Barrett JC, Larionov V (2003) Cloning of human centromeres by transformation-associated recombination in yeast and generation of functional human artificial chromosomes. Nucleic Acids Res 31:922–934
25. Kouprina N, Pavlicek A, Noskov VN, Solomon S, Otstot J, Isaacs W, Carpten JD, Trent JM, Barrett JC, Jurka J, Larionov V (2005) Dynamic structure of the SPANX gene cluster mapped to the prostate cancer susceptibility locus HPCX at Xq27. Genome Res 15:1477–1486
26. Kim JH, Leem SH, Sunwoo Y, Kouprina N (2003) Separation of long-range human *TERT* gene haplotypes by transformation-associated recombination cloning in yeast. Oncogene 22:2452–2456
27. Kouprina N, Lee N, Pavlicek A, Samoshkin A, Kim JH, Lee HS, Varma S, Reinhold WC, Otstot J, Solomon G, Davis S, Meltzer PS, Schleutker J, Larionov V (2012) Exclusion of the 750-kb genetically unstable region at Xq27 as a candidate locus for prostate malignancy in HPCX1-linked families. Genes Chromosomes Cancer 51:933–948
28. Noskov V, Leem SH, Solomon G, Mullokandov M, Chae J-Y, Yoon Y-H, Shin Y-S, Kouprina N, Larionov V (2003) A novel strategy for analysis of gene homologs and segmental genome duplications. J Mol Evol 56:702–710
29. Kouprina N, Mullokandov M, Rogozin I, Collins K, Solomon G, Risinger J, Koonin E, Barrett JC, Larionov V (2004) The *SPANX* gene family of cancer-testis specific antigens: rapid evolution, an unusual case of positive selection and amplification in African great apes and hominids. Proc Natl Acad Sci U S A 101:3077–3082
30. Kouprina N, Pavlicek A, Mochida GH, Solomon G, Gersch W, Yoon YH, Collura R, Ruvolo M, Barrett JC, Woods CG, Walsh CA, Jurka J, Larionov V (2004) Accelerated evolution of the *ASPM* gene controlling brain size begins prior to human brain expansion. PLoS Biol 2:653–663
31. Pavlicek A, Noskov N, Kouprina N, Barrett JC, Jurka J, Larionov V (2004) Evolution of the tumor suppressor *BRCA1* locus in primates: implications for cancer predisposition. Hum Mol Genet 13:1–15
32. Leem SH, Kouprina N, Grimwood J, Kim JH, Mullokandov M, Yoon YH, Chae JY, Morgan

J, Lucas S, Richardson P, Detter D, Glavina T, Rubin E, Barrett JC, Larionov V (2004) Closing the gaps on human chromosome 19 revealed genes with a high density of repetitive tandemly arrayed elements. Genome Res 14:239–246

33. Kouprina N, Leem SH, Solomon G, Ly A, Koriabine M, Otstot J, Pak E, Dutra A, Zhao S, Barrett JC, Larionov V (2003) Segments missing from the draft human genome sequence can be isolated by TAR cloning in yeast. EMBO Rep 4:257–262

34. Kouprina N, Earnshaw WC, Masumoto H, Larionov V (2013) A new generation of human artificial chromosomes for functional genomics and gene therapy. Cell Mol Life Sci 70:1135–1148

35. Kazuki Y, Oshimura M (2011) Human artificial chromosomes for gene delivery and the development of animal models. Mol Ther 19: 1591–1601

36. Nakano M, Cardinale S, Noskov VN, Gassman R, Vagnarelli P, Kandels-Lewis S, Earnshaw WC, Larionov V, Masumoto H (2008) Inactivation of a human kinetochore by a specific targeting of chromatin modifiers. Dev Cell 14:507–522

37. Noskov VV, Chuang RY, Leem SH, Larionov V, Kouprina N (2011) Isolation of circular yeast artificial chromosomes for synthetic biology and functional genomics studies. Nat Protoc 6:89–96

38. McCormick MK, Shero JH, Cheung MC, Kan YW, Hieter PA, Antonarakis SE (1989) Construction of human chromosome 21-specific yeast artificial chromosomes. Proc Natl Acad Sci U S A 86:9991–9995

39. Theis JF, Newlon CS (1997) The *ARS309* chromosomal replicator of *Saccharomyces cerevisiae* depends on an exceptional *ARS* consensus sequence. Proc Natl Acad Sci U S A 94: 10786–10791

Chapter 2

Recombineering Linear BACs

Qingwen Chen and Kumaran Narayanan

Abstract

Recombineering is a powerful genetic engineering technique based on homologous recombination that can be used to accurately modify DNA independent of its sequence or size. One novel application of recombineering is the assembly of linear BACs in *E. coli* that can replicate autonomously as linear plasmids. A circular BAC is inserted with a short telomeric sequence from phage N15, which is subsequently cut and rejoined by the phage protelomerase enzyme to generate a linear BAC with terminal hairpin telomeres. Telomere-capped linear BACs are protected against exonuclease attack both in vitro and in vivo in *E. coli* cells and can replicate stably. Here we describe step-by-step protocols to linearize any BAC clone by recombineering, including inserting and screening for presence of the N15 telomeric sequence, linearizing BACs in vivo in *E. coli*, extracting linear BACs, and verifying the presence of hairpin telomere structures. Linear BACs may be useful for functional expression of genomic loci in cells, maintenance of linear viral genomes in their natural conformation, and for constructing innovative artificial chromosome structures for applications in mammalian and plant cells.

Key words Linear BAC, Recombineering, *E. coli*, Genomic DNA, Chromosome, Phage N15, Plasmid

1 Introduction

BACs are important resources for the functional study of mammalian genes because of their capacity to clone several hundred kilobases of DNA, which can accommodate large chromosomal segments containing complete genes and all their natural regulatory signals. However, all the currently available BAC clones are in circular form, while mammalian genes are naturally carried on linear chromosomes. The ability to assemble linear BACs may provide a useful model system for eukaryotic studies because they will more closely represent the structure of natural mammalian chromosomes. While linear artificial chromosomes constructs based on truncation of actual human chromosomes are already available, the techniques required to build such linear vectors are extremely tedious and not applicable to BAC clones isolated from existing libraries [1, 2].

Kumaran Narayanan (ed.), *Bacterial Artificial Chromosomes*, Methods in Molecular Biology, vol. 1227,
DOI 10.1007/978-1-4939-1652-8_2, © Springer Science+Business Media New York 2015

In the microbial world, although it is more common to find DNA maintained in circular form there are exceptions where linear chromosomes and plasmids can be found, e.g., *Borrelia*, *Streptomyces*, *Agrobacterium*, and a number of gram-negative bacteria [3–6]. In *E. coli*, the prophage of the lambda-like N15 phage replicates as a linear plasmid with telomeres. The N15 core telomeric sequence (*telRL*) is a 56 bp site that is contained within a larger 310 bp DNA sequence known as the telomerase occupancy site (*tos*) [6–8]. The *tos* site is the target for the viral protelomerase enzyme TelN to process into left (*tel*L) and right (*tel*R) telomeres, enabling its prophage to replicate as a linear plasmid. We recently adapted the phage N15 system to assemble linear BACs using recombineering technology in *E. coli* [9–12]. In this technique, a DNA fragment containing the phage N15 telomeric (*tos*) sequence is inserted into a BAC followed by its resolution in vivo in *E. coli* into separate left (*tel*L) and right (*tel*R) telomeres by the phage protelomerase enzyme TelN, resulting in a self-replicating linear BAC [9].

Specifically, a *tos* sequence is amplified along with a selectable zeocin resistance marker (Zeo) (Fig. 1a) as a PCR cassette and inserted into a target BAC via homologous recombination (Fig. 1b) [11]. By design, primers TosZeoF2 and TosZeoR2 that are used to amplify the *tos*-Zeo cassette (Fig. 1a) contain homology sequences that target this DNA into a site centered on the unique *Sfi*I restriction site located within the vector section of many BAC clones (Fig. 1b), leaving its genomic insert intact.

The resulting recombinant BAC (in this chapter any BAC clone containing the recombined *tos*-Zeo cassette is referred to as "BAC-*tos*") remains circular at this stage and carries the *tos* sequence as well as the zeocin marker within its backbone (Fig. 1b) [9]. After recombination, zeo^r colonies are screened by PCR to verify presence of the intact *tos* sequence (Fig. 1c). Primers Sfi1 and Sfi2, positioned outside the junction where the recombination occurs, amplify the *tos*-Zeo insertion, generating a 1.5 kb PCR product.

To linearize the BAC-*tos* in vivo, it is transferred into a $telN^+$ *E. coli* strain that constitutively expresses active TelN enzyme for resolution of the *tos* sequence by the protelomerase, generating the terminal *telL* and *tel*R telomeres (Fig. 1d) [9]. The linear BAC-tos can be column extracted for further downstream applications (Fig. 1e). The presence of telomeres on the linear BAC-*tos* can be verified by treatment with RecBCD (exonuclease V), which hydrolyzes DNA with open ends. Linear BAC-*tos* with terminal telomeres will be covalently closed and therefore resist RecBCD digestion (Fig. 1f) [9].

The linearization protocols in this chapter are described using a 100-kb BAC (BAC4396) that contains the entire human beta-globin locus [9, 13]. But the same principles can be applied to any BAC clone. The biggest advantage of linearizing BACs by recombineering is that it is not limited by size or sequence

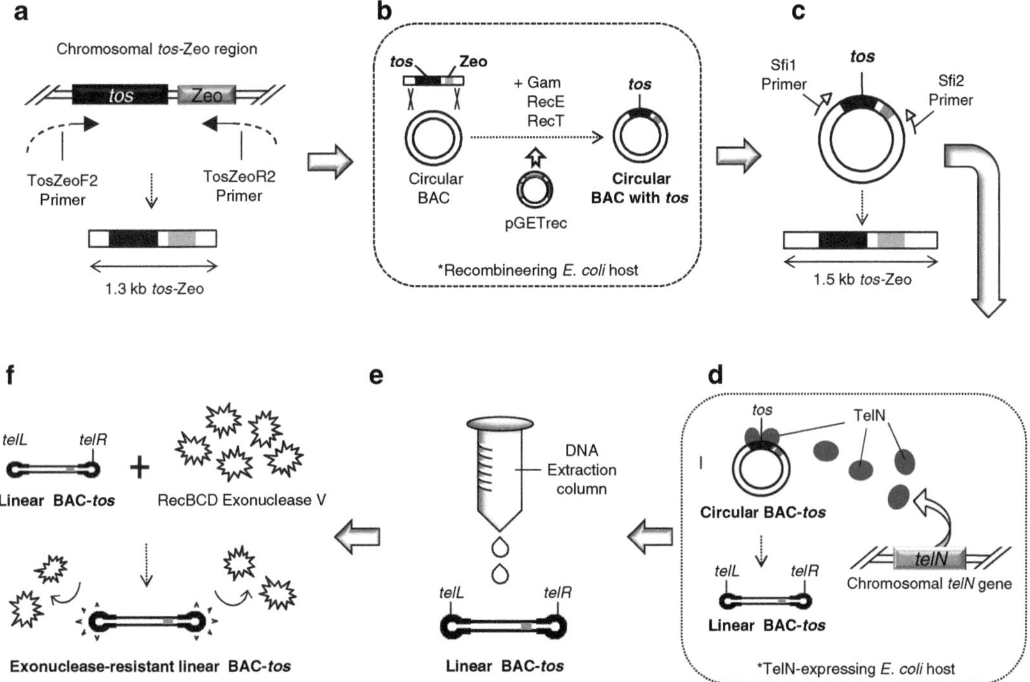

Fig. 1 Overview of BAC linearization. (**a**) Synthesizing *tos*-Zeo cassette for recombineering. A *tos*-Zeo cassette is amplified by PCR from a chromosomal *tos*-Zeo region, using primers TosZeoF2 and TosZeoR2. The resulting 1.3 kb *tos*-Zeo cassette contains the phage N15 telomeric sequence *tos* and a Zeocin resistance gene, with overhangs that complement DNA sequence at the target site on the circular BAC. (**b**) Recombineering the *tos-Zeo* sequence into BACs. First, the *tos*-Zeo cassette is transformed into an *E. coli* host carrying the target circular BAC. Expression of the recombineering enzymes (Gam, RecE, RecT) from plasmid pGETrec then induces insertion of the *tos*-Zeo cassette into the circular BAC by recombineering [11]. The precise insertion of *tos*-Zeo cassette into target BAC results in a circular BAC containing the *tos* sequence. The BAC remains circular at this stage. (**c**) PCR screening for recombinants. Zeo^R colonies that arise after recombineering are screened by PCR for the presence of the integrated *tos*-Zeo cassette. Primers Sfi1 and Sfi2 are located flanking the recombinant junction, and produce a 1.5 kb PCR product when the *tos*-Zeo cassette is correctly targeted into a BAC. (**d**) In vivo linearization of BAC-*tos*. The circular BAC-*tos* is transformed into a *telN*+ *E. coli* constitutively expressing the TelN protein, resolving the *tos* sequence into individual left (*telL*) and right (*telR*) telomeres and generating a linear BAC-*tos* [9]. (**e**) Extraction of linear BAC-*tos*. Linearized BAC-*tos* DNA is then column extracted for downstream applications. (**f**) Detecting presence of telomeres. To test for the integrity of telomeric ends formed on the linearized BAC-*tos*, the DNA is incubated with RecBCD exonuclease V that degrades open-ended linear DNA. As the telomeric ends on TelN-linearized BAC-*tos* are covalently closed, they will be resistant to exonuclease attack. Diagram is not drawn to scale

composition. As a result, virtually any BAC may be linearized through the incorporation of a *tos* site, followed by telomere resolution in a TelN-expressing host [9]. Linear BACs assembled using this method may be useful for functional expression of genomic loci in cells, maintenance of linear viral genomes in their natural conformation, and for constructing innovative artificial chromosome structures for applications in mammalian and plant cells.

2 Materials

2.1 Bacterial Strains and Plasmids

1. Plasmid pGETrec [11]: This plasmid is ampicillin resistant and contains the *recE*, *recT*, and *gam* genes that are under the control of an arabinose inducible promoter, P_{BAD} [14].
2. Target BAC: BAC4396 [13]. This 100-kb BAC contains the entire human beta-globin gene cluster on a pBeloBAC11 backbone [13].
3. Bacterial host for recombineering: *Escherichia coli* DH10B. DH10B is F⁻ *mcrA* Δ(*mrr-hsd*RMS-*mcr*BC) φ80dlacZΔM15 Δ*lac*X74 *deo*R *rec*A1 *end*A1 *ara*Δ139 Δ(*ara*, *leu*)7697 *gal*U *gal*K λ⁻ *rps*L *nup*G [15]. DH10B is routinely used as the recombineering host because it is commonly used for cloning and functional studies of large BACs [16, 17]. Therefore by directly performing recombineering in this 'working' strain there will not be a need to isolate and transform the newly recombineered BAC into a different strain for downstream experiments.
4. The tos-*Zeo* carrying strain *asd*⁻ *zeo*ʳ *tos*⁺ *cm*ʳ DH10B [9]: This strain serves as the template to amplify the 1.3 kb *tos*-Zeo PCR cassette used for recombineering into BACs (Fig. 1a). It carries a 519 bp N15 genomic sequence that includes the 310 bp *tos* (N15 positions 24,471–24,989 bp; Gen-Bank Accession No. NC_001901) [9] (*see* **Note 1**).
5. TelN-expressing host, *telN*⁺ DH10B (full genotype: *asd*⁻ *zeo*ʳ *telN*⁺ DH10B [9]): This strain is auxotrophic for diaminopimelic acid (DAP), and must be grown under zeocin selection and supplemented with 5 mM of DAP in nutrient-rich Brain Heart Infusion (BHI) broth or agar (*see* **Note 2**).

2.2 Enzymes

1. PCR enzyme pack, e.g., Long PCR Enzyme Mix from Thermo Scientific (*see* **Note 3**).
2. RecBCD (Exonuclease V): Epicentre RecBCD Enzyme Pack, including ATP.
3. Restriction enzymes and their respective buffers, e.g., *Dpn*I and SuRE/Cut buffer A from Roche.

2.3 DNA Isolation and Purification Kits

1. Plasmid extraction kit, e.g., QIAprep® Spin Miniprep Kit by Qiagen Inc.
2. DNA Gel Extraction Kit, e.g., QIAquick® Gel Extraction Kit by Qiagen Inc.
3. DNA extraction kit for linear DNA: NucleoBond® Xtra Midi/Maxi Kit (Macherey-Nagel Inc.) (*see* **Note 4**).

2.4 Primers

1. Primers for recombineering the tos-*Zeo* cassette into BAC4396:

   TosZeoF2: 5′-GAA ACC TGT CGT GCC AGC TGC ATT AAT GAA TCG GCC AAC GCG AAC CCC TTG CGG CCG CCC ttc ccc cgt ttc taa gtc tc-3′

   TosZeoR2: 5′-GCG GAT GAA TGG CAG AAA TTC GAT GAT AAG CTG TCA AAC ATG AGA ATT GGT CGA CGG CCC gca ata aac aag ttt cga gg-3′

   These primers have 60 nt sequence at the 5′ end (in upper case letters) that is homologous to a sequence centered around the *Sfi*I restriction site on the vector section of BAC4396, and a 20 nt sequence at the 3′ end (in lower case letters) that is complementary to the *tos*-Zeo region of the *asd*$^-$ *zeo*^{r} *tos*$^+$ *cm*^{r} DH10B strain (*see* Subheading 2.1) (*see* **Note 5**). Using the genomic DNA of *asd*$^-$ *zeo*^{r} *tos*$^+$ *cm*^{r} DH10B strain as the PCR template, the amplified *tos*-Zeo cassette will contain flanking sequences that are complementary to the BAC4396 (*see* **Note 6**).

2. Primers for screening BAC4396-tos clones that are generated after recombineering with *tos*-Zeo:

   Sfi1: 5′-aaa gtg taa agc ctg ggg tg-3′

   Sfi2: 5′-aac agt act gcg atg agt gg-3′

   These primers anneal just outside the recombineering target site on BAC4396, amplifying across the recombineering junctions. Recombinant BAC4396-tos clones will give a 1.5 kb product while non-recombinant BAC4396 clones without the 1.3 kb tos-*Zeo* cassette insertion will give a 300 bp product, enabling detection of the recombination event rapidly using PCR screening.

2.5 Antibiotics, Reagents, and Chemicals

1. Diaminopimelic acid (DAP), 100 mM: Dissolve 9.5 g DAP in 50 mL of distilled water in a clean beaker. Heat solution up to 50 °C while stirring. When powder has completely dissolved, cool to room temperature before filter-sterilizing.
2. Antibiotics:

No.	Selective antibiotic	Working concentration (μg/mL)	Diluent	Preparation
1	Ampicillin	100	Water	Filter-sterilized
2	Chloramphenicol	12.5	Ethanol	Not sterilized
3	Streptomycin	20	Water	Filter-sterilized
4	Zeocin	25	Water	Purchased from Invitrogen Inc. and used directly

3. Brain Heart Infusion (BHI) media, e.g., BHI agar or broth from BD Inc.
4. TE buffer for dissolving and eluting DNA: 10 mM Tris–HCl, pH 8.0, 1 mM EDTA.
5. 10 % (w/v) L-(+)-arabinose for inducing pGETrec: Dissolve 5 g of dry L-(+)-arabinose in 50 mL of sterile water, filter with 0.2 μm syringe filter under sterile condition. Prepare freshly before recombineering. Keep at 4 °C until used.
6. Ice-cold 10 % (v/v) glycerol for preparing electrocompetent cells: Mix 100 mL of glycerol with 900 mL of water and autoclave. Cool at 4 °C overnight, then chill at −20 °C for 1–2 h before use (*see* **Note** 7).
7. 3 M Sodium acetate for DNA precipitation, pH 5.2: Dissolve 24.6 g of sodium acetate anhydrous in 80 mL of distilled water. Adjust pH to 5.2 with glacial acetic acid. Top up solution to 100 mL final volume and autoclave before use (*see* **Note 8**).
8. Absolute ethanol (100 % v/v) for DNA precipitation: Chill at −20 °C at least an hour before use.
9. 70 % ethanol (v/v) for DNA precipitation: Mix 700 mL of ethanol with 300 mL of distilled water.
10. Isopropanol for DNA extraction: Use at room temperature.

2.6 Specialized Equipment and Consumables

1. Electroporation unit: Gene Pulser Xcell Microbial System (Bio-Rad Inc.).
2. Gene Pulser® 0.2 cm electrode gap cuvette (Bio-Rad Inc.). Cool at −20 °C about 1 h before use (*see* **Note 9**).
3. Pulsed field gel electrophoresis (PFGE) system, e.g., CHEF-DR III System (Bio-Rad Inc.).

3 Methods

3.1 PCR Amplifying and Purifying the tos-Zeo Cassette

The main objective of this set of PCR is to amplify the *tos* sequence as part of a *tos*-Zeo cassette, and insert it into a chosen BAC clone (Fig. 1a). High fidelity DNA polymerase with proofreading activity should be used in order to ensure accurate amplification of the targeting cassette.

1. Grow strain *asd*$^-$ *zeo*^{r} *tos*$^+$ *cm*^{r} DH10B [9] in 5 mL of BHI broth (BHI is updated in Subheading 2.5) supplemented with 5 mM of DAP and zeocin at 30 °C with 150 rpm shaking for 30–48 h.
2. Use 1 μL aliquot of the *asd*$^-$ *zeo*^{r} *tos*$^+$ *cm*^{r} DH10B culture as template to obtain DNA. Prepare the following mixture in a PCR tube for each reaction (*see* **Note 10**).

No.	Component	Volume (μL)
1	10× Long PCR Buffer with 15 mM $MgCl_2$	5
2	10 mM dNTPs (Thermo Scientific)	1
3	TosZeoF2 Primer (0.5 μg/μL)	1
4	TosZeoR2 Primer (0.5 μg/μL)	1
5	*asd⁻ zeo^r tos^+ cm^r* DH10B culture	1
6	Long PCR Enzyme Mix (5 U/μL)	0.5
7	Nuclease-free water	40.5
	Total	50

3. Subject the mixture to the following cycling condition:

Step	Temperature (°C)	Duration (s)	
1	95	120	
2	95	30	35 cycles
3	45	30	
4	68	120	
5	68	600	
6	4	Forever	

4. Electrophorese the PCR product in a 1 % (w/v) agarose gel (*see* **Note 11**).
5. Stain the gel with ethidium bromide or other nucleic acid stains, and visualize gel with a UV light box to confirm that a 1.3 kb band is present (*see* **Note 12**). Carefully excise the 1.3 kb band into clean microcentrifuge tubes with a clean, sharp scalpel.
6. Extract the DNA from the excised agarose using a gel extraction kit according to the manufacturer's instructions, e.g., QIAquick® Gel Extraction Kit by Qiagen Inc.
7. If using QIAquick® Gel Extraction Kit by Qiagen Inc., weigh the gel slices into a 2 mL microcentrifuge tube. By putting maximum 400 mg of gel into each tube, add 3 volumes of Buffer QG to the gel e.g., for every 100 mg of gel 300 μL of Buffer QG is added.
8. Incubate the gel and QG buffer mixture at 50 °C for 10 min or until gel has completely dissolved. Vortex the mixture every 2–3 min to help dissolve the gel.
9. Add 1 volume of isopropanol to the mixture, then mix by inverting the tube.
10. Apply the mixture from each tube into one QIAquick column, then centrifuge at 15,000 × *g* for 1 min at room temperature. Discard flow-through.

11. Add 500 μL of Buffer QG to column and repeat centrifuge at 15,000 × *g* for 1 min at room temperature. Discard flow-through.
12. Add 750 μL of Buffer PE to column. Incubate column for 2–5 min, then centrifuge at 15,000 × *g* for 1 min at room temperature.
13. Discard flow-through, then repeat centrifuge at 15,000 × *g* for 1 min at room temperature to remove residual Buffer PE.
14. To elute the PCR product, transfer column to a clean collection tube, then add 50 μL of distilled water (pH 8.5) into the center of the column. Let the column stand for 1 min before centrifuging at 15,000 × *g* for 1 min at room temperature.
15. To remove potential contaminating bacterial genomic DNA of the *asd*$^-$ *zeo*^{r} *tos*$^+$ *cm*^{r} DH10B strain, the purified PCR product is digested with *Dpn*I. *Dpn*I is able to specifically digest methylated bacterial DNA while leaving the newly synthesized PCR product intact, which is unmethylated. Set up the *Dpn*I digestion as follows:

No.	Component	Volume (μL)
1	10× SuRE/Cut buffer A	6
2	Purified PCR product in water	50
3	*Dpn*I (10 U/μL)	3
4	Nuclease-free water	1
	Total	60

16. Precipitate the PCR product by mixing the *DpnI-digested* DNA with 50 μL of distilled water (pH 8.5), 10 μL of 3 M sodium acetate (pH 5.2), and 250 μL of −20 °C absolute ethanol.
17. Shake the mixture vigorously and incubate at −80 °C for 30 min.
18. Centrifuge at 15,000 × *g* for 15 min at room temperature.
19. Wash the pellet with 250 μL of 70 % ethanol and centrifuge at 15,000 × *g* for 5 min at room temperature.
20. Remove the remaining supernatant with a micropipette tip and let the pellet air-dry for 1–3 min.
21. Resuspend the pellet in 5–10 μL of TE buffer.

3.2 Transferring the Recombineering Plasmid pGETrec into DH10B (BAC4396)

Transforming plasmid pGETrec into an *E. coli* strain introduces the recombineering machinery into this host [11]. Therefore, in order to introduce recombineering ability into DH10B (BAC4396), i.e., DH10B strain carrying the human beta-globin

BAC4396, pGETrec is transformed into this host. This step generates strain DH10B (pGETrec, BAC4396), where both plasmid pGETrec and BAC4396 coexist in a host that is proficient to carry out recombineering (Fig. 1b) (*see* **Note 13**).

Perform all procedures under sterile condition unless otherwise specified. Keep cells cool on ice at all times. Precool all plasticware (centrifuge bottles, microcentrifuge tubes, pipette tips, serological pipettes, etc.) and glassware at −20 °C for a few hours. These materials should be kept at −20 °C up until immediately before they come into contact with bacterial cells (*see* **Note 14**).

1. Grow a 5 mL overnight starter culture of DH10B (BAC4396) in LB at 220 rpm and 37 °C, supplemented with chloramphenicol.
2. Inoculate a fresh 250 mL of LB broth containing chloramphenicol with the DH10B (BAC4396) starter culture.
3. Grow the culture in a shaker at 220 rpm and 37 °C until it reaches exponential stage, specifically at OD_{600} of 0.55–0.65 (*see* **Note 15**).
4. Harvest the cells by aliquoting the culture into a precooled sterile centrifuge bottle.
5. Incubate the culture at −20 °C for 20 min, swirling the bottle gently at 3-min intervals.
6. Centrifuge the bottle at 6,800 × *g*, at −2 °C, for 12 min.
7. Sit the bottle on ice, and carefully pour off the broth, leaving the pellet intact at the bottom of the bottle.
8. Gently, but thoroughly, resuspend the cell pellet in 250 mL of ice-cold 10 % (v/v) glycerol using a precooled 10 mL serological pipette (*see* **Note 16**).
9. Centrifuge the resuspended culture at 6,800 × *g*, at −2 °C, for 12 min.
10. Repeat the glycerol washing steps with 250 mL of ice-cold 10 % (v/v) glycerol (*see* **steps 7–9**) twice.
11. Carefully pour off all supernatant, leaving the pellet at the bottom.
12. Leave the bottle upright on ice for 2–3 min so that any residual 10 % (v/v) glycerol from the sides is collected at the bottom.
13. Resuspend the pellet in the residual glycerol using a precooled 1 mL micropipette tip with a wide bore (*see* **Note 17**).
14. Collect the entire volume of resuspended cells into one chilled microcentrifuge tube before dispensing 45 μL of aliquots each into individual chilled microcentrifuge tubes (*see* **Note 18**).
15. Using a precooled pipette tip, add 1–2 μL (approximately 0.01–0.05 μg) of pGETrec plasmid into a tube containing

45 μL of the DH10B (BAC4396) electrocompetent cells. Transfer the cells and DNA mixture into a prechilled 0.2 cm electrode gap cuvette; keep cuvette on ice until it is ready for electroporation.

16. To estimate transformation efficiency, perform a control transformation reaction by adding a clean preparation of 10 pg of pUC19 control plasmid DNA (e.g., obtained from Invitrogen Inc.) into one tube of electrocompetent cells (*see* **Note 19**).
17. Transfer the cells and control plasmid mixture into a prechilled 0.2 cm electrode gap cuvette; keep cuvette on ice until it is ready for electroporation.
18. Just before electroporation, firmly tap the cuvette on a flat surface, e.g., the lab bench twice to bring all cells down to the bottom.
19. Wipe the cuvette dry before placing it into the electroporation cuvette chamber (*see* **Note 20**).
20. Set the electroporation unit to the following setting: 2.5 kV, 200 Ω, and 25 μF.
21. Press the pulse button located on the pulse controller unit. As soon as a beep sounds, release the pulse button (*see* **Note 21**).
22. Immediately remove the cuvette from chamber and add 1 mL of LB broth without any antibiotic (*see* **Note 22**).
23. Using a glass Pasteur pipette transfer the cells from the cuvette into a 14 mL round bottom snap-cap tube (*see* **Note 23**).
24. Grow the cells for 60 min at 220 rpm, 37 °C. This will help the cells recover from the electroporation treatment.
25. At the end of the recovery incubation, serially dilute transformed cells and plate them onto LB agar plates containing chloramphenicol and ampicillin.
26. Incubate plates at 37 °C for overnight.
27. Pick 15–20 *zeo*^{r} *amp*^{r} colonies [i.e., putative DH10B (pGET-rec, BAC4396) colonies] onto a single LB agar plate containing chloramphenicol and ampicillin with a sterile tooth pick, and grow the plate at 37 °C overnight.
28. Pick and grow each colony from the plate in LB broth, supplemented with chloramphenicol and ampicillin, at 220 rpm, 37 °C, overnight (*see* **Note 24**).

3.3 Recombineering tos-Zeo Cassette into BAC4396

The main objective of this recombineering step is to introduce the *tos* sequence into the circular BAC4396 by insertion of the *tos*-Zeo cassette (Fig. 1b). The *tos*-Zeo cassette is electroporated into DH10B (pGETrec, BAC4396) (*see* Subheading 3.2) and, once plasmid pGETrec is induced to express the recombineering genes, will insert into the target site on BAC4396 (Fig. 1b).

All procedures listed here should be performed under sterile condition unless otherwise specified. Keep cells cool on ice at all times. Precool all plasticware (centrifuge bottles, microcentrifuge tubes, pipette tips, serological pipettes, etc.) and glassware at −20 °C for at least a few hours until immediately before they come into contact with bacterial cells (*see* **Note 14**).

1. Grow a 5 mL starter culture of DH10B (pGETrec, BAC4396) in LB at 150 rpm and 30 °C for 30–48 h, supplemented with ampicillin and chloramphenicol (*see* **Notes 1** and **25**).
2. Inoculate a fresh 250 mL of LB broth containing ampicillin and chloramphenicol with the starter culture of DH10B (pGETrec, BAC4396) culture.
3. Grow the culture in a shaker at 150 rpm and 30 °C until it reaches exponential stage, specifically at OD_{600} of 0.45–0.50, before the culture is induced in the next step (*see* **Note 26**).
4. To induce the expression of the recombineering genes *recE*, *recT*, and *gam* on plasmid pGETrec, supplement the culture with L-(+)-arabinose to a final concentration of 0.2 % (w/v) under sterile condition (*see* **Note 27**).
5. Return the culture to grow at 150 rpm and 30 °C, until OD_{600} reaches between 0.55 and 0.65 (*see* **Note 28**).
6. Harvest cells by aliquoting the culture into a precooled sterile centrifuge bottle.
7. Incubate the culture at −20 °C for 20 min, swirling the bottle gently at 3-min intervals.
8. Centrifuge the bottle at 6,800 × *g*, at −2 °C, for 12 min.
9. Sit the bottle on ice, and carefully pour off the broth, leaving the pellet intact at the bottom of the bottle.
10. Gently but thoroughly resuspend the cell pellet in 250 mL of ice-cold 10 % (v/v) glycerol using a precooled 10 mL serological pipette.
11. Centrifuge the resuspended culture at 6,800 × *g*, at −2 °C, for 12 min.
12. Repeat the glycerol washing steps with 250 mL of ice-cold 10 % (v/v) glycerol (*see* **steps 9–11**) twice.
13. Sit the bottle on ice, and carefully remove all supernatant with a precooled serological pipette.
14. Leave the bottle upright on ice for 2–3 min, so that residual 10 % (v/v) glycerol is collected at the bottom.
15. Resuspend the pellet in the residual glycerol using a precooled 1 mL micropipette tip with a wide bore.
16. Collect the entire volume of resuspended cells into one chilled microcentrifuge tube, before dispensing 45 μL of aliquots each into individual chilled microcentrifuge tubes.

17. Using a precooled pipette tip, add 1–2 μL of precooled DNA in TE buffer containing approximately 0.3–1.0 μg of purified *tos*-Zeo PCR cassette (*see* Subheading 3.1) into a tube of the DH10B (pGETrec, BAC4396) electrocompetent cells. Transfer the cells and DNA mixture into a prechilled 0.2 cm electrode gap cuvette; keep cuvette on ice until it is ready for electroporation.
18. To estimate transformation efficiency, perform a control transformation reaction by adding a clean preparation of 10 pg of pZeoSV2 (+) control plasmid DNA into one tube of electrocompetent cells (*see* **Notes 29** and **30**).
19. Transfer the cells and DNA mixture into a prechilled 0.2 cm electrode gap cuvette; keep cuvette on ice until it is ready for electroporation.
20. Just before electroporation, firmly tap the cuvette on a flat surface, e.g., the lab bench twice to bring all cells down to the bottom.
21. Wipe the cuvette dry before placing it into the electroporation cuvette chamber.
22. Set the electroporation unit to the following setting: 2.5 kV, 200 Ω, and 25 μF.
23. Press the pulse button located on the pulse controller unit. As soon as a beep sounds, release the pulse button (*see* **Note 21**).
24. Immediately remove the cuvette from chamber and add 1 mL of BHI broth without any antibiotic added.
25. Using a glass Pasteur pipette transfer cells from the cuvette into a 14 mL round bottom snap-cap tube, and grow them for 120 min at 150 rpm, 30 °C. This will help the cells recover from the electroporation treatment (*see* **Note 31**).
26. At the end of recovery incubation, serially dilute transformed cells and plate onto LB agar plates containing chloramphenicol and zeocin (*see* **Note 32**).
27. Incubate plates at 30 °C for at least 20 h.
28. Pick 15–20 *zeo*^{r} *cm*^{r} colonies [i.e., putative DH10B (BAC4396-tos) colonies], spot them onto a single LB agar plate containing chloramphenicol and zeocin with a sterile tooth pick and grow the plate at 30 °C for 24–48 h.
29. Pick and grow each colony from the plate in LB broth, supplemented with chloramphenicol and zeocin, at 150 rpm, 30 °C for 30–48 h (*see* **Note 24**).

3.4 PCR Screening of BAC-tos

Perform PCR screening using primers Sfi1 and Sfi2 (Fig. 1c) on the cultures grown from Subheading 3.3, **step 29** to identify correctly inserted *tos*-Zeo in individual *zeo*^{r} *cm*^{r} colonies, i.e., putative DH10B (BAC4396-tos) clones. Set up the PCR according to the instructions provided by the enzyme manufacturer; optimization may be necessary for a different enzyme (*see* **Note 33**).

1. Grow 5 mL each of the *zeo*^{r} *cm*^{r} colonies at 150 rpm at 30 °C for 30–48 h in LB broth supplemented with chloramphenicol and zeocin.
2. Mix the following components in a PCR tube for each reaction:

No.	Component	Volume (μL)
1	10× Long PCR Buffer with 15 mM $MgCl_2$	5
2	10 mM dNTPs (Thermo Scientific)	1
3	Sfi1 Primer (0.5 μg/μL)	1
4	Sfi2 Primer (0.5 μg/μL)	1
5	DH10B (BAC4396-tos) culture	1
6	Long PCR Enzyme Mix (5 U/μL)	0.5
7	Nuclease-free water	40.5
	Total	50

3. Subject the reaction to the following cycling condition:

Step	Temperature (°C)	Duration (s)	
1	95	120	
2	95	30	30 cycles
3	55	30	
4	68	120	
5	68	600	
6	4	Forever	

4. Electrophorese the resulting PCR product in a 1 % (w/v) agarose gel. Positive recombinants with the *tos*-Zeo cassette correctly integrated, i.e., BAC4396-tos, will produce a 1.5 kb band, while non-recombinants without *tos*-Zeo cassette integration (or with *tos*-Zeo integration at other unintended sites) will produce a 300 bp product (Fig. 1c) (*see* **Note 34**).

3.5 Separating Recombinant BAC-tos from pGETrec Plasmid

In this section, the newly recombineered BAC4396-tos, which is generated in Subheading 3.3 and validated in Subheading 3.4, is isolated from the recombineering host and re-transformed into freshly made DH10B competent cells (*see* **Note 32**).

1. Purify circular BAC4396-tos from DH10B (BAC4396-tos) clones (Subheading 3.4), which have been previously verified

by sequencing, using standard plasmid extraction kit according to manufacturer's instructions, e.g., QIAprep Spin Miniprep Kit by Qiagen Inc.

2. If using QIAprep Spin Miniprep Kit, pellet down DH10B (BAC4396-tos) which has been grown in 5 mL of LB broth for 20–24 h at 30 °C and 150 rpm by centrifuging at 6,800 × *g*, room temperature for 5 min. Remove all supernatant by pipetting.
3. Resuspend the pellet in 250 μL of Buffer P1 containing RNase A, then transfer cell suspension to a microcentrifuge tube.
4. Add 250 μL of Buffer P2, invert tube to mix suspension thoroughly.
5. Add 350 μL of Buffer N3, then immediately invert tube to mix.
6. Centrifuge sample for 10 min at 15,000 × *g*, room temperature for 10 min.
7. Carefully pipette out supernatant from tube without removing the white debris, then load into a QIAprep column.
8. Wash the column by adding 500 μL of Buffer PB and centrifuging at 15,000 × *g*, room temperature for 1 min. Discard flow-through.
9. Add 750 μL of Buffer PE into column and centrifuge at 15,000 × *g*, room temperature for 1 min. Discard flow-through.
10. Repeat centrifugation at 15,000 × *g*, room temperature for 1 min to remove residual Buffer PE.
11. To elute the DNA, transfer column to a clean collection tube, then add 50 μL of TE buffer to the column. Let column stand for 1 min, then centrifuge at 15,000 × *g*, room temperature for 1 min.
12. Digest the eluted DNA with *Dpn*I, followed by precipitation as described in Subheading 3.1, **steps 15–21**.
13. Quantify the precipitated BAC4396-tos DNA by spectrophotometer, and set up a restriction digestion reaction with a small DNA aliquot:

No.	Component	Volume (μL)
1	10× SuRE/Cut buffer A	2
2	~100 ng BAC4396-tos, diluted in water	17
3	*Nar*I (10 U/μL)	1
	Total	20

14. Visualize the digested BAC4396-tos by pulsed-field electrophoresis (PFGE) to verify the intactness of the isolated BAC4396-tos.
15. Grow a 5 mL starter culture of DH10B in LB at 150 rpm for 30–48 h at 30 °C, supplemented with streptomycin.
16. Inoculate a fresh 250 mL of LB broth containing streptomycin with the DH10B starter culture.
17. Grow the culture in a 30 °C shaker at 150 rpm until it reaches an OD_{600} of 0.55–0.65 (*see* **Note 15**).
18. Harvest cells by aliquoting the culture into a precooled sterile centrifuge bottle.
19. Incubate the culture at −20 °C for 20 min, swirling the bottle gently at 3-min intervals.
20. Centrifuge the bottle at 6,800 × *g*, −2 °C, for 12 min.
21. Sit the bottle on ice, and carefully remove the broth, leaving the pellet intact at the bottom of the bottle.
22. Gently but thoroughly resuspend the cell pellet in 250 mL of ice-cold 10 % (v/v) glycerol using a precooled 10 mL serological pipette.
23. Centrifuge the resuspended culture at 6,800 × *g*, at −2 °C, for 12 min.
24. Repeat the glycerol washing steps with 250 mL of ice-cold 10 % (v/v) glycerol (*see* **steps 21–23**) twice.
25. Sit the bottle on ice, and carefully remove all supernatant with a precooled serological pipette.
26. Leave the bottle upright on ice for 2–3 min, so that residual 10 % (v/v) glycerol collects at the bottom.
27. Resuspend the pellet in the residual glycerol using a precooled 1 mL micropipette tip with a wide bore.
28. Collect the entire volume of resuspended cells into one chilled microcentrifuge tube, before dispensing 45 μL of aliquots each into individual chilled microcentrifuge tubes.
29. Using a precooled wide bore pipette tip, add 1–2 μL of BAC4396-tos DNA (100–200 ng) in TE buffer into a tube containing a 45 μL aliquot of DH10B electrocompetent cells. Transfer the cells and DNA mixture into a prechilled 0.2 cm electrode gap cuvette; keep cuvette on ice until it is ready for electroporation.
30. Perform a control transformation reaction by adding a clean preparation of 10 pg of pZeoSV2 (+) control plasmid DNA into one tube of electrocompetent cells.
31. Transfer the cells and control DNA mixture into a prechilled 0.2 cm electrode gap cuvette; keep cuvette on ice until it is ready for electroporation.

32. Just before electroporation, firmly tap the cuvette on a flat surface, e.g., the lab bench twice to bring all cells down to the bottom.
33. Wipe the cuvette dry before placing it into the electroporation cuvette chamber.
34. Set the electroporation unit to the following setting: 2.5 kV, 200 Ω and 25 μF.
35. Press the pulse button located on the pulse controller unit. As soon as a beep sounds, release the pulse button (*see* **Note 21**).
36. Immediately remove the cuvette from chamber and add 1 mL of LB broth without adding any antibiotic.
37. Using a glass Pasteur pipette transfer cells from the cuvette into a 14 mL round bottom snap-cap tube, and grow them for 120 min at 150 rpm and 30 °C to help the cells recover from the electroporation treatment.
38. At the end of recovery incubation, serially dilute transformed cells and plate onto LB agar plates containing chloramphenicol and zeocin.
39. Incubate plates at 30 °C for 16–48 h.
40. Pick 15–20 *zeo*^{r} *cm*^{r} colonies [i.e., putative DH10B (BAC4396-tos) colonies] onto a single LB agar plate containing chloramphenicol and zeocin with a sterile tooth pick and grow the plate at 30 °C for 24–48 h.
41. Pick and grow each colony from the plate in LB broth, supplemented with chloramphenicol and zeocin, at 150 rpm, 30 °C for 36–48 h (*see* **Note 24**).

3.6 In Vivo Linearization of BAC-tos

In this section circular BAC4396-tos DNA is introduced into a TelN-expressing *E. coli* host cell (*telN*$^+$ *E. coli*) (Fig. 1d) to resolve the *tos* sequence into telomeres in vivo, linearizing the BAC. All steps prior to electroporation are performed at room temperature unless otherwise stated. During electroporation, care must be taken to keep the cells chilled as much as possible to maximize their viability and thence transformation efficiency.

1. Isolate circular BAC4396-tos from PCR verified DH10B (BAC4396-tos) clones with standard plasmid extraction kit, e.g., QIAprep Spin eluting the DNA in 50 μL of TE buffer (*see* Subheading 3.5, **steps 2–11**). Run a diluted sample by PFGE to confirm identity and integrity of isolated DNA (*see* **Note 35**).
2. Once verified, precipitate the DNA. Mix eluted DNA with 50 μL of distilled water (pH 8.5), 10 μL of 3 M sodium acetate (pH 5.2) and 250 μL of −20 °C absolute ethanol.

3. Shake the mixture vigorously and incubate at −80 °C for 30 min (*see* **Note 36**).
4. Centrifuge at 15,000 × *g* for 15 min at room temperature.
5. Wash the pellet with 250 μL of 70 % ethanol and centrifuge at 15,000 × *g* for 5 min at room temperature.
6. Remove the remaining supernatant with a micropipette tip and let the pellet air-dry for 5–10 min.
7. Resuspend the pellet in 5–10 μL of TE buffer.
8. Evaluate the quality of precipitated DNA by PFGE.
9. Grow the *telN*+ DH10B strain in 5 mL of LB broth supplemented with streptomycin, zeocin, and 5 mM of DAP at 30 °C with 150 rpm shaking for 30–48 h.
10. Dilute the 5 mL *telN*+ DH10B culture into 250 mL of LB broth containing streptomycin, zeocin and 5 mM of DAP.
11. Grow the diluted culture in a 30 °C shaker at 150 rpm until the culture reaches an OD_{600} of 0.55–0.65 (*see* **Note 15**).
12. Harvest the cells by aliquoting the culture into a precooled sterile centrifuge bottle.
13. Incubate the culture at −20 °C for 20 min, swirling the bottle gently at 3-min intervals.
14. Centrifuge the bottle at 6,800 × *g*, at −2 °C, for 12 min.
15. Sit the bottle on ice, and carefully remove the broth, leaving the pellet intact at the bottom of the bottle.
16. Gently but thoroughly resuspend the cell pellet in 250 mL of ice-cold 10 % (v/v) glycerol using a precooled 10 mL serological pipette.
17. Centrifuge the resuspended culture at 6,800 × *g*, at −2 °C, for 12 min.
18. Repeat the glycerol washing steps with 250 mL of ice-cold 10 % (v/v) glycerol (*see* **steps 15–17**) twice.
19. Sit the bottle on ice and carefully remove all supernatant with a precooled serological pipette.
20. Leave the bottle upright on ice for 2–3 min, so that residual 10 % (v/v) glycerol is collected at the bottom.
21. Resuspend the pellet in the residual glycerol using a precooled 1 mL micropipette tip with a wide bore.
22. Collect the entire volume of resuspended cells into one chilled microcentrifuge tube, before dispensing 45 μL aliquots each into individual chilled microcentrifuge tubes.
23. Electroporate the *telN*+ DH10B cells with 0.1–0.5 μg of the precipitated BAC4396-tos DNA. For control, electroporate

10 pg of pUC19 plasmid following the protocol described in Subheading 3.2 (*see* **steps 15–23**).

24. Grow the cells at 30 °C, 150 rpm for 120 min for recovery.
25. At the end of recovery growth, serially dilute cells and plate them onto BHI agar plates containing chloramphenicol, streptomycin, zeocin and 5 mM of DAP. Incubate plates at 30 °C for 24–48 h.
26. Pick individual transformed *telN*+ DH10B (BAC4396-tos) colonies onto a fresh BHI agar plate containing chloramphenicol, zeocin and 5 mM of DAP with a sterile tooth pick and incubate at 30 °C until colonies arise (30–48 h).
27. Grow each clone in BHI broth supplemented with chloramphenicol, zeocin and 5 mM of DAP at 150 rpm, 30 °C, for 30–48 h until they reach late-log growth (*see* **Notes 1** and **37**).

3.7 Extraction of Linear BAC-tos

Linear BAC-*tos* is extracted from *telN*+ DH10B (BAC4396-tos) to check for the identity of the newly recombineered BAC-*tos* and to obtain purified linear DNA for downstream applications (Fig. 1e). To obtain large quantity of intact linear BAC4369-tos DNA, we use the Macherey-Nagel's NucleoBond® Xtra Midi or Maxi Kit (*see* **Note 38**).

The procedures described below are based on the protocol provided by the manufacturer for purification of low-copy plasmid using NucleoBond® Xtra Midi Kit, with minor modifications. Perform all steps at room temperature unless otherwise specified.

1. Grow 5 mL of *telN*+ DH10B (BAC4396-tos) starter culture in LB broth containing chloramphenicol, zeocin and 5 mM of DAP for 36–48 h at 30 °C with 150 rpm shaking.
2. Dilute the 5 mL starter culture into 200 mL of LB broth supplemented with chloramphenicol, zeocin and 5 mM of DAP, and grow for 36–48 h at 30 °C with 150 rpm shaking.
3. Warm the Elution Buffer ELU to 50 °C.
4. Harvest the cultures into a centrifuge bottle and centrifuge at 6,000 × *g*, 4 °C for 15 min.
5. Carefully pour off the supernatant and completely resuspend the pellet in 16 mL of Resuspension Buffer RES + RNase A with a serological pipette.
6. Add in 16 mL of Lysis Buffer LYS and invert the bottle gently five times (*see* **Note 39**). Incubate at room temperature for 5 min.
7. While the suspension is incubating, equilibrate a NucleoBond® Xtra Column and its inserted column filter by directly pipetting 12 mL of Equilibration Buffer EQU onto the rim of the column filter, making sure that the entire column is wet. Let the column empty by gravity flow (*see* **Note 40**).

8. Add 16 mL of Neutralization Buffer NEU to the suspension. Immediately invert the bottle 10–15 times (*see* **Note 41**).
9. Gently invert the suspension again three times, and without delay carefully pour the content into the equilibrated column. Continue to refill the column until all of the suspension has been loaded. Let the column empty by gravity flow.
10. Wash the column by directly pipetting 5 mL of Equilibration Buffer EQU onto the rim of the column filter. Let the column empty by gravity flow.
11. Remove and discard the column filter (*see* **Note 42**).
12. Wash the column by adding 8 mL of Wash Buffer WASH into the column. Let the column empty by gravity flow.
13. Remove the flow-through collection tray and place a sterile centrifuge tube of 10 mL or bigger capacity directly underneath the column (*see* **Note 43**).
14. Elute the BAC DNA by adding 5 mL of pre-warmed Elution Buffer ELU (*see* **step 3**), 1 mL at a time, into the column (*see* **Note 44**). The eluate should flow directly into the centrifuge tube.
15. Add 3.5 mL of room-temperature isopropanol to the eluted DNA. Secure the tube tightly and vortex vigorously for 10 s. Incubate the mixture at room temperature for 2 min.
16. Centrifuge the mixture at 17,000 × *g* for 45 min at room temperature (*see* **Note 45**).
17. Carefully remove the supernatant with a pipette tip while avoiding the pellet (*see* **Note 46**).
18. Wash the pellet with 2 mL of room-temperature 70 % (v/v) ethanol. Centrifuge at 17,000 × *g* for 5 min at room temperature.
19. Carefully remove the supernatant with a pipette tip and briefly centrifuge for a few seconds again to bring down the residual ethanol. Remove the last remaining residual ethanol with a micropipette tip.
20. Dry the pellet at room temperature for 5–10 min (*see* **Note 47**).
21. Gently resuspend the pellet in 100–150 μL of TE buffer, and incubate in a 37 °C water bath with constant agitation for 30–60 min to dissolve the pellet (*see* **Note 48**).
22. Determine the DNA yield using a spectrophotometer and check the quality by electrophoresing a small aliquot.
23. Subject an aliquot of the DNA to restriction digestion and analyze the restriction patterns by PFGE (Fig. 2). For single-site digestion, e.g., by *Nar*I (Fig. 2a), two bands are observed for linearized BAC4396-tos (lanes 5–7), while only one band will be present for circular BAC4396-tos (lane 8). For restriction

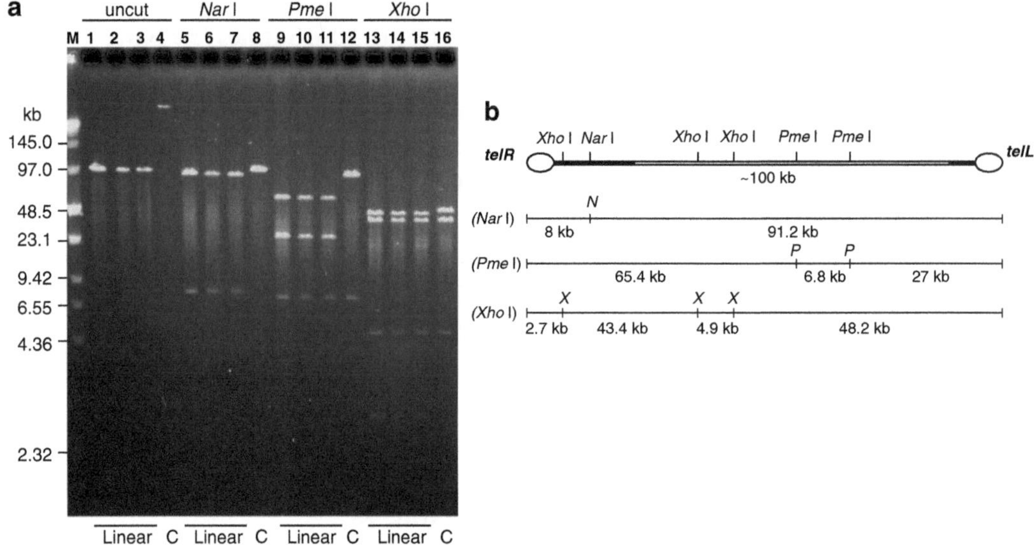

Fig. 2 In vivo linearization of BAC-*tos*. (**a**) Pulsed-field gel electrophoresis analysis of three independent clones of the 100 kb BAC4396-tos DNA, linearized in vivo, compared with the parent circular BAC4396-tos. *Lane M*, low range PFG marker (New England Biolabs Inc.); *lanes 1–3*, uncut TelN-linearized 100 kb BAC4396-tos; *lanes 5–7*, *9–11*, and *13–15* show TelN-linearized BAC4396-tos digested with *Nar*I, *Pme*I, and *Xho*I, respectively. *Lane 4*, uncut circular 100 kb BAC4396-tos; *lanes 8*, *12*, and *16*, circular 100 kb BAC4396-tos digested with *Nar*I, *Pme*I, and *Xho*I, respectively. The conformation of the DNA is indicated as either Linear or circular (labeled 'C'). (**b**) Predicted *Nar*I, *Pme*I, and *Xho*I restriction fragments for linear BAC4396-tos. Both **a** and **b** are reprinted from [9] with permission from Elsevier

digestions with more than one cut site, e.g., by *Pme*I and *Xho*I (Fig. 2b), an additional band is observed for linearized BAC4396-tos (lanes 9–11, and lanes 13–15) compared to its circular counterpart (lanes 12 and 16), hence verifying the linear conformation of BAC4396-tos.

3.8 Detecting the Presence of Telomeres on Linear BAC-tos

To detect the presence of *telL* and *telR* telomeres on the linear BAC4396-tos, a RecBCD assay is performed [9] (Fig. 1f). The presence of telomeres will enable BAC4396-tos to resist RecBCD enzyme digestion and remain intact (*see* **Note 49**). Perform all steps at room temperature unless otherwise stated.

1. Mix 300 ng of linear BAC4396-tos, diluted in 20.5 μL of distilled water, with 2.5 μL of 10× Reaction Buffer, 1 μL of 25 mM ATP and 1 μL of 10 U/μL RecBCD.
2. Incubate the mixture at 37 °C for 45 min with gentle mixing every 15–20 min.
3. Perform the same reaction on intact circular BAC4396-tos (positive control for RecBCD resistance) and restriction enzyme-linearized circular BAC4396-tos (positive control for RecBCD exonuclease activity).

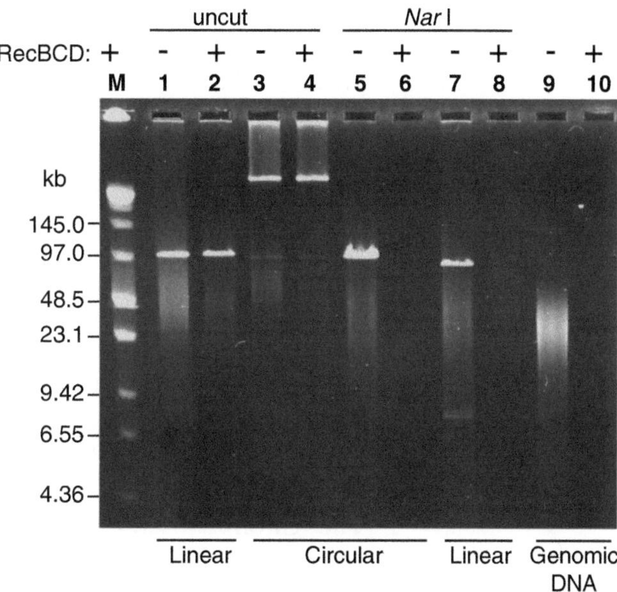

Fig. 3 Detection of hairpin telomeres on BAC-*tos* ends by RecBCD digestion. Linear BAC4396-tos with telomeres is resistant to RecBCD exonuclease digestion. Linear BAC4396-tos is treated in vitro with RecBCD and resolved using pulsed-field gel electrophoresis. *Lane M*, low range PFG marker (New England Biolabs Inc.); *lanes 1–2*, uncut TelN-linearized BAC4396-tos; *lanes 3–4*, uncut circular BAC4396-tos; *lanes 5–6*, circular BAC4396-tos digested with *NarI*; *lanes 7–8*, TelN-linearized BAC4396-tos digested with *NarI*; *lanes 9–10*, total genomic DNA from HT1080 cells. Lanes marked '+' denote DNA treated with RecBCD exonuclease, while lanes marked '–' denote control DNA not treated with RecBCD exonuclease prior to electrophoresis. Reprinted from [9] with permission from Elsevier

4. Examine the samples by PFGE (Fig. 3). BAC4396-tos linearized by TelN will be resistant to RecBCD attack and travel as an intact 100 kb band (lane 2), demonstrating presence of the protective telomeres. Circular BAC4396-tos will similarly be resistant due to the absence of open ends, and travel as a supercoiled band (lane 4). When the same circular and linear BAC4396-tos are first linearized by restriction enzyme *NarI* to expose their DNA ends (lanes 5 and 7), both samples will be destroyed by subsequent RecBCD digestion (lanes 6 and 8).

4 Notes

1. This *asd*$^{-}$ *zeo*^{r} *tos*$^{+}$ *cm*^{r} DH10B strain as well as other strains containing *tos* DNA or that will be inserted with *tos* DNA, should be grown at less vigorous conditions in order to prevent rearrangement of the N15 telomeric sequence [9, 18].

All cultures on plates should be incubated at 30 °C for at least 20–24 h. The exact timeline can be judged by looking for the presence of well grown colonies on the plate. All liquid culture should be grown at 30 °C with shaking at 150 rpm for 30–48 h. For liquid culture, it is important to determine the exact time it needs to grow until it reaches late log phase. An OD_{600} reading should be taken at the end of the growth in order to ensure the culture density is at the late log phase. For the DH10B host, we always aim for OD_{600} of 1.0–1.2. This is the reading we commonly get at the late log phase for this strain. We recommend taking the OD_{600} reading when the strains are grown for the first time in the lab and ensuring they are given enough time to reach a value of 1.0–1.2. Once this is worked out, all subsequent growth of the strain can be based on this timeline.

2. The chromosomally integrated TelN gene is constitutively expressed in this *telN*$^+$ DH10B strain and it is used for in vivo resolution of *tos*-containing BACs into linear BACs (Fig. 1d) [9].

3. For amplification of the 1.3 kb *tos*-Zeo cassette, we strongly recommend using long-template PCR enzymes that possess proofreading activity because the sequences of the targeting cassette, especially the *tos* element, need to be faultless in order for TelN to process it during resolution. However, for routine screening of recombinants by PCR, where a size difference between recombinants and non-recombinant is used to detect a recombineering event, regular DNA Taq Polymerase without proofreading activity, e.g., DNA Taq Polymerase by Invitrogen Inc., may be used.

4. The Resuspension Buffer RES should be kept at 4 °C after addition of RNase.

5. When designing long primers for recombineering, we recommend ordering a high-quality HPLC-purification step in order to quality control for full-length products.

6. The targeted region around the *Sfi*I site can be found on the backbone of BAC clones from many BAC libraries. Therefore TosZeoF2 and TosZeoR2 may be used as universal primers for targeting the *tos*-Zeo cassette into BAC clones.

7. The 10 % (v/v) glycerol needs to be ice-cold to preserve viability of the electrocompetent cells in order to maximize recombineering efficiency.

8. When working with glacial acetic acid, work in a fume hood with safety goggles and protective gloves.

9. The cuvettes should be chilled before electroporation. This will help maximize transformation and/or recombineering efficiency.

10. We typically prepare eight or more reactions and pool them together to obtain enough PCR product for purification.
11. We like to make an elongated well on the gel by taping several wells on the well comb together during agarose gel preparation. After PCR the contents of all the tubes are loaded slowly into this long well. This way the DNA band will migrate in one long well and can be subsequently cut out and purified cleanly without having to deal with intervening agarose fragments between wells.
12. Always wear protective gear and goggles when working under UV light.
13. Plasmid pGETrec [11] is a 6.6 kb high-copy plasmid that carries the recombineering genes, *recE*, *recT*, and *gam*. Because of its small size calcium chloride transformation [19, 20] can also be used for transformation of this plasmid into the *E. coli* host.
14. Exposing cells to room temperature will severely reduce transformation efficiency and the resulting number of positive clones.
15. The time required to reach OD_{600} of 0.55–0.65 may vary between cultures and should be monitored closely. To keep track of the growth, start monitoring the culture early. About 2 h after inoculation, start taking the OD_{600} reading approximately every 30 min. As the OD_{600} begins to approach 0.4, take more frequent readings, e.g., every 10 min, or even 5 min, judging on the rate of growth (i.e., for faster growing cultures, take readings at shorter intervals). As soon as the reading is anywhere within this range, stop the growth by aliquoting the culture into a precooled sterile centrifuge bottle and place it on ice for 15–20 min, then proceed to make the competent cells. If, however, the culture overgrows past OD_{600} 0.65, it should be discarded and a fresh culture started.
16. Glycerol, placed at −20 °C before use, could partially freeze and form ice crystals. Do not pour crystallized glycerol into the bottle and resuspend the cells. If glycerol is found partially frozen, swirl it under running tap water for a few minutes to thaw the solution completely. It should still remain extremely cold after this quick thawing process. Pour 250 mL into the bottle containing the cell pellet and resuspend the pellet by gently pipetting it up and down. This resuspension step should be carried out fast, within 1–2 min, and the bottle returned to the centrifuge for the following step.
17. Typically 1 mL or less of cells is collected from this residual glycerol.
18. Approximately 10–20 tubes of 45 μL aliquots can be routinely obtained. For each electroporation one 45 μL tube is used. The remaining electrocompetent cells may be stored at −80 °C at this point for future use.

19. The choice of control plasmid DNA differs for each transformation. For this electroporation, pUC19 is used as the control plasmid. This is because pUC19 confers resistance to ampicillin, of which the parental cells DH10B (BAC4396) will be susceptible to. Therefore DH10B (BAC4396) transformed with pUC19 will be selected on LB plates containing both chloramphenicol and ampicillin, with BAC4396 giving resistance to chloramphenicol and pUC19 to ampicillin.
20. Drying the cuvette's surface reduces the occurrence of electrical arcing, which could significantly reduce the viability of the cells. Avoid touching the aluminum electrodes because this could raise the temperature of the mixture and reduce electroporation efficiency.
21. If a "pop" sound is heard, arcing has occurred. Discard this electroporation reaction and repeat using another batch of competent cells, DNA, and cuvette. Plating these "popped" cells is unlikely to give any recombinants.
22. Adding antibiotic at this recovery stage will dramatically reduce cell viability, effectively ruining the experiment.
23. It is not necessary to sterilize the Pasteur pipettes as long as they are clean. A bulb should be used at the end of the pipette to gently suck the cells up and down in the cuvette. After electroporation the cells will be very fragile and therefore harsh pipetting should be avoided as it will lyse the cells.
24. For storage purpose, thoroughly mix 750 μL of the grown culture with 250 μL of 50 % (v/v) glycerol/LB medium by vigorous shaking. Flash-freeze the tube in a −80 °C isopropanol bath and store the stock culture at −80 °C. To maintain a working stock, streak each colony onto a LB agar plate containing the appropriate antibiotics, and re-streak every fortnight to maintain stability of the strain.
25. In our example we use BAC4396. For a specific BAC clone that is to be used, it is necessary to first transform pGETrec into the DH10B strain containing this BAC clone to generate the recombineering strain, DH10B (pGETrec, BAC clone) (*see* Subheading 3.2).
26. It is important to not overgrow the culture past OD_{600} of 0.50, as overgrown cultures will result in limited recombinants. In order to accurately keep track of the growth, start monitoring the culture early.
27. Even after the culture is removed from the incubator it will continue growing. Therefore it is very important to work fast. Remove the culture from the incubator, take the OD_{600} reading, and when the growth target is reached, induce with arabinose and immediately return the culture into the incubator.

28. The time required to reach OD_{600} of 0.55–0.65 may vary between cultures and should be monitored closely. A reading may be taken 20–30 min after induction in order to judge how fast the culture is growing. It is common to find the growth rate slowing down a little after induction, an effect of the bacterial cells being forced to overproduce the recombineering proteins from plasmid pGETrec [12, 18]. The most important thing is to stop the growth of the culture as it moves into the window between OD_{600} of 0.55–0.65. As soon as the reading is anywhere within this range, stop the growth by aliquoting the culture into a precooled sterile centrifuge bottle and place it on ice for 15–20 min and proceed to make the competent cells. If, however, the culture overgrows past OD_{600} 0.65, it should be discarded and a fresh culture started. Overgrown cultures will suffer from diminished recombineering proficiency and the experiment is unlikely to work.
29. For this electroporation, we typically transform the zeocin-resistant plasmid pZeoSV2(+) (Invitrogen Inc.) and select on LB plates containing zeocin, ampicillin, and chloramphenicol. Since DH10B (pGETrec, BAC4396) will already be ampicillin (and chloramphenicol) resistant, and will naturally grow on LB plates containing this ampicillin, the control plasmid must be resistant to an antibiotic other than ampicillin and chloramphenicol—hence, zeocin is used.
30. In our hands, a transformation efficiency of at least 1×10^8 CFU/μg is achieved routinely and consistently. Efficiencies lower than this could result in very few recombinants, or none at all. Therefore we strongly recommend fine-tuning the competent cell preparation to reach an efficiency of at least 1×10^8 CFU/μg before proceeding. The key to achieving this is to (1) work quickly during preparation of the competent cells, and (2) ensure that all instructions for using precooled materials and reagents are followed closely.
31. For the recovery stage, the cells should be grown at the slower condition of 150 rpm and 30 °C (as opposed to the standard 220 rpm and 37 °C) in order to promote the stability of the repeat-containing *tos* sequence [9, 18]. Because of the slower growth condition, the cells are incubated for 120 min, much longer time than usually done for electroporation. This longer incubation helps the cells undergo recombination as efficient as cells grown at standard conditions.
32. We have often found pGETrec plasmid persisting in the transformed cells even if ampicillin selection is not added. But at this stage, elimination of pGETrec is not a priority. The first step is to identify the correct recombinant containing the *tos-Zeo* cassette by PCR screening, described in Subheading 3.4.

Only after determining the correct recombinants by PCR screening, the modified BAC will be moved away from pGET-rec by a re-transformation step in Subheading 3.5 into freshly prepared DH10B competent cells [11].

33. We use the Long PCR Enzyme Mix from Thermo Scientific Inc., but any other long PCR kit should be suitable because the PCR is only performed for screening purpose.

34. It is strongly recommended to confirm that the *tos*-Zeo cassette is correctly recombined, i.e., by sequencing several PCR positive clones and by restriction digestion checking, before proceeding any further with these modified BACs.

35. We routinely digest the BAC4396-tos with a single cutting restriction enzyme, e.g., *Nar*I before electrophoresis [9]. However, for a chosen BAC any restriction enzyme that can verify its intactness can be used. Once you are satisfied that the BAC is intact and not rearranged it can be transferred into the telN$^+$ *E. coli* strain (Fig. 1d).

36. Vortexing large BAC DNA can damage it. Therefore shake the tube to mix.

37. For storage purpose, thoroughly mix 750 μL of the grown culture with 250 μL of 50 % (v/v) glycerol/BHI medium by vigorous shaking. Flash-freeze the tube in a –80 °C isopropanol bath and store this stock culture at –80 °C. To maintain a working stock, streak each colony onto a BHI agar plate containing chloramphenicol, streptomycin, zeocin, and 5 mM of DAP, and re-streak every fortnight to maintain viability of the DAP-auxotrophic clones.

38. We have used conventional plasmid extraction kit, e.g., QIAprep® Spin Miniprep Kit (Qiagen Inc.) to successfully isolate intact linear BAC-*tos* for routine plasmid checking. Nevertheless we find the NucleoBond® kits to produce the highest yield of linear DNA in excellent purity (260 nm:280 nm ratio > 1.80). Using the NucleoBond® Xtra Midi kit we typically achieve a yield of around 50 μg of linear DNA from 200 mL of culture.

39. Avoid vortexing to mix as this will generate contaminating chromosomal DNA.

40. Place a collection tray at the bottom to collect unwanted flow-through. The buffer will not completely run dry from the column; as long as the column is not dripping anymore it is ready for use.

41. Make sure the bottle is no more than two-third full so that the suspension can be mixed homogeneously.

42. Make sure the off-white precipitate in the column filter does not get into the column or it may clog the column and reduce the purification yield and/or quality.

43. Be sure to use high quality tubes that can withstand very high speed centrifugation of up to 17,000 × *g* for at least 45 min. We regularly use the Nalgene™ Oak Ridge High-Speed Polypropylene Copolymer (PPCO) 10 mL centrifuge tubes.
44. Adding 1 mL aliquots of the Elution Buffer ELU gives a higher yield compared to directly adding 5 mL in one go.
45. It will be a good idea to mark the edge of the tube that faces outward in the centrifuge rotor to indicate the position of the pellet, so that the translucent pellet will not be disturbed or lost during subsequent washing steps.
46. Be very careful when removing supernatant. The pellet is translucent to the naked eye and can easily dislodge from the bottom of the tube.
47. Do not over dry the pellet or it will take a long time to dissolve.
48. Dissolving the pellet could take up to 2–3 h depending on the size of pellet and how dry it is.
49. The protocol described here uses RecBCD Exonuclease V from Epicentre Inc. Different suppliers may have different composition of buffer and enzyme concentration, therefore perform the RecBCD digestion according to the manufacturer's instructions. We have also tested a RecBCD Exonuclease V pack from New England BioLabs Inc., which produces similar results.

Acknowledgements

The authors are grateful to Nikolai Ravin for providing N15 reagents and to Sek-Chuen Chow for support and encouragement. Q.C. is grateful to Monash University Malaysia for a HDR Scholarship. This work was partly funded by a Fundamental Research Grant Scheme FRGS/1/2011/ST/MUSM/02/2 from the Ministry of Higher Education Malaysia to K.N.

References

1. Kazuki Y, Hoshiya H, Takiguchi M et al (2011) Refined human artificial chromosome vectors for gene therapy and animal transgenesis. Gene Ther 18:384–393
2. Kakeda M, Nagata K, Osawa K et al (2011) A new chromosome 14-based human artificial chromosome (HAC) vector system for efficient transgene expression in human primary cells. Biochem Biophys Res Commun 415:439–444
3. Ferdows MS, Barbour AG (1989) Megabase-sized linear DNA in the bacterium Borrelia burgdorferi, the Lyme disease agent. Proc Natl Acad Sci U S A 86:5969–5973
4. Allardet-Servent A, Michaux-Charachon S, Jumas-Bilak E et al (1993) Presence of one linear and one circular chromosome in the Agrobacterium tumefaciens C58 genome. J Bacteriol 175:7869–7874

5. Lezhava A, Mizukami T, Kajitani T et al (1995) Physical map of the linear chromosome of Streptomyces griseus. J Bacteriol 177: 6492–6498
6. Hertwig S (2007) Linear plasmids and prophages in gram-negative bacteria. In: Meinhardt F, Klassen R (eds) Microbial linear plasmids. Springer, Berlin
7. Deneke J, Ziegelin G, Lurz R et al (2000) The protelomerase of temperate Escherichia coli phage N15 has cleaving-joining activity. Proc Natl Acad Sci U S A 97:7721–7726
8. Ravin NV, Strakhova TS, Kuprianov VV (2001) The protelomerase of the phage-plasmid N15 is responsible for its maintenance in linear form. J Mol Biol 312:899–906
9. Ooi YS, Warburton PE, Ravin NV et al (2008) Recombineering linear DNA that replicate stably in E. coli. Plasmid 59:63–71
10. Narayanan K, Chen Q (2011) Bacterial artificial chromosome mutagenesis using recombineering. J Biomed Biotechnol 2011:971296
11. Narayanan K, Williamson R, Zhang Y et al (1999) Efficient and precise engineering of a 200 kb beta-globin human/bacterial artificial chromosome in E. coli DH10B using an inducible homologous recombination system. Gene Ther 6:442–447
12. Narayanan K, Sim EU, Ravin NV et al (2009) Recombination between linear double-stranded DNA substrates in vivo. Anal Biochem 387: 139–141
13. Kaufman RM, Pham CT, Ley TJ (1999) Transgenic analysis of a 100-kb human beta-globin cluster-containing DNA fragment propagated as a bacterial artificial chromosome. Blood 94:3178–3184
14. Guzman LM, Belin D, Carson MJ et al (1995) Tight regulation, modulation, and high-level expression by vectors containing the arabinose PBAD promoter. J Bacteriol 177:4121–4130
15. Grant SG, Jessee J, Bloom FR et al (1990) Differential plasmid rescue from transgenic mouse DNAs into Escherichia coli methylation-restriction mutants. Proc Natl Acad Sci U S A 87:4645–4649
16. Narayanan K, Warburton PE (2003) DNA modification and functional delivery into human cells using Escherichia coli DH10B. Nucleic Acids Res 31:e51
17. Osoegawa K, Woon PY, Zhao B et al (1998) An improved approach for construction of bacterial artificial chromosome libraries. Genomics 52:1–8
18. Narayanan K (2008) Intact recombineering of highly repetitive DNA requires reduced induction of recombination enzymes and improved host viability. Anal Biochem 375:394–396
19. Hanahan D (1983) Studies on transformation of Escherichia coli with plasmids. J Mol Biol 166:557–580
20. Inoue H, Nojima H, Okayama H (1990) High efficiency transformation of Escherichia coli with plasmids. Gene 96:23–28

Chapter 3

BAC Sequencing Using Pooled Methods

Christopher A. Saski, F. Alex Feltus, Laxmi Parida, and Niina Haiminen

Abstract

Shotgun sequencing and assembly of a large, complex genome can be both expensive and challenging to accurately reconstruct the true genome sequence. Repetitive DNA arrays, paralogous sequences, polyploidy, and heterozygosity are main factors that plague de novo genome sequencing projects that typically result in highly fragmented assemblies and are difficult to extract biological meaning. Targeted, sub-genomic sequencing offers complexity reduction by removing distal segments of the genome and a systematic mechanism for exploring prioritized genomic content through BAC sequencing. If one isolates and sequences the genome fraction that encodes the relevant biological information, then it is possible to reduce overall sequencing costs and efforts that target a genomic segment. This chapter describes the sub-genome assembly protocol for an organism based upon a BAC tiling path derived from a genome-scale physical map or from fine mapping using BACs to target sub-genomic regions. Methods that are described include BAC isolation and mapping, DNA sequencing, and sequence assembly.

Key words Bacterial artificial chromosomes, High-throughput DNA sequencing, Sub-genome assembly, Assembly strategy, Quality evaluation

1 Introduction

High-throughput sequencing technologies allow for the high-resolution measurement of an organism's genome, epigenome, and transcriptome state [e.g., genotyping by sequencing [1], ENCODE project [2]]. The utility of deep sequencing, however, is limited by the quality of a reference genome sequence to which sequence tags can be mapped. If a reference genome is unavailable, one must be constructed, often a de novo assembly of the data that is then used in alignment. Clearly, this circular procedure opens the possibility for propagation of assembly errors in downstream analyses. As complete genome sequencing has become more common and efforts have expanded to include large and complex eukaryotes, even highly heterozygous out-crossers and polyploids, clever strategies and design must be focused on sub-genomic segments (e.g., BACs), rather than current shotgun approaches. By reducing

Kumaran Narayanan (ed.), *Bacterial Artificial Chromosomes*, Methods in Molecular Biology, vol. 1227,
DOI 10.1007/978-1-4939-1652-8_3,

the complexity of the genome into a set of manageable BAC-size pieces, BAC clones can be multiplexed without indexing and largely reconstructed through de novo assembly.

This chapter describes the sub-genome assembly protocol for an organism based upon a BAC tiling path derived from a genome-scale physical map or from fine mapping using BACs to target sub-genomic regions. Several examples of this genome reduction strategy have been published [3–6]. Sequencing and bioinformatics technology is constantly changing. Here, high-throughput sequencing data derived from the Illumina and Roche/454 platforms are discussed using the Celera wgs-assembler [7]. Certainly, other sequencing platforms and assemblers will be successful in the assembly of high-quality sub-genome reference scaffolds.

This chapter starts with methods for BAC isolation and mapping: (Subheading 3.1) DNA preparation for sequencing, (Subheading 3.2) High Information Content Fingerprinting (HICF), (Subheading 3.3) DNA fingerprint map construction, and (Subheading 3.4) BAC Minimum Tiling Path (MTP) selection. This is followed with protocols for sequencing: (Subheading 3.5) high-throughput sequence acquisition, and (Subheading 3.6) sequence data preprocessing methods, and concluded with strategies for sequence assembly: (Subheading 3.7) scaffold assembly with the Celera assembler, (Subheading 3.8) simulation-based approach for selecting optimal sequencing strategy and read coverage, and (Subheading 3.9) computing assembly quality scores to validate genome assembly.

2 Materials

2.1 DNA Isolation

1. 250 mL flasks.
2. 50 mL centrifuge tubes (Nalgene).
3. Fixed-angle centrifuge (e.g., Beckman Coulter Avanti J-E).
4. Swing-bucket centrifuge (e.g., Beckman Coulter Avanti J-E).
5. Plate Shaker.
6. 1.2 mL microfuge tubes.
7. Isopropanol.
8. 100 % Ethanol.
9. Phenol.
10. $T_{10}E_1$:10 mM Tris–HCl, pH 8.0, 1 mM Ethylenediaminetetraacetic acid (EDTA), pH 8.0.
11. $T_{10}E_{50}$:10 mM Tris, pH 8.0, 50 mM Ethylenediaminetetraacetic acid (EDTA) pH 8.0.
12. 7.5 M NH_4OAc.

13. 2XYT Broth.
14. Chloramphenicol: Stock concentration, 100 μg/mL, working concentration,12.5 μg/mL.
15. Incubator shaker.
16. 0.5 M Ethylenediaminetetraacetic acid (EDTA), pH 8.0.
17. 4 N Sodium Hydroxide (NaOH).
18. 10 % Sodium dodecyl sulfate (SDS).
19. 7.5 M Potassium acetate (KOAc).
20. 100 % Glacial Acetic acid.
21. Isopropyl β-D-1-thihogalactopyranoside (IPTG).
22. 5-bromo-4-chloro-3-indolyl-β-D-galactopyranoside (X-gal).
23. Ribonuclease A.
24. Tape Plate Sealer (e.g., Fischer BioReagents).
25. Filter/receiver plates (Pall Corporation).
26. SNaPShot Multiplex Kit (Life Technologies).
27. 3730xl DNA Analyzer (Life Technologies).

2.2 Software

1. *wgs-Assembler:* Powerful and general purpose genome assembler, *v7.0.* Available at http://wgs-assembler.sourceforge.net *under the GNU General Public License* [7, 8].
2. *Trimmomatic:* Versatile Illumina sequence trimmer, *v0.30.* Available at http://www.usadellab.org/cms/?page=trimmomatic [9, 10].
3. *Bowtie2:* Fast short read aligner, *v2.0.0-beta6.* Available at http://bowtie-bio.sourceforge.net/bowtie2/index.shtml [11, 12].
4. *Samtools:* Software tools for manipulating SAM-BAM alignment files, *v0.1.18 (r982:295*). Available at http://samtools.sourceforge.net [13, 14].

3 Methods

3.1 High Information Content Fingerprinting (HICF) and MTP Selection

3.1.1 DNA Isolation

1. Inoculate BAC clones into 1.2 mL of 2XYT media supplemented with 12.5 μg/mL of chloramphenicol and incubate at 37 °C for 18 h.
2. Spin culture at 2,101 ×*g* for 10 min at 4 °C.
3. Decant media and drain upside down on paper towels for 5 min.
4. Add 100 μL of Solution 1 (10 mM EDTA, pH 8.0 supplemented with RNase).
5. Shake at 250 rpm at room temperature on a plate shaker.

6. Add 200 μL of Solution 2 (0.2 N NaOH and 1 % w/v SDS).
7. Immediately add 150 μL of Solution 3 (7.5 M NH_4OAc), seal with tape plate sealer and mix well by inversion 10 times.
8. Centrifuge at 3,000 × *g* for 20 min at 4 °C.
9. Transfer supernatant to filter/receiver plate.
10. Centrifuge 2,101 × *g* for 5 min at 15 °C.
11. Discard filter and add 270 μL of room temperature isopropanol, seal with tape plate sealer and mix 5–7 times by inversion.
12. Centrifuge 2,101 × *g* for 30 min at 15 °C.
13. Decant and let drain upside down for 1 min on paper towels.
14. Add 450 μL of 70 % ethanol and centrifuge at 2,101 × *g* for 10 min at 15 °C.
15. Decant ethanol and drain upside down for 1 min and then dry in sterile laminar flow hood for 20 min at room temperature.
16. Reconstitute DNA in 30 μL of ultrapure ddH_2O (Gibco) overnight at 4 °C.

3.1.2 Digestion and Labeling

1. Purified BAC DNA is subject to the digestion with 4 type-II restriction enzymes according to the reaction cocktail in Table 1 and incubated for 2 h at 37 °C.
2. After digestion, label restricted overhang compliments with four different fluorophores, each A,G,C,T with the SNapshot® multiplex kit (Life Technologies) using the reaction mixture in Table 2.
3. Incubate the reaction for 1 h at 65 °C in a thermal cycler.

Table 1
1× Digestion cocktail for one 96-well plate

Reagent	Amount (μL)
10× Buffer 2	206
100× BSA	20.6
Rnase (Ambion)	20.6
Banl	10.3
Hin DIII	10.3
Nhe I	20.6
Xho I	10.3
PvuII	20.6
DDH_2O	1,226
Final volume	1,545.3

Table 2
1× Labeling cocktail for one 96-well plate

Reagent	Amount (μL)
Snapshot	25.2
10× NEB buffer [2]	126
33 mL Tris–HCI pH 9.0	441
Final volume	595.2

4. After incubation, purify the labeled fragment products by adding 65 μL of a 1:10 solution of (7.5 M NH_4OAc:100 % ETOH stored at −20 °C).
5. Incubate 12 min on ice and then spin at 3,000 × *g* for 20 min at 15 °C in a swing-bucket centrifuge.
6. Decant, and drain upside down on paper towels for 5 min.
7. Wash with 200 μL of 70 % ETOH, spin at 3,000 × g for 10 min at 15 °C in a swing-bucket centrifuge.
8. Decant and dry for 15 min.
9. Load on 3730xl DNA Analyzer (Applied Biosystems) following the manufacture's recommended protocols, including the addition of an internal size standard for each clone (LIZ1200, Applied Biosystems) and a 50 cm capillary array.

3.2 Fingerprint Map Construction

3.2.1 Preparation of Digitized BAC Fingerprints and BAC Contig Assembly

1. Following capillary electrophoresis, upload raw trace files (*.fsa) into GeneMapper® (Applied Biosystems) for sizing quality analysis.
2. Upload the output of GeneMapper sizing quality files into FPminer (Bioinforsoft). Edit fingerprints for background noise and vector bands using a color threshold of 120 for red, green, blue, and yellow; and 100 for the size standard (orange).
3. Remove bands occurring in more than 10 % of the restriction profiles (vector and abundant repeats) using the global compare clones function.
4. Assign each clone a digitized fingerprint and output (*.sizes) for assembly with Fingerprint Contigs software [15].
5. Use the following parameters, found on the configure menu, for setting up the first build using a 1,200 bp size standard and generating a fragment profile using a 50 cM capillary array: Fast Sulston radio button checked, (input genome size in kb), (input BAC insert size—average—in kb), define the band size by (genome size(kb)/average BAC insert size)/average number of fragments per clone from FPminer, Gel length = 36,000, contig display page size = 3,000, HICF radio button highlighted.

6. Conduct the initial (base) assembly on the "FPC main analysis" menu with the following parameters: Tolerance = 3, Cutoff = 1e-^{60}, Best contig of=40, click "Build Contigs (Runs Kill first)."
7. Remove questionable "Q" clones after the initial contig assembly using the DQ'er function, where if>=10 % contigs are removed of questionable clones.
8. Assess the contig ends for overlap at lower Sulston cutoffs by lowering the cutoff through an iterative process and analyzing the output for normal (relatively even) clone distribution and estimated genome size, with a cutoff no higher than 10–25. Set the cutoff, with the Auto Merge/Add radio button highlighted, FromEnd = at least (30 % of the average BAC insert size), Match = at least 2, and the Ends → Ends button selected.
9. Add singletons by selecting clones on the main FPC menu, click search commands, highlight singletons, and then return to the FPC Main Analysis menu and highlight Ends Only, Auto Merge/Add, and click KeySet → FPC.

3.3 Minimum Tiling Path (MTP) Selection

3.3.1 Selecting a Reduced Set of Overlapping BACs for Sequencing

1. On the main FPC window, select the MTP button is to bring up the main MTP selection menu.
2. Use parameters for finding overlapping pairs: Min FPC Overlap: 30, Max FPC Overlap: 50, FromEnd: 30 % of the average BAC insert size, Use Fingerprints radio button highlighted—Min Shared Bands: 14.
3. After overlapping pairs have been identified, click Pick MTP clones button with the "All contigs" and "Give preference to large clones" radio buttons highlighted.

3.4 BAC DNA Preparation for Sequencing

3.4.1 Single Colony Isolation and Culture Preparation

Streak BAC cultures for single colony isolation onto semi-solid LB agar Petri dishes containing 12.5 μg/mL chloramphenicol supplemented with isopropyl β-D-1-thihogalactopyranoside (IPTG) and 5-bromo-4-chloro-3-indolyl-β-D-galactopyranoside (X-gal) and incubate for 18 h at 37 °C.

3.4.2 DNA Isolation

1. Select a single colony, prepare preculture by inoculating a 1.2 mL culture of 2xYT broth (Fisher BioReagents) supplemented with 12.5 μg/mL of chloramphenicol. Incubate the preculture for 4 h at 37 °C, shaking at 250 rpm.
2. After the 4 h preculture, transfer 50 μL to a 50 mL flask containing 45 mL of 2xYT broth (Fisher BioReagents) supplemented with 12.5 μg/mL of chloramphenicol, incubate for 16 h (this is a critical step, *see* **Note 1**) at 37 °C at 250 rpm.
3. Transfer cultures to 42 mL oak ridge centrifuge tubes (Nalgene®) and centrifuge at 9,690 × *g*.

4. Decant cultures and drain upside down on laboratory towels for 5 min, and place on ice.
5. With a 5 mL pipette, add 2 mL of Solution 1 (10 mM EDTA, pH 8.0) to each cell pellet.
6. Tightly cap each tube and vortex vigorously to completely disrupt all cell aggregates until a homogenous solution remains, and incubate at room temperature for 5 min.
7. This is a critical step (*see* **Note 2**). Add 4 mL of Solution 2 (0.2 N NaOH and 1 % w/v SDS) slowly down the side of the tube being careful not to create any mixing or agitation. Gently replace the tubes on ice.
8. Immediately add 3 mL of solution 3 (7.5 M NH_4OAC, stored at −20 °C) slowly down the side of the tube being careful not to create any mixing or agitation. Carefully place the tubes in a fixed-angle centrifuge, and centrifuge for 15 min at 19,837 × *g*, 4 °C.
9. Transfer the supernatant to a new tube and centrifuge for 15 min at 19,837 × *g*, 4 °C to remove residual cellular debris.
10. Transfer the supernatant to a new tube and add 9 mL of isopropanyl (room temp) and mix gently by inverting 15–20 times.
11. Centrifuge 15 min at 19,837 × *g*, 4 °C, decant supernatant and drain upside down on paper towels for 5 min.
12. Dissolve pellet with 2.25 mL of $T_{10}E_{50}$ (10 mM Tris–HCl pH 8.0 and 50 mM EDTA pH 8.0) and incubate at room temperature for 15 min.
13. (Stopping point, *see* **Note 3**) Add 1.15 mL of 7.5 M (KOAc) and mix by tapping gently and freeze at −80 °C for at least 30 min. This is a safe stopping point for an indefinite period of time.
14. Thaw tubes at room temperature and mix by tapping after they are completely thawed and centrifuge at 5,054 × *g* for 10 min at 4 °C.
15. Transfer supernatant to a clean tube and add 8.5 mL of 100 % ETOH and mix well by inversion 15–20 times and centrifuge at 19,837 × *g*, 4 °C.
16. Decant and drain tubes upside down for 5 min on paper towels.
17. This is a critical step (*see* **Note 4**). Dry pellet in laminar flow hood for 12 min and dissolve pellet in 700 μL of $T_{10}E_{50}$ (10 mM Tris–HCl pH 7.5 and 50 mM EDTA pH 8.0) and incubate at room temperature for 15 min.
18. Transfer supernatant (DNA suspension) to 1.2 mL microfuge tubes and add 600 μL of phenol (Fisher BioReagents) and vortex for 10 s to mix well.
19. Centrifuge tubes at 20,000 × *g* in a benchtop microfuge at 4 °C for 5 min.

20. This is a critical step (*see* **Note 5**). Transfer the top layer carefully avoiding the interphase to a new microfuge tube and add 0.1 volume of 3 M NaOAc pH 5.3 and 3 volumes of 100 % ETOH and mix well by vortexing.
21. Centrifuge tubes at 20,000 × *g* in a benchtop microfuge 4 °C for 30 min.
22. Decant supernatant and wash pellet with 70 % ETOH and spin in a benchtop microfuge 4 °C for 15 min.
23. Decant supernatant, dry pellet, and resuspend in 40 μL of $T_{10}E_1$ (10 mM Tris–HCl pH 8.0 and 1 mM EDTA pH 8.0).

3.5 High-Throughput Sequencing

3.5.1 Sequence Pooled BAC Roche/454 Libraries as 3,000 bp Mate-Pairs at 50–100× Depth

Pool targeted BACs for sequencing accordingly by adding equi-molar DNA amounts (as determined by fluorimetry) to a final DNA mass of 5 μg. Produce Roche 454 DNA sequencing libraries and sequence at a reputable facility using 3,000 bp mate-pair library preparation techniques and Titanium (Roche) sequencing chemistry.

3.5.2 Sequence Pooled BAC Illumina Libraries as Linear and 3,000 bp Mate-Pairs 50–100× Depth

Alternatively, sequence BACs using Illumina technology. Pool BAC DNA as described in Subheading 3.5.1 in equi-molar ratios and use it as template for two library types, linear and mate-pair, respectively. For linear DNA libraries, fragment 500 ng of template DNA using fragmentase (New England Biolabs) to an average fragment length of 500–800 bp, as measured by a Bioanalyzer (Agilent). Prepare linear DNAseq libraries by either the NEBnext® (New England Biolabs) or the Truseq (Illumina) kits, according to the manufacturer's instructions. Construct mate pair library with the Nextera mate pair kit (Illumina) using the gel plus option, selecting for fragments ~3,000 bp. Collect DNA sequence using a 250 × 250 bp paired-end MiSeq run.

3.6 Sequence Data Preprocessing

3.6.1 Preprocessing Roche/454 Reads for Assembly with the Celera wgs-Assembler

1. Convert raw mate pair SFF files to wgs-assembler FRG format using the sffToCA tool (Parameter example: "-trim chop -clear 454 –linker titanium -insertsize 3000 300 -libraryname lib1 -output lib1.frg lib1.sff").
2. Convert SFF files to FASTA format (sequence and quality) with sff_extract.py [16] while cutting with titanium linker sequence.
3. Mask FASTA formatted sequences contaminant sequences (e.g., BAC vector and E .coli) with cross_match (Parameter example: "-minmatch 10 -minscore 20 –screen").
4. Using a text processing script, sort reads into true mate-pair (MP) and unpaired (UP) libraries.
5. Remove duplicate and short reads as per info from SFFtoCA log.
6. Convert cleaned FASTA formatted files into FRG format using convert-fasta-to-v2.pl (Parameter example: -454 -l WBS1R1_MP_ALL -s lib1.fa -q lib1.qual -m lib1.list -mean 3000 -stddev 300 > lib1.frg).

3.6.2 Preprocessing Illumina Paired Reads for Assembly with the Celera wgs-Assembler

1. Construct an adaptor for trimmomatic [10]. Include all relevant Illumina adaptor, PCR, and other sequences as well as vector splice sites with flanking DNA around 100 bp (forward and reverse complement) (Parameter example: "ILLUMINACLIP: adaptor_file.fa: 2:40:15 LEADING:3 TRAILING:6 SLIDINGWINDOW:4:15 MINLEN:36").
2. Align trimmed reads to a potential contaminant file using bowtie2 [17].
3. Using samtools and a text processing script, filter reads that match contaminant with a MAQ score >30.
4. Convert cleaned FASTQ files into FRG format using fastqToCA (Parameter example: "-insertsize 3000 300 -libraryname lib1 -technology illumina -type illumina -innie -mates left.fastq right.fastq > lib1.frg").

3.7 Contig Assembly with the Celera wgs-Assembler

1. Use the runCA script to assemble contigs and scaffolds (Optional parameter settings: "overlapper = mer obtOverlapper = mer ovlOverlapper = mer unitigger = bog doToggle = 1").
2. BLAST-align BAC-end sequences (BES) to scaffolds. Filter low-quality hits (e.g., Percent Identity <98 %; alignment length >300 bp).
3. If known, use FPC MTP order and BES orientation to order and orient scaffolds. Use this information to construct a reference pseudomolecule.

3.8 Selecting Sequencing Coverage, BAC Pool Sizes and Assembly Parameters by Applying Computer Simulations and Assembly Quality Scores Utilizing a Related Sequenced Organism

1. Select a closely related sequenced genome from database of sequenced organisms, e.g., NCBI genome database [18].
2. Select simulated BAC sizes and a minimum tiling path of BACs for pooling. These may be obtained from published data, such as [19].

3.8.1 Generate Simulated Sequencing Reads Using a Similar Genome Sequence, Realistic Read Lengths, Quality Values and Error Rates

3. Select fragment length distributions D reflecting available sequencing technology (a separate distribution for each technology, e.g., 454 single and Illumina paired). The fragment length distribution may be obtained from published sequencing experiments, such as [20], or from the respective sequencing technology providers.
4. Select quality value Q and error E distributions reflecting available sequencing technology (a separate distribution for each technology, e.g., 454 single and Illumina paired). The distributions may be estimated from findings in literature, such as [21], or obtained from the respective sequencing technology providers.
5. Select model for read placement P on target sequence (uniform or nonuniform, a separate distribution for each technology). A window-based approach is applied, where each BAC is divided into windows of 100 bp and assigned a read density from max{N(5,1),0} (where N stands for the normal distribution), and finally the densities are scaled to sum to 1 and used as probabilities of assigning a read in the windows [4].

6. Generate 100× single-end sequencing coverage by randomly selecting (N*100)/L fragment start positions (where N is genome length and L is average read length), according to the read placement model P [22]. Add sequencing errors and quality values according to distributions E and Q.
7. Generate 100× paired sequencing coverage by randomly selecting (N*100)/(½L) fragment start positions with fragment length randomly selected from D (where N is genome length, L is average read length, D is fragment length distribution), according to the read placement model P [22]. Add sequencing errors and quality values according to distributions E and Q.

3.8.2 Assemble Simulated Reads Using Different Sequencing Coverage, BAC Pool Sizes, Assembly Algorithms, and Compute Assembly Quality Scores

1. Select various BAC pool sizes such as 3, 5, and 10 Mbp.
2. Select various levels of coverage, such as 50× single + 50× paired, 100× single, 100× paired.
3. Select subsets of simulated reads for each BAC pool size and sequencing coverage combination (optionally, select several subsets for each combination and average the results).
4. Assemble each combination of BAC pool size and read coverage, as detailed in Subheading 3.7.

3.8.3 Compute Assembly Quality Scores

1. Find alignments of assembled contigs to the reference sequences, using BLASTN [23] with default parameters, and discarding any matches shorter than 1 kbp.
2. Process the matches into a table T with following characteristics: Let reference R be length n, and assembly A of length m, then for each position i, where i = 1,…,n, the value of T[i] is the coordinate of the closest position j in the assembly, T[i] = argminj{| i − j |}, such that there is a match between R[i] and A[j]. A match on the reverse strand of the assembly indicates an inversion and the value of T[i] becomes -j. If there does not exist any match for position i, then T[i] = 0.
3. Compute scores for relocation, inversion, redundancy, match, coverage based on the table T [4].
   (a) Relocation RL = 1 − #d/p, where #d is the number of disagreements and p = (x2 − x) is the number of sampled pairs (e.g., with x = 10,000). The score is computed from a random subset of x reference positions for the sake of computational efficiency. When the order of two sampled points, a and b, and their values do not agree, e.g., T[a] > T[b] but a < b, the number of disagreements, #d, is increased for both a and b.
   (b) Inversion $I = 1 - \sum 1_i/l$, where $1_i = 1$ when T[i] < 0 and is 0 otherwise.
   (c) Redundancy RD = #u/m, where #u is the number of references positions where exactly one assembled position matches.

(d) Match $M = 1/2 \ (\sum_s(|u_s|/n)^2 + \sum_t(|v_t|/n)^2)$, with $\Sigma_s u_s + \sum_t v_t = n$ where u are matching segments ($T \neq 0$) and s are gap segments ($T = 0$).

(e) Coverage $C = \sum l_i/n$, where $l_i = 1$ when $T[i] \neq 0$ and is 0 otherwise.

3.8.4 Select the Sequencing Coverage, BAC Pool Size, Assembly Algorithm That Yields the Best Assembly Quality Scores

1. Sum the individual quality scores for each assembly, and select the highest scoring assembly A (*see* **Note 6**).
2. Use the BAC pool size, sequencing coverage, and assembly algorithm that generated A, in the sequencing and assembly steps in Subheadings 3.4–3.7.

3.8.5 Selecting Sequencing Coverage, BAC Pool Sizes and Assembly Parameters by Applying Assembly Quality Scores Utilizing a Known Sequenced Region of the Genome

1. Obtain the reference sequence of a genome region by, e.g., Sanger sequencing
2. Obtain high-throughput sequencing reads for the same genome region, as detailed in Subheading 3.5, *see* [3].
3. Assemble subsets of the high-throughput reads, as detailed in Subheadings 3.6 and 3.7, with various BAC pool sizes and sequencing coverage as detailed in Subheading 3.8.2.
4. Compute the quality scores for each assembly as detailed in Subheading 3.8.3.
5. Select the sequencing coverage (and assembly algorithm, parameters) that yields the highest quality scores (*see* **Note 6**).

3.9 Assessing the Quality of the Final Assembly by Computing Assembly Quality Scores Against Known Sequenced Regions of the Genome

1. Obtain accurate reference sequence for known subsequences of the target genome, e.g., from Sanger sequencing.
2. Use the methods described in Subheading 3.8.3 to compute the assembly quality scores of pooled BAC assembly against the known reference sequence.
3. Average over several known regions' scores to obtain estimate of pooled BAC assembly quality in terms of the assembly quality scores.

4 Notes

1. Incubation time is critical to maintain the cellular culture in exponential growth phase. Incubation times longer than 16 h can lead to increased amounts of secondary metabolites and host DNA.
2. This step lyses the cellular suspension and technique is critical in keeping host DNA (E. coli) shearing and co-precipitation at a minimum. It is the physical forces and agitation of the suspension that caused the host chromosome to co-precipitate with the BAC plasmid. Move promptly to the next step.

3. This is a safe stopping point. The supernatant can be safely stored at −80 °C for an indefinite period of time.
4. At this stage, the pellet is moderately attached to the tube. Over-drying will cause the DNA to be resistant to reconstitution. It is also critical to use the T10E50 pH 8.0 solution as DNA is better reconstituted in a somewhat alkaline conditions.
5. When transferring the top layer, be sure to keep the pipette tip away from the interphase completely. Trace amounts of phenol will inhibit all downstream reactions.
6. In practice, determine the parameters of acceptable sequencing cost and time, and choose the highest scoring assembly within those parameters.

References

1. Elshire RJ, Glaubitz JC, Sun Q, Poland JA, Kawamoto K, Buckler ES, Mitchell SE (2011) A robust, simple genotyping-by-sequencing (GBS) approach for high diversity species. Plos One 6:e19379
2. Dunham I, Kundaje A, Aldred SF, Collins PJ, Davis CA, Doyle F, Epstein CB, Frietze S, Harrow J, Kaul R et al (2012) An integrated encyclopedia of DNA elements in the human genome. Nature 489:57–74
3. Feltus FA, Saski CA, Mockaitis K, Haiminen N, Parida L, Smith Z, Ford J, Staton ME, Ficklin SP, Blackmon BP et al (2011) Sequencing of a QTL-rich region of the *Theobroma cacao* genome using pooled BACs and the identification of trait specific candidate genes. BMC Genomics 12:379
4. Haiminen N, Feltus FA, Parida L (2011) Assessing pooled BAC and whole genome shotgun strategies for assembly of complex genomes. BMC Genomics 12:194
5. Gonzalez VM, Benjak A, Henaff EM, Mir G, Casacuberta JM, Garcia-Mas J, Puigdomenech P (2010) Sequencing of 6.7 Mb of the melon genome using a BAC pooling strategy. BMC Plant Biol 10:246
6. Steuernagel B, Taudien S, Gundlach H, Seidel M, Ariyadasa R, Schulte D, Petzold A, Felder M, Graner A, Scholz U et al (2009) De novo 454 sequencing of barcoded BAC pools for comprehensive gene survey and genome analysis in the complex genome of barley. BMC Genomics 10:547
7. Miller JR, Delcher AL, Koren S, Venter E, Walenz BP, Brownley A, Johnson J, Li K, Mobarry C, Sutton G (2008) Aggressive assembly of pyrosequencing reads with mates. Bioinformatics 24:2818–2824
8. wgs-assembler (2013) http://sourceforge.net/apps/mediawiki/wgs-assembler/. Accessed 31 Oct 2013
9. Lohse M, Bolger AM, Nagel A, Fernie AR, Lunn JE, Stitt M, Usadel B (2012) RobiNA: a user-friendly, integrated software solution for RNA-seq-based transcriptomics. Nucleic Acids Res 40:W622–W627
10. Trimmomatic. (2013) www.usadellab.org/cms/?page=trimmomatic
11. Langmead B, Salzberg SL (2012) Fast gapped-read alignment with Bowtie 2. Nat Methods 9:357–359
12. bowtie2 (2013) http://bowtie-bio.sourceforge.net/bowtie2/index.shtml. Accessed 31 Oct 2013
13. Li H, Handsaker B, Wysoker A, Fennell T, Ruan J, Homer N, Marth G, Abecasis G, Durbin R, Proc GPD (2009) The sequence alignment/map format and SAMtools. Bioinformatics 25:2078–2079
14. samtools (2013) http://samtools.sourceforge.net/. Accessed 31 Oct 2013
15. Soderlund C, Humphray S, Dunham A, French L (2000) Contigs built with fingerprints, markers, and FPC V4.7. Genome Res 10:1772–1787
16. sff_extract (2013) https://github.com/JoseBlanca/seq_crumbs. Accessed 31 Oct 2013
17. Langmead B, Trapnell C, Pop M, Salzberg SL (2009) Ultrafast and memory-efficient alignment of short DNA sequences to the human genome. Genome Biol 10:R25
18. National Center for Biotechnology Information (2002) The NCBI handbook [Internet]. The reference sequence (RefSeq) project, chapter 18. National Library of Medicine (US), National Center for Biotechnology Information, Bethesda, MD, http://www.ncbi.nlm.nih.gov/books/NBK21091/

19. Ouyang S, Zhu W, Hamilton J, Lin H, Campbell M, Childs K, Thibaud-Nissen F, Malek RL, Lee Y, Zheng L, Orvis J, Haas B, Wortman J, Buell CR (2007) The TIGR rice genome annotation resource: improvements and new features. Nucleic Acid Res 35(Database Issue):D846–D851
20. Van Nieuwerburgh F, Thompson RC, Ledesma J, Deforce D, Gaasterland T, Ordoukhanian P, Head SR (2012) Illumina mate-paired DNA sequencing-library preparation using Cre-Lox recombination. Nucleic Acids Res 40:e24
21. Luo C, Tsementzi D, Kyrpides N, Read T, Konstantinidis KT (2012) Direct comparisons of Illumina vs. Roche 454 sequencing technologies on the same microbial community DNA sample. PLoS One 7:e30087
22. Haiminen N, Kuhn DN, Parida L, Rigoutsos I (2011) Evaluation of methods for de novo genome assembly from high-throughput sequencing reads reveals dependencies that affect the quality of the results. PLoS One 6:e24182
23. Altschul SF, Gish W, Miller W, Myers EW, Lipman DJ (1990) Basic local alignment search tool. J Mol Biol 215:403–410

Part II

Modification of BACs

Chapter 4

Making BAC Transgene Constructs with Lambda-Red Recombineering System for Transgenic Animals or Cell Lines

Scott Holmes, Suzanne Lyman, Jen-Kang Hsu, and JrGang Cheng

Abstract

The genomic DNA libraries based on Bacteria Artificial Chromosomes (BAC) are the foundation of whole genomic mapping, sequencing, and annotation for many species like mice and humans. With their large insert size, BACs harbor the gene-of-interest and nearby transcriptional regulatory elements necessary to direct the expression of the gene-of-interest in a temporal and cell-type specific manner. When replacing a gene-of-interest with a transgene in vivo, the transgene can be expressed with the same patterns and machinery as that of the endogenous gene. This chapter describes in detail a method of using lambda-red recombineering to make BAC transgene constructs with the integration of a transgene into a designated location within a BAC. As the final BAC construct will be used for transfection in cell lines or making transgenic animals, specific considerations with BAC transgenes such as genotyping, BAC coverage and integrity as well as quality of BAC DNA will be addressed. Not only does this approach provide a practical and effective way to modify large DNA constructs, the same recombineering principles can apply to smaller high copy plasmids as well as to chromosome engineering.

Key words BAC, Recombineering, Lambda *red*, Transgene construct, Transgenic animal, In vivo expression

1 Introduction

Taking advantage of the extensive gene-flanking sequence of the BAC that likely contains key regulatory components, BAC transgene constructs have been used to direct the expression of foreign genes for the past two decades [1–3]. Not only can BAC transgenes in animals maintain expression as the targeted gene in a tissue-specific manner, but they also maintain expression at physiological levels due to low copy number that mirrors natural genome organization [1, 2]. In addition, BACs can be used to make point mutations or transgene insertions and to deliver the effect of such mutations into cell lines, plants, or animals [4–6]. Using BACs to make transgenic animals has gained popularity, as it requires less effort

Kumaran Narayanan (ed.), *Bacterial Artificial Chromosomes*, Methods in Molecular Biology, vol. 1227, DOI 10.1007/978-1-4939-1652-8_4, © Springer Science+Business Media New York 2015

and time than that of the Knock In approach. Large-scale generation of BAC transgenic mice such as in the GENSAT project [1] has proved that BAC transgenes show less integration position effects and dosage artifacts when compared with traditional transgenic manifestation.

Using recombination-mediated DNA engineering (recombineering), one can manipulate plasmid DNA of different sizes with ease [4, 7, 8]. Standard recombineering with inducible λ-*red* uses two arms that flank the transgenes to integrate the transgene into target DNA via homologous recombination (a positive selection marker (SM) is included in the transgene) [8]. This also is called linear transformation as linear DNA for targeting can be integrated into plasmids or chromosomes and it is selectable. Due to the recombination efficiency of λ-*red* operon, which can be either provided from hosts or plasmids, one can modify a single copy gene in bacteria chromosomes with homologous arms as short as 30–50 bp [9].

The methods in this chapter describe the rationale behind designing a BAC transgene construct and the steps needed to insert foreign DNA within a targeted locus on the BAC. Specific attention is given to choosing a BAC for targeting [10, 11]. Two methods are applied to make BAC recombinogenic (targetable) by providing inducible recombination machinery via plasmid or host. Several approaches are then provided for making the targeting cassettes (TC) by flanking a transgene and selection marker with homologous arms. Furthermore, a description of how to employ interchangeable selection markers to produce a modified BAC for transgenic animals or cell cultures is included. Though making transgenic animals with BACs introduced via pronuclear injection does not require eukaryotic selection markers, it is recommended to include selection pressure with cell cultures to increase transfection efficiency.

1.1 Overview of BAC Transgene Design

A well-planned design and thorough execution are essential for generating a BAC transgene (BAC-tg) that can meet the desired goal. As any modification in the BAC transgene construct is hard to correct once introduced inside genomic DNA, appropriate selection of a BAC clone, transgene, and insertion site merits careful attention.

1.1.1 Choosing the Target Gene

Choose a suitable target gene-of-interest that can direct the expression of the transgene in a manner similar to endogenous expression. Most of the time, the gene of interest is predetermined, especially for an individual laboratory that is familiar with the resources regarding one's own projects. Initially, the expression profile of the gene-of-interest is based on the expression studies of RNA blots or microarrays. Currently, multiple resources such as gene chip results, atlas of gene expression based on in situ hybridization, and RNA deep sequencing have demonstrated the expression of individual transcripts. GENSAT, or similar BAC transgenic projects can also provide a guideline for the expected expression of a specific BAC clone [1, 2, 4]. Additional regulatory mechanisms such as Tet on/off or recombinase-mediated activation/suppression can be incorporated within the transgene to achieve a better temporal or spatial resolution [12–14].

1.1.2 Finding a Specific BAC and Confirming Coverage

For those BAC libraries that map to their respective genomic DNA annotations, a specific BAC can be identified through the accession number of a targeted gene with a genome browser such as NCBI, UCSC or Ensembl (*see* **Notes 1** and **2**). Other libraries require high-density arrays screening with gene-specific probes. For the mouse and human BAC, "Mitocheck BAC Finder" server http://mitocheck.org/ is a convenient way to find gene-specific BACs and the DNA sequence flanking the start and stop codons. A smaller BAC is easier to handle, yet it is vital to ensure complete coverage of the gene-of-interest and nearby regulatory elements (*see* **Note 3**). Nearby genes or chromosome markers (CpG islands, p300 binding sites, DNase hyper-sensitive locations) can be used to predict a transcription unit. Most genome browser websites provide updated information regarding gene structure. Once a gene-specific BAC clone is located, its ID address can be identified and the BAC purchased through vendors such as the BACPAC Resources Center, Children's Hospital Oakland Research Institute (CHORI).

Once a BAC clone (Fig. 1a) is acquired, the coverage of genomic DNA by the BAC needs to be checked by PCR reactions to ensure it contains targeting sites as well as enough upstream and downstream sequence for genotyping and for covering the transcriptional unit (*see* **Note 4**). This eliminates possible contamination and rearranged BAC clones. To determine the orientation of the BAC insert within the backbone vector, T7 or SP6 primers can be used against the 5′ and 3′ end sequences of the BAC insert.

1.1.3 Gene Annotation

Information about gene structures is needed in order to select the target site. Special attention is needed for identifying the protein coding genes. Note the presence of initiation codons (ATG), ATG tri-nucleotide prior to the initiation codons, exons, alternative ATG sites, and non-ATG initiation codons. Also, the 5′ untranslated region (UTR) should span from the transcription start site(s) to the translation start site(s), and the last coding exon should be a stop codon with a 3′ UTR and a poly A signal(s). In addition, the gene expression might involve alternative splicing.

DNA sequence information, at least for the target sites and the outside sequences of homologous arms, is needed for recombineering. Repetitive elements, which can complicate PCR and targeting reactions, need to be noted and avoided in the homologous arm. When a repeat-containing homologous arm is used for targeting, off-target reactions may be generated, resulting in undesired

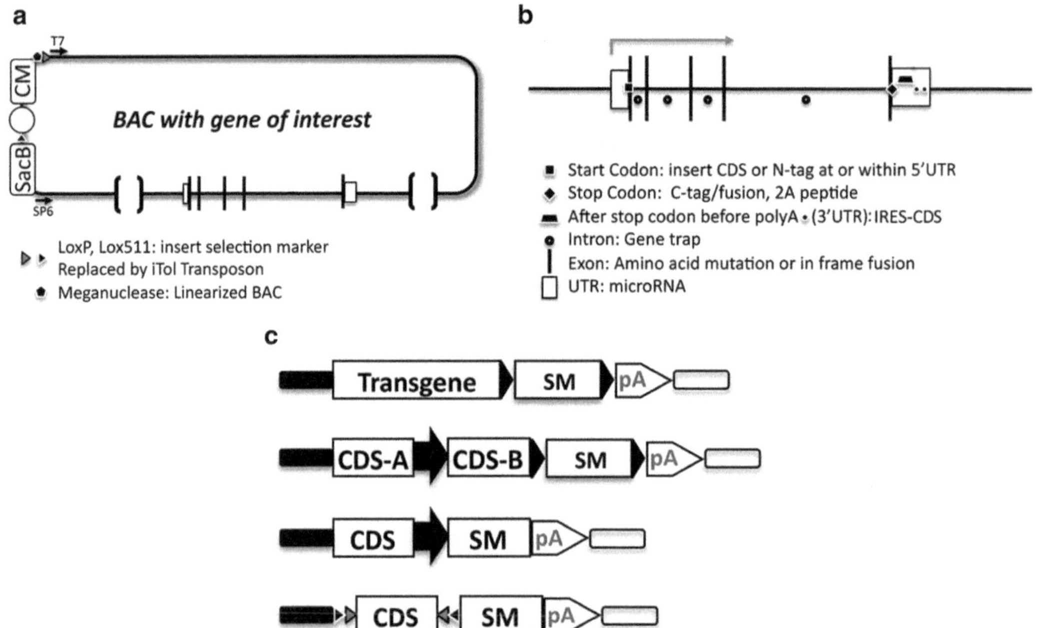

Fig. 1 (**a**) Schematic overview of a bacterial artificial chromosome (BAC) clone. BACs are large, low-copy (1–2 per bacterial cell) vectors based on the *E. coli* fertility plasmid (F Factor) and contain a large insert of up to 300 kb of genomic DNA from a mouse or other organism. To cover the whole genome of an organism like a mouse, BACs with different and sometimes overlapping inserts are stored in libraries (*see* **Note 1**). (**b**) The insertion site in the gene of interest on the BAC. Note the transcription start site is at the beginning of the 5′ untranslated region (UTR). The coding sequence (CDS) of an inserted protein runs from the start codon (ATG) to the stop codon. (**c**) Transgene components within targeting cassettes. The transgene is inserted into a gene of interest contained in the BAC. The transgene can contain any functional DNA fragment or coding sequence (CDS) like fluorescent proteins, recombinases, or toxins. Selection markers (SM) must be included to select for the BAC in the bacterial host and in cell culture. Common selection markers include Neomycin, Hygromycin, and Zeocin. loxP or Frt (*black triangles*) can be used to remove the selection marker. IRES or F2A elements (*black arrows*) inserted between CDSs allow expression of multiple CDSs on the same mRNA transcript. FLEX cassettes (*grey triangles*) allow CRE-dependent activation of any enclosed CDS. Homologous arms marked by black and white boxed franking transgene

recombination byproducts. A GC-rich area, occasionally associated with start codon, can potentially complicate PCR reactions and cloning. For model organisms whose whole genome sequences have been annotated, it is just a matter of downloading the DNA sequence around the targeting sites, the entire transcriptional unit of the targeted gene, or even whole coverage of BAC. DNA Software such as MacVector, Vector NTI or many available freeware programs can be used to manage DNA data as long as they can handle the Genebank format.

1.1.4 Insertion Site

The BAC transgene utilizes the endogenous gene transcription machinery to transcribe and translate the transgene (for peptide transgenes). To ensure the transgene expression replicates that of the endogenous gene it is inserted into, some decisions need to be determined on where and how to insert the transgene into the selected gene-of-interest (*see* Fig. 1b) (*see* **Note 5**). Because the regulatory components can be found upstream of the gene (promoter), within the intron of the gene, or after the gene (enhancer), a general rule is to insert the transgene such that it produces minimal alteration of the targeted gene. While a post-transcriptional interaction such as RNA export or RNA stability transgene could be beneficial, most transgene designs only tap into transcriptional regulation. However, when fusing a transgene to an endogenous gene to yield a fusion protein, the translational regulation and modification of endogenous proteins needs to be considered in addition to any transcriptional and post-transcriptional regulation.

Some characteristics for the insertion site are as follows (*see* Fig. 1b):

1. Insertion into the 5′ UTR should be between the transcriptional start site and the translational start sites. Make note of ATG sequences upstream of the initiation codon.
2. Initiation codon insertion/replacement can be an N-terminal fusion protein if skipping the stop codon of the transgene and both CDSs are in frame. There should be a stop codon replacement only for C-terminal fusion [15].
3. The 3′ UTR can be used to insert transgenes with an Internal Ribosome Entry Site (IRES) between the stop codon and polyA signals [16].
4. The insertion site can be in an intron to yield an intragenic fusion as that of a gene trap, which both require a RNA splicing acceptor [2].
5. For an N-terminal tag, the tag should be in frame without a stop codon (notice signal peptide). For a C-terminal tag, the tag should be in frame before the stop codon (notice C-terminal modification). For an intragenic fusion, the transgene should be in frame for both N- and C-terminal.
6. For a fusion protein, post-translational processing should be considered including signal peptides and peptide end modifications.

1.1.5 Choosing and Designing the Transgene

Choose a transgene that meets the project's requirements (*see* Fig. 1c). It may be any functional DNA fragment that contains coding proteins or functional RNAs. Common transgenes include tracers (fluorescence proteins, or enzymes, such as galactosidases or phosphatases), recombinase (Cre, Flp) (*see* **Note 6**), trans-acting regulatory factors (Gal4, tTA/rtTA), or toxins (e.g., diphtheria toxin). Pay attention to species-specific considerations such as codon usage and translational regulation when expressing a bacterial gene in eukaryotic cells. When expressed as messenger RNA, a Kozak sequence must be included as part of the transgene.

It can be part of the endogenous sequence or included with the targeting cassette. A good example of a Kozak sequence is gc/tcgccaccATG, where ATG is the start codon of the gene-of-interest. A polyA signal could be included in the transgene to stabilize its expression. When inserting the transgene into the C-terminal or 3′UTR of the targeted gene, endogenous polyA signals will suffice. Occasionally, an endogenous or artificial intron may increase mRNA exportation from the nucleus [17]. Other (optional) characteristics may be included. For example, viral components such as WRPE (Woodchuck hepatitis virus post-transcriptional regulatory element) can increase transcription and translation efficiency when added in-between CDS and poly A site [18]. S/MAR (scaffold/matrix attachment region) can be included in the BAC or within its transcription unit to maintain the BAC without integrating into chromosome [19]. Functional RNA, such as microRNA and microRNA sponge [20], can be expressed within the transcription unit of a BAC. As for siRNA, which requires DNA polymerase III promoter (Human H1 or U6 promoter) for expression, it can be combined with recombinase-dependent activation, such as the removal of LoxP floxed STOP cassette with Cre, to achieve tissue-specific knock down. The transgene can become inducible by Cre-dependent expression for a loxP-flanked and removable STOP cassette or a Tet-on/off with tetracycline-repressor tTA/rtTA and Tet Response Element (TRE) [12, 14].

Recombinase-mediated cassette exchange (RMCE) can be used to create a floxed transgene and alter the gene orientation (FLEX construct) [21]. A tag should be beneficial to be included in the transgene. Examples of tagging include: cellular localization tagging to direct expression of the transgene in specific organelles; an immuno-tag for purification or staining; or a fluorescence tag as a tracker of protein of the target gene [15]. Genes in eukaryotic cells are monocistronic, so it is necessary to add the transgene in front of the targeted gene to be expressed. To make multiple gene products from a single transgene construct, the following components can be included: (a) internal ribosomal entry site (IRES) to direct the expression of two transgenes or a single transgene after a stop codon in a targeted gene (wild-type IRES from encephalomyocarditis virus (EMCV) works well, and the IRES sequence from mammalian mRNA can also serve as a splicing donor) [22]; (b) 2a peptides-self cleavage peptide with viral origin (such as Foot and Mouth Disease virus) can be used to link multiple CDS to form a single protein [23].

1.1.6 Selection Markers

A selection marker such as an antibiotic-resistant gene is used for positive selection when manipulating DNA in bacteria or in cell lines (*see* **Note 7**). Common selection markers that can be used for both bacteria and eukaryotic cells include neomycin/kanamycin, hygromycin, and zeocin. Dual promoters containing both prokaryotic and eukaryotic promoters allow the flexibility of selection marker, allowing the vector to be transferred between bacteria and eukaryotic cells [9]. To avoid any interference with a transgene expression, it is preferable if the selection marker is removable, e.g., either with Frt or loxP flanking antibiotic-resistant genes or residing within an intron that can be removed by RNA splicing [2, 9]. The location of the selection markers can be between the CDS and the polyA signal of the transgene (preferred), upstream of the 3′ homologous arm (right arm), or downstream of the 5′ homologous arm (left arm) in the event that insertion of an N-terminal tag is necessary. The selection markers may also serve as a STOP cassette to control transgene expression. Additional selection can be added into the BAC vector by using a Cre/LoxP reaction or by targeting to select in eukaryotic cells. When combined with transposons such as iTol2 in BAC backbone, one may increase BAC integration into chromosome [5].

2 Materials

Prepare all solutions using ultrapure water (prepared by purifying deionized water to attain a sensitivity of 18 MΩ cm at 25 °C) and analytical grade reagents. Sterilization either by autoclave or filtration (0.22 μm filter) is needed. Prepare and store all reagents at room temperature (unless indicated otherwise). Diligently follow all waste disposal regulations when disposing waste materials such as bacteria culture, medium, gel staining reagents.

2.1 Cloning Reagents

1. Phusion high-fidelity polymerase (NEB).
2. pGEM® T Easy kit (Promega).
3. QIAquick gel extraction kit (Qiagen).
4. DH5a, Stb3 *E. coli* strain (if cloning a difficult fragment).
5. QIAquick miniprep kit (Qiagen).
6. L-arabinose (10 %).
7. Isopropylthio-β-D-galactoside (IPTG: 1 M, 20 ml/plate).
8. 5-Bromo-4-chloro-3-indolyl-β-D-galactopyranoside (X-gal: 50 mg/ml in dimethylformamide, 40 ml/plate).
9. Low salt LB: 1 % BactoTryptone, 0.5 % Bacto yeast extract, 0.5 % NaCl.
10. Antibiotics: 20 μg/ml Chloramphenicol (CM, stock: 34 mg/ml in ethanol); 100 μg/ml Ampicillin (Amp, stock: 50 mg/ml in water); 25 μg/ml Kanamycin (Kan, stock: 50 mg/ml in water); 25 μg/ml Zeocin (Zeo, stock: 100 mg/ml in water); 50 μg/ml Hygromycin (Hyg, stock: 100 mg/ml in water).

11. SOC 0.2 %(W/V) Bacto-Tryptone, 0.05 %(W/V) Bacto-Yeast Extract, 10 mM NaCl, 2.5 mM KCl, 10 mM $MgCl_2$ and $MgSO_4$, 20 mM glucose.

2.2 Recombineering Reagents (See Notes 8 and 9)

1. pKD46 (BAD-λ-red) plasmid (*see* **Note 10**), which contains λ-Red operon under the control of arabinose-inducible promoter [7] (From CGSC: E. coli Genetic Stock Center).
2. DY380 (ts-λ-CI857-red) cells contain heat inducible λ-red in bacteria genome. Can be substituted with SW102 cells (DY380, *galK*) (From the National Cancer Institute, Frederick, MD).
3. DH10B cells (BAC host) (Invitrogen).
4. EL250 cells (DY380 with pBAD-Flpe insert at TetR, to be used to manipulate Frt sites flanking selection marker with arabinose induction). Can be substituted with SW105 cells (EL250, *galK*) (From the National Cancer Institute, Frederick, MD).
5. EL350 cells (DY380 with pBAD-Cre insert at TetR, to be used to manipulate loxP sites flanking selection marker with arabinose induction). Can be substituted with SW105 cells (EL350, *galK*) (From the National Cancer Institute, Frederick, MD).
6. Primer design and usage depend on individual BAC, insertion sites, and transgene to be inserted. One usually requires a 5′ and 3′ genotyping primer pair to check upstream and downstream insertion; one primer pair checks against the BAC and another one against the transgene. The primers checking against the BAC sequence should be outside of homologous arms (*see* **Note 11**).
7. Electroporation Instrument (such as Electroporator from Eppendorf) with 1 mm separated gap cuvette (BioRad).

3 Methods

3.1 Preparing the BAC

It is necessary to purify BAC DNA in order to move BAC between hosts, clean up BAC for use as a PCR template, and map the BAC with enzyme digestion and electrophoresis.

3.1.1 Isolating BAC DNA

Use an alkaline lysis protocol to isolate BACs from bacteria. Specifically, we use buffers P1, P2, and N3 from the Qiagen Miniprep kit.

1. Inoculate BAC-containing bacteria with 3 ml LB (12.5 μg/ml CM).
2. Incubate at 32 °C overnight on an orbital shaker (225 rpm).
3. Pour 2 ml overnight culture into a 2 ml round bottom tube.
4. Centrifuge at 21,130×*g*, 30 s.

5. Pour off supernatant and invert on rack to drain.
6. Tap on paper towel to remove any droplets.
7. Add 250 μl P1 and resuspend pellet completely.
8. Add 250 μl P2 and mix by inverting tube several times to improve lysis (*see* **Note 12**).
9. Add 350 μl N3 and mix by inverting tube several times.
10. Centrifuge at 21,130 × *g* for 4 min.
11. Transfer 800 ml of the supernatant to a 1.7 ml microcentrifuge tube.
12. Centrifuge at 21,130 × *g* for 4 more min.
13. Transfer supernatant to new microcentrifuge tube and add 0.5 ml isopropanol.
14. Mix well by inverting.
15. Centrifuge with hinge down (tube hinge facing the center of the rotor) at 21,130 × *g* for 15 min.
16. Carefully pour off supernatant.
17. Add 1 ml 70 % ethanol.
18. Spin at with hinge up (tube hinge facing away from the center of the rotor) at 21,130 × *g* for 10 min (this shifts the pellet, removing any remaining salt).
19. Carefully aspirate the supernatant, removing any droplets from the walls of the tube (*see* **Note 13**).
20. Invert tube on a rack and let air dry for 10 min (*see* **Note 14**).
21. Resuspend pellet in 20 μl H_2O with 10 mM Tris–HCl (pH 8) (*see* **Note 15**).

3.1.2 Preparation of Electrocompetent Cells

The procedures described below are optimized for DH10B cells, but can be adapted to other *E. coli* strains by adjusting the voltage setting during electroporation.

1. Inoculate 5 ml LB media with frozen DH10B cells stock and grow overnight. With standard transformation using electrocompetent or chemically competent cells, diluted overnight culture (1/20 dilution with LB medium) needs to be grown for 3 h to reach OD ~0.6 during the log grow phase [24]. However, overnight cultures (OD > 1.0) are preferred for BAC transformation, as high-density cells are easier to be transformed with large constructs such as BAC. In our lab, we have compared the transformation efficiency based on OD. Both OD ~0.6 and overnight culture can be used. However, overnight culture shows better consistency of having transformed colonies when using freshly prepared BAC miniprep as DNA source.

2. Pellet the bacteria culture with centrifugation (12.5 ml in a 50 ml conical tube, centrifuge at 1,157×*g* for 5 min at 4 °C, and transfer into a 2 ml tube after resuspension. If using less than 5 ml of overnight saturated culture, pellet all into a 2 ml round bottom tube with a bench-top centrifuge at room temperature).
3. Pour off supernatant.
4. Resuspend pellet in ice-cool 1 ml H_2O and add 1 ml more to bring volume to 2 ml (*see* **Note 16**).
5. Centrifuge in a microcentrifuge at 16,100×*g* for 30 s.
6. Pour off supernatant.
7. Repeat same procedure (**steps 3–5**) two more times.
8. After the last wash, remove as much H_2O as possible by pipetting out.
9. Pipet 50 μl H_2O into the tube and stir gently to resuspend.

3.1.3 BAC Transformation

1. Grow bacteria (DH10B and its derivatives—DY380, EL350) in 5 ml LB overnight at 32 °C with shaking.
2. Prepare electrocompetent cells (*see* Subheading 3.1.2).
3. Grow BAC, BAC-tg with its host in 3 ml LB overnight at 32 °C with shaking (*see* **Note 17**).
4. Make BAC miniprep (*see* Subheading 3.1.1).
5. Add 2 μl freshly prepared BAC into freshly prepared competent cells.
6. Pipet 50 μl electrocompetent cells into microcentrifuge tube containing 100–300 ng DNA to be transformed (*see* **Note 18**).
7. Pipet the cells and DNA into a 1 mm gap electroporation cuvette (1 mm separated gap cuvette such as BIORAD).
8. Wipe off sides of cuvette and tap on bench to remove any air bubbles.
9. Electroporate at 1,660 V (time constant should be between 3.8 and 5.4 ms) (*see* **Note 19**) (All Voltage based on the Electroporator 2510 from Eppendorf).
10. Pipet 300 μl SOC into electroporation cuvette.
11. Transfer SOC plus transformed cells back into the same tube that had contained the DNA.
12. Incubate at 32 °C with shaking for 1 h. Cells are ready to be plated onto an agar plate.
13. After incubation with SOC, plate all of culture on agar plate containing 20 μg/ml CM (and other antibiotics).
14. Incubate overnight at 32 °C.

3.1.4 PCR with BAC

1. BAC confirmation and BAC-tg genotyping require standard PCR reactions. For BAC sequencing or PCR cloning high-fidelity (HF) PCR enzymes should be used in order to minimize errors.
2. BAC cultures should serve as the PCR template.
3. For colony screening, inoculate 12–24 colonies with 50 μl LB with proper antibiotic and then incubate them for 3–4 h at 32 °C with shaking.
4. Follow the PCR protocol for standard Taq polymerase (*see* Table 1) or HF Taq polymerase (*see* Table 2). Conduct PCR with the appropriate primers.
5. Run 10 μl of standard PCR (5 μl HF-PCR) on an agarose gel to verify the PCR product.
6. For cloning of PCR reactions that yield unique band, the fragment can be purified using a gel extraction kit, e.g., QIAquick kit. With multiple bands, use gel purification to isolate a band with the correct size.

3.2 BAC Targeting

BAC targeting includes multiple steps: (1) Making BAC recombinogenic (*see* Subheadings 3.2.1 to 3.2.3). (2) Preparing targeting cassette (*see* Subheadings 3.3 to 3.4.3). (3) Targeting (*see* Subheading 3.5). (4) Removing of SM (*see* Subheading 3.6) (Optional) (*see* Fig. 2).

Table 1
PCR cycle protocol for standard Taq polymerase (25 μl total volume)

Cycle number	Denaturation	Annealing	Extension	Final
1	240 s @ 94 °C	–	–	–
2–35	30 s @ 94 °C	30 s @ 58 °C	60 s per kb @ 72 °C	–
36	–	–	420 s @ 72 °C	–
Hold	–	–	–	25 °C thereafter

Table 2
PCR cycle protocol for high-fidelity (phusion) Taq polymerase (50 μl total volume)

Cycle number	Denaturation	Annealing	Extension	Final
1	120 s @ 94 °C	–	–	–
2–35	30 s @ 98 °C	15 s @ 62 °C	15 s per kb @ 72 °C	–
36	–	–	420 s @ 72 °C	–
Hold	–	–	–	25 °C thereafter

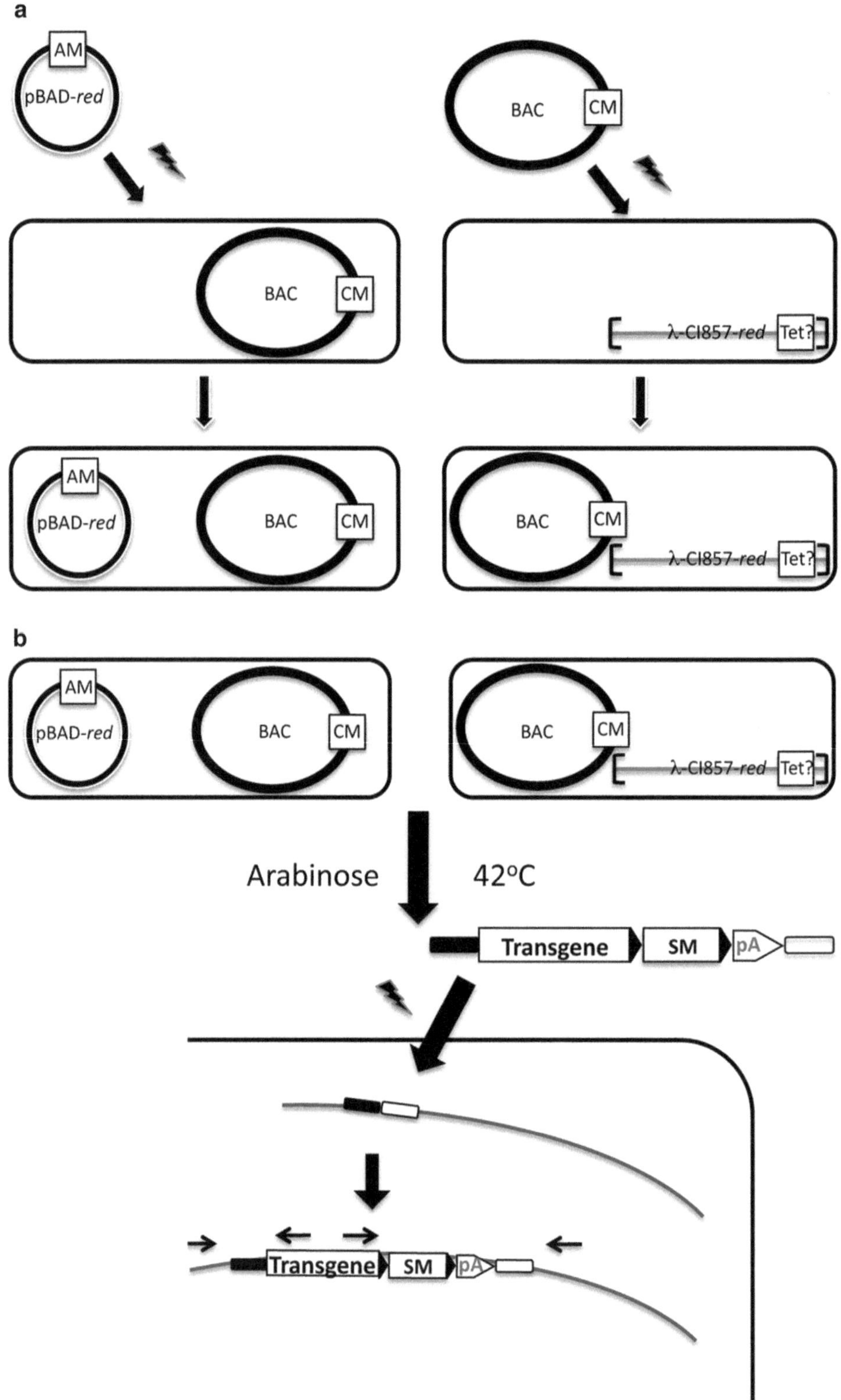

Fig. 2 (**a**) Making the recombinogenic host. Two strategies are shown to make a recombinogenic host. (*Left side*) The first strategy introduces the plasmid pBAD-red (pKD46) that contains the recombinogenic machinery (lambda-red) into a host already containing the BAC (lightning represents electroporation). This strategy is easier as it depends on antibiotic selection and plasmids are imported more efficiently than larger BAC vectors.

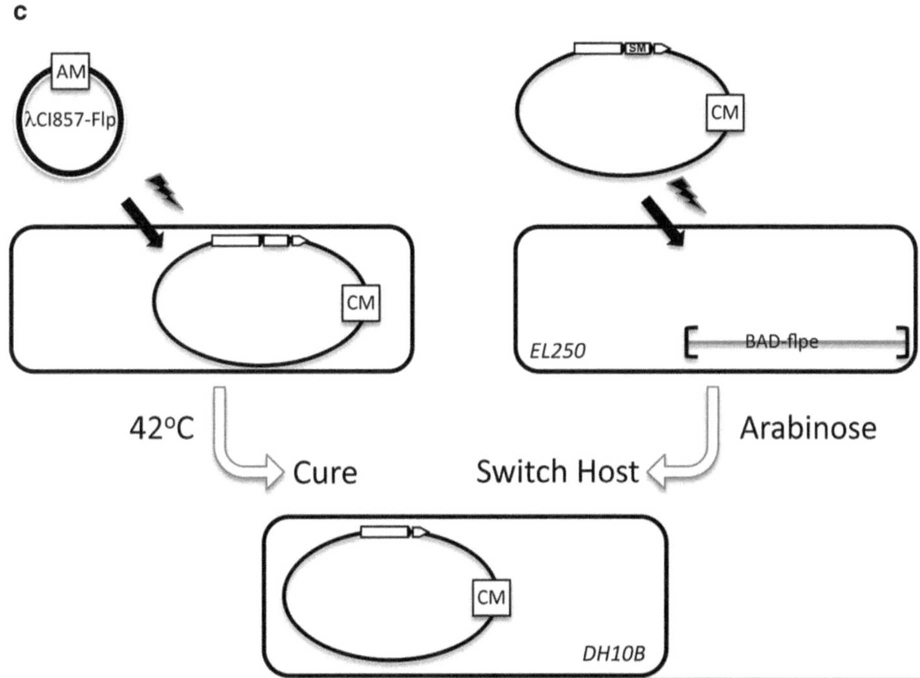

Fig. 2 (continued) (*Right side*) The second strategy introduces the BAC into a bacteria host that contains the lambda-red machinery in the bacterial chromosome. The second strategy typically is hindered by the low transformation efficiency of the large, low-copy BAC. Note the two selection markers described are Chloramphenicol (CM) and Ampicillin (AM). (**b**) Recombining the targeting cassette into a BAC clone. The recombinogenic host contains both the BAC and the lambda-red machinery, either on a plasmid (*Left side*) or integrated as a fragment within the *E. coli* chromosome (*Right side*). Arabinose (for plasmid) or high temperatures (for chromosome) is used to induce the lambda-red machinery. Electroporation is used to introduce the targeting cassette. Genotyping PCR is used to confirm integration. (**c**) Removal of floxed selection marker. Two approaches are shown to remove an FRT-flanked selection marker; the plasmid-based approach (*Left side*) introduces Flpe recombinase as a plasmid (the plasmid can be removed by curing) and the bacteria-based approach (*Left side*) uses EL250 with Flpe on the *E. coli* chromosome. Note that switching hosts is necessary to remove the Flpe-machinery

3.2.1 Making BAC Recombinogenic

In order to use recombineering to modify the BAC DNA, the host bacteria must contain both inducible λ-*red* components together with the DNA target and then induced to become recombinogenic. Targeted plasmid such as BACs can be transformed into DY380 for modification with the λ-*red* components (*see* Subheading 3.2.3) (*see* Fig. 2a). Alternatively, pKD46 can be transformed as a plasmid into the host of the BACs with the condition that pKD46 (Amp+) and target plasmid have different replication origin or different antibiotic resistance. It is usually not an issue as pKD46 (Amp+, ts101 ori) can be introduced into a host containing a BAC/PAC (Cm/Kan, F ori). However, issues may arise when using modified GENSAT BACs (Cm+, Amp+) or other common Amp+ cloning vectors due to the conflict of sharing the same antibiotic resistant gene. In this case, use DY380 or pKD46-like plasmid with different selection markers such as pET (Tet+) (*see* Subheading 3.2.2).

3.2.2 Transform BAC-Containing Bacteria with the pBAD-λ-Red (pKD46) Plasmid

1. Pipet 1 ml of overnight BAC culture into 2 ml round bottom tube.
2. Prepare electrocompetent cells (*see* Subheading 3.1.2).
3. Transfer the 50 μl of electrocompetent cells to a microcentrifuge tube containing 0.1 μg pKD46 plasmid.
4. Perform electroporation.
5. Pipet 300 μl SOC into cuvette, pipetting up and down twice and transfer into same tube that contained the plasmid.
6. Incubate at 32 °C with shaking for 1 h.
7. Spread 45 μl on an agar plate containing 12.5 μg/ml CM and 100 μg/ml Amp.
8. Incubate at 32 °C overnight.
9. Grow four colonies in 3 ml LB with Amp/Cm and pick one or two to perform targeting.

3.2.3 Transform λ-Red-Containing DY380 Cells with the Target BAC

1. Pipet 4 ml of overnight DY380 culture into 2 ml round bottom tube.
2. Prepare electrocompetent cells (*see* Subheading 3.1.2).
3. Prepare BAC (*see* Subheading 3.1.1).
4. Perform electroporation with BAC and DY380 cells (*see* Subheading 3.1.3 **steps 6–14**).
5. Pipet 300 μl SOC into cuvette, pipetting up and down twice and transfer into same tube that incubate at 32 °C with shaking for 1 h.
6. Spread all SOC on an agar plate containing 12.5 μg/ml CM and 25 μg/ml tetracycline.
7. Incubate at 32 °C overnight.
8. Grow four colonies in 3 ml LB with Tet/Cm and pick one or two to perform targeting.

3.3 Making the Targeting Cassette

The cassette contains two homologous arms (*see* **Notes 20–24**) that flank the transgene and SM. The 5′ homologous arm (left arm) and 3′ arm (right arm) can be between 30 and 200 bp. Longer arms have better specificity and targeting efficiency than the shorter arms. Shorter arms (<75 bp) can be conveniently added on by PCR, but the PCR amplification efficiency decrease with longer template. When DNA synthesis is not cost feasible, assembling these longer arms requires some cloning steps.

3.3.1 Assemble Homologous Arms by Cloning

If a unique restriction site already exists for cloning the transgene in between the arms, then primers LA-5 (5′ of left arm) and RA-3 (3′ of right arm) can be designed to amplify an approximately 200–400 bp PCR product which is then TA cloned into pGEM-TEasy (or any TA cloning vector) (*see* Subheading 3.3.2 and then 3.3.4).

Alternatively, if no restriction site is available, overlapping PCRs can be used to amplify two 100–200 bp regions with primers LA-5 and LA-3 to synthesize the left arm and primers RA-5 and RA-3 to make the right arm. Primers RA-5 and LA-3 should be located between primers LA-5 and RA-3 without generating PCR products that share overlapping sequences. Unique restriction enzyme sites are incorporated into the 3′ side of the primer LA-3 and the 5′ side of the primer RA-5 such that the restriction sites are complementary when overlapped (*see* Subheading 3.3.3). The transgene with selection marker can be added in-between homologous arms to make the targeting vector (*see* Subheading 3.3.5). The targeting vector with locus-specific homologous arms flanking the transgene and selection marker can then be excised to generate the targeting cassette for recombineering (*see* Subheading 3.4).

3.3.2 Amplification of PCR Products Containing an Existing and Unique Restriction Enzyme Site

1. Set up a high-fidelity PCR reaction using Phusion following the manufacturer's guidelines. Use 1 μl of the BAC overnight culture as the template and amplify with primer L (Forward) and primer R (Reverse).
2. Check 5 μl of the PCR on a 1 % agarose gel to make sure there is a product of the correct size (*see* **Note 25**).
3. Perform a restriction digest using the enzyme within PCR products to make certain the restriction enzyme site is there as expected.
4. To add a dA to the PCR ends for TA cloning, add 0.1 μl Taq polymerase to the PCR and set in the PCR machine at 72 °C for 10 min.
5. Run all PCR products on a 1 % agarose gel containing SYBR Safe DNA gel stain.
6. Using a clean razor blade, excise the band from the gel while viewing under blue light.
7. Isolate the DNA using the QIAquick gel extraction kit.
8. Quantitate the DNA using a spectrophotometer.
9. Continue as TA cloning (*see* Subheading 3.3.4).
10. The resulting TA cloned plasmid with PCR products, which are generated by primer L and primer R, can be called generally L/R plasmid and specified with the name of the target gene.

3.3.3 Generation of Cloning Site by Overlapping PCR

1. Synthesize primers LA-5 and LA-3 (left arm) and RA-5 and RA-3 (right arm). Incorporate a few restriction enzyme sites in LA-3 and RA-5 primers that are noncutters both in homologous arms also in the cloning vector (i.e., pGEMT-Easy). These restriction sites could possibly allow subsequent directional cloning of the transgene by using unique enzymes to remove the transgene from its vector. If this is not possible, the transgene can be blunted after digestion to clone into a blunt site in-between arms.

Table 3
High-fidelity, overlapping PCR reaction

Number of cycles	Time	Temperature (°C)
1 Cycle	30 s	98
5 Cycles	10 s	98
	15 s	52
	15 s	72
1 Cycle	5 min (Add primers during this 5 min period)	72
35 Cycles	10 s	98
	15 s	60
	15 s	72
1 Cycle	7 min	72
1 Cycle	10 min	4

2. Set up a high-fidelity PCR using Phusion to amplify left (LA) and right (RA) arms [25].
3. Check 3 μl of the PCR on a 1 % agarose gel to make sure products of the expected sizes are present.
4. Run all PCR products on a SYBR safe gel, excise fragments, and isolate using QIAquick gel extraction kit.
5. Quantify the DNA using a spectrophotometer. Dilute the fragment with the lowest concentration 1:10 and then dilute all other fragments such that their concentrations are the same in order to optimize the template concentration for subsequent PCR reactions.
6. Set up a high-fidelity overlapping PCR reaction using 1 μl of each diluted fragment (*see* Table 3). For the first 5 cycles, no primers are added. This step combines both arms through overlapping sequence (hence the term overlapping PCR) and DNA extension reactions. The annealing temperature should be set at the Tm of the overlapping area (e.g., 52 °C for a 16 bp overlap). After 5 cycles, resume standard HF-PCR reaction.
7. After confirming a correct LR-PCR product by size and by restriction enzyme digestion, follow the same procedure as **steps 4–8** of Subheading 3.3.2 and then proceed to TA cloning (*see* Subheading 3.3.4).

3.3.4 TA Cloning

1. Set up a ligation to TA cloning vector using a 3:1 ratio of insert:vector and following the instructions of pGEM-T Easy cloning (Promega) or Topo-TA cloning (Invitrogen).
2. Transform DH5α cells (NEB) or Top10 (Invitrogen) chemically competent cells with 42 °C heat shock for 30 s and add 300 μl SOC after the transformation and incubate for 1 h at 37 °C.
3. Prepare an Amp agar plate to set up blue/white screening of colonies by spreading 20 μl 0.1 M IPTG and 40 μl 20 mM X-gal onto the plate.
4. Spread 125 μl of the transformation onto the plate and incubate at 37 °C overnight.
5. Pick 12–24 white colonies into 50 μl LB containing Amp in a 96-well-plate and incubate with shaking for 3 h.
6. PCR screen using primers L and R.
7. Grow four positive colonies in 3 ml LB containing Amp at 37 °C overnight.
8. Prepare miniprep DNA using the QIAspin miniprep kit and check insert with suitable restriction enzyme digestion.

3.3.5 Preparation of Targeting Vector

1. Digest the L/R plasmid DNA (*see* Subheading 3.3.2, **step 10**) with the unique restriction enzyme within the homologous region amplified from the targeted BAC. Treat linearized DNA fragment with calf alkaline phosphatase (CIP) to reduce self-ligation.
2. Harvest the transgene and selection marker from the respective plasmids with enzyme digestion or PCR amplification. Enzyme sites or PCR primers depend upon the plasmid used in a particular project (*see* Subheadings 1.1.5 and 1.1.6).
3. In order to blunt one or both ends after the digest, treat the reaction with Klenow fragment. The Klenow reaction is as follows: add 1 μl dNTP mix (10 mM each) and 1 μl Klenow, incubate at room temperature for 30 min, and then heat inactivate at 75 °C for 20 min.
4. Run the digest on a SYBR safe gel, excise the band, and purify the product using the QIAquick gel extraction kit (*see* **Note 26**).
5. Quantitate both DNA fragment using a spectrophotometer.
6. Set up a ligation reaction using a rapid ligation kit, e.g., Roche. Use 25 ng of vector and 3:1 ratio of insert:vector. Bring the volume to 9 μl with H_2O. Add 10 μl of 2× ligase buffer and 1 μl of ligase. Incubate the ligation reaction at room temperature for a minimum of 5 min.
7. Transform chemically DH5α competent cells and add 300 μl SOC after the transformation and incubate for 1 h at 37 °C.

8. Plate on agar plate with appropriate antibiotics.
9. Incubate overnight at 37 °C.
10. Pick 24 colonies in 50 μl LB containing appropriate antibiotics in 96-well plate and incubate at 37 °C for 3 h.
11. PCR screen the colonies using the following two reactions: primer LA-5 and a reverse primer near N-terminal of transgene, and primer RA-5 with primer near C-terminal of transgene with primer R. This ensures that transgene is flanked with homologous arms.
12. Grow four positive colonies in 3 ml LB plus antibiotics overnight at 37 °C.
13. Isolate plasmid DNA using the QIAspin miniprep kit.
14. Digest the plasmids with enzymes that will indicate proper construction. Additionally, PCR both ends with primer pairs L/tgR and tgF/R to confirm the integrity of targeting cassette (*see* **Note 27**).

3.4 Preparing the Targeting Cassette

Once the targeting cassette (which includes the homologous arms flanking the transgene and the selectable marker) is cloned, the cassette can be excised by enzyme digestion or via HF-PCR with primers L/R. Enzyme excision is an easy way to harvest the targeting cassette without a size limitation of targeting cassette, but the technique requires enzyme sites that can remove the intact targeting cassette. PCR not only can increase the concentration of the targeting cassette when plasmid DNA is limited, but it can also generate the targeting cassette when there are no enzymes sites that can be used to excise the targeting cassette.

3.4.1 Preparing the Targeting Cassette by Enzyme Excision

1. Set up a restriction enzyme digest using 14–17 μl of the targeting cassette vector in a 20 μl reaction to release the targeting cassette with the sites that flank targeting cassette (usually NotI sites in the pGEMTEasy vector). Additional enzymes, which only digest vector backbone, can be included to remove the vector backbone and reduce background from any residual vector after targeting.
2. Run the digest on a SYBR safe gel, excise the fragment, and gel purify.
3. Run 1/5 of purified product to confirm the size of band and quality of DNA.
4. Longer arms increase targeting efficiency, and as few as 10 ng (which can be seen on the gel) can be effective in targeting to a BAC.

3.4.2 Preparing the Targeting Cassette by PCR

1. Set up a high-fidelity PCR reaction using Primer L/R with a template either the digested and purified targeting cassette from Subheading 3.4.1 or a 1:100 dilution of the targeting

cassette vector (1 μl, ~2 ng). Follow the PCR program guidelines for the Phusion PCR enzyme and use an extension time of 15 s per kb.

2. Check 3 μl on a gel to make sure the correctly sizes product is present.
3. Digest the PCR reaction with 2 μl DpnI and 2 μl vector backbone-specific enzymes(s) to remove the DNA template. The digestion reduces background products after targeting.
4. Run all PCR products on a SYBR safe gel, excise the band, and gel purify. Depend on the PCR efficiency, usually 100–500 ng targeting cassette can be acquired.
5. Depending on the length of homologous arms, 10–200 ng of targeting cassette is needed for one targeting reaction.

3.4.3 Preparing the Targeting Cassette by Using Long Primers

When the combined DNA size of the transgene and selection marker is short, one can skip homologous arms cloning steps by introducing a pair of 50 bp homology arms in the PCR primers to amplify the transgene with the selection marker. This alternative method enables making targeting cassette with one single PCR reaction, and is suitable for limited-size transgene.

1. Synthesize primers LA and RA. These primers are approximately 70 mers with 50 bases homologous to the BAC at the insertion site and 18–25 bases (Tm 50–60 °C) homologous to the transgene. 50 mer each of homologous arms is standard. It can be as few as 30 mer which results in reduced targeting efficiency.
2. Set up a high-fidelity PCR reaction using 1 μl of approximately 2 ng/μl transgene vector as the template.
3. Follow HF-PCR for extension guidelines and **steps 2–4** of Subheading 3.4.2 for preparation of the targeting cassette.

3.5 Recombining the Targeting Cassette into a BAC Clone (See Fig. 2b)

1. Inoculate four tubes containing 3 ml LB with 20 μg/ml CM and 100 μg/ml Amp with individual colonies from the BAC/pKD46 plate from Subheading 3.2.3. Grow overnight with shaking at 32 °C (With BAC/DY380 grow in 20 μg/ml CM and 25 μg/ml Tet).
2. Pipet 625 μl of overnight culture into 12.5 ml LB in a 50 cc conical tube.
3. Add 125 μl 10 % L-arabinose to a final concentration of 0.1 %.
4. Grow at 32 °C with shaking for 3 h (With BAC/DY380 after 3 h at 32 °C, shift to 42 °C for 15 min, and cool down quickly with ice slurry).
5. Pipet 1 ml of culture into 1 cm cuvette and check OD_{600} nm with spectophotometer. Usually it takes 3 h to reach OD_{600} around 0.6.

6. Centrifuge remaining culture at 1,157×*g* for 5 min at 4 °C.
7. Prepare electrocompetent cells (*see* Subheading 3.1.2).
8. Electroporate the cells (*see* **steps 6–14** of Subheading 3.1.3) with the targeting cassette (*see* **Note 28**).
9. After incubation in SOC, plate 100 μl on an agar plate containing 20 μg/ml CM and the antibiotic for the selection marker (e.g., Kan (25 μg/ml) for a neo cassette).
10. Incubate at 32 °C overnight.
11. Screen 24 colonies with PCR using 5′ and 3′ genotyping primers.
12. Grow four positive colonies in 3 ml LB containing μg/ml CM and the selection marker antibiotic.
13. PCR BAC culture (1 μl) with both 5′ and 3′ genotyping primers (*see* Subheading 3.1.4).
14. Prepare BAC minipreps from the cultures with positive PCR result (*see* Subheading 3.1.1).
15. Digest 8 μl with SpeI to check quality of BAC. The average fragment of mouse (or human) genomic DNA after SpeI digestion is about 9 (8) kb. This digest should generate multiple distinguishable bands for the BAC, whereas genomic DNA will produce a smear because of the presence of too many sites for this enzyme. Other enzymes with smaller average fragment, such as XbaI or PstI, can be used when SpeI digestion only yields large DNA fragments.

3.6 Removal of Selection Marker (See Fig. 2c)

1. Grow EL250 cells in 5 ml LB and overnight at 32 °C with shaking if using Flpe (Frt sites) or EL350 cells if using Cre recombinase (loxP sites) (*see* **Note 29**).
2. Add 0.1 % Arabinose to induce the expression of Flpe (Cre) for 1 h at 32 °C.
3. Prepare electrocompetent cells (*see* Subheading 3.1.2).
4. Electroporate (*see* **steps 6–14** of Subheading 3.1.3) 2 μl freshly prepared BAC (20–40 ng).
5. After incubation with SOC, plate all of the culture on an agar plate containing 20 μg/ml CM.
6. Incubate overnight at 32 °C (*see* **Note 30**).
7. Pick 24 colonies in wells containing 50 μl 20 μg/ml CM and another duplicate set of 24 wells with only the selection marker antibiotic (*see* **Note 31**).
8. Incubate for 3 h or until growth is visible at 32 °C with shaking.
9. PCR screen with 3′ genotyping primers.
10. Choose four positive colonies that not only displayed negative growth in the selection marker antibiotic, but also yielded

smaller PCR band, indicated lost of selection market. Grow them in 3 ml LB containing 20 μg/ml CM.

11. Prepare BAC minipreps (*see* Subheading 3.1.1).
12. PCR with both 5′ and 3′ genotyping primers.
13. Transfer BAC to its original host (DH10B) (*see* Subheading 3.1.3).
14. After quality control, BAC transgene constructs can be used for cell transfection and embryo injection (*see* **Notes 34** and **35**).

4 Notes

1. A common BAC library is organized as 384-well format, so any BAC clone can be identified with its own ID address <Library code - plate # well number> [26]. A few BAC libraries are the basis of whole genomic DNA sequencing. Thus, some clones have been sequenced completely. Additional clones within the same library, as well as clones in different libraries but of the same species or different strains, have been mapped back to their particular genomic locus by using a T7 or SP6 primer to cover the sequences at the ends of the BAC in question [26].
2. Different BACs can originate from different libraries, and these libraries may contain different genomic DNA and sometimes different vector backbones. When genomic DNA comes from eukaryotic cells, DNA elements such as repeats, GpC islands, or AT stretches may cause some instability inside of bacteria. Some libraries are CM-resistant (BAC) and some are Kan-resistant (PAC) based on the original vector. Some GENSAT clones are both CM- and Amp-resistant since the targeting vector sequence was inserted during targeting and not resolved. Details for each BAC library about the source of DNA, cloning vectors, and cloning sites can be found with the CHORI web link (https://bacpac.chori.org/home.htm).
3. Due to their size, most BACs contain not only the targeted gene but also nearby genes. If these genes are intact, gain-of-function mutations of those genes are expected in transgenic animals. However, when the gene has no dosage-dependent effect or the expression level is tightly regulated, such gain-of-function mutations may or may not pose an issue. If an unexpected phenotype occurred with BAC transgene and expression of nearby genes is suspected, one can either find a BAC with just the right coverage or remove nearby genes in the BAC by targeting (using the present protocol) or BAC trimming [27].
4. Without completely sequencing the DNA of the BAC insert, BAC-end sequencings have been used to map inserts of genomic DNA back to the location on a chromosome. This resulted in BACs with similar coverage that may contain

differences due to strain variation or DNA recombination. Such variations are rare in a pure mouse strain, but it is worth checking the quality of a BAC before and after each manipulation. There is also the chance of cross-contamination between wells, so it is recommended to check individual colonies by PCR reaction with the primers of the gene-of-interest(s) to confirm the identity of a specific BAC.

5. Not all loci are compatible with a selected transgene. Thus, flexibility in removing part of a targeted gene or relocation to a different site is needed if troubleshooting cannot help. GC-rich areas are problematic during targeting. Not only are such areas more difficult to target and genotype (the PCR reactions must be modified), but also the associated PCR reactions are prone to make truncations in DNA even when using high-fidelity enzymes.
6. The popular mouse C57/black 6 BAC library (RP23/24) is constructed in the vector pBACe3.6/pTARBAC [28]. Both vectors contain the T7/Sp6 polymerase binding site for BAC-end sequencing or PCR, a homing enzyme site (PI-SceI) for linearization of the BAC, as well as SacB for counter-selection. Additional features such as the presence of loxP and loxP511 might cause complications when using these BACs to express Cre recombinase. However, with a RMCE reaction, a selection marker with dual promoters can be inserted in those loxP sites and selected in the eukaryotic cell lines [11].
7. The purpose of a selection marker is to increase the ability of acquiring the proper-targeted construct. It is advantageous to use a "removable" cassette in order to keep, delete, or exchange a selection cassette depending on the experimental design. FRT and loxP flanking cassettes can take advantage of the RMCE method to be modified with different selection methods (Neo, Hygro, and Zeo) [21].
8. Always start a bacterial culture from a frozen stock. It is essential to maintain the recombinogenic property of the plasmids and the DY380 (EL250, 350) bacteria.
9. Make glycerol stocks of recombinogenic bacteria by the following steps. Streak bacteria on an agar plate. Pick 6–12 individual colony into 1 ml LB, and after a short incubation around 30 min divide the culture into two halves. Grow one half in 32 °C and the other half in 42 °C overnight. Make glycerol stocks with the duplicate culture that grow at 32 °C but die completely at 42 °C. Constant expression of λ-red reduced genomic DNA stability and cell viability. Those cells can survive O/N induction of λ-red indicated the lost of recombination property.
10. Obtain a large preparation of functional pKD46 in −20 °C from the original source [7] or after its λ-red function has been

assayed. To reduce freezing and thawing, keep a small aliquot in a cold room.

11. Proper genotyping assays in conjunction with BAC preparation and digestion is necessary to monitor the fabrication of a BAC transgene throughout targeting, removal of a selection marker, and switching a host when using DY380 and EL250.
12. If the protocol yields no BAC DNA, retest each buffer by using them for smaller plasmid miniprep. Check if the bacteria lysis is completed. A completed lysis solution changes from cloudy to transparent.
13. Be careful of the pellet when discarding the supernatant via pouring or suction. The pellet may be hard to see but there should always be enough debris to form a white visible pellet. The volume of the starting culture may be increased up to threefold if necessary and combined into a single 2 ml tube. Half of the BAC miniprep may be digested with enzymes like *SpeI* or *XbaI* and separated on a 1 % agarose gel to confirm the presence of BACs. The digested BAC will show as multiple distinct bands as average fragment size of genomic DNA after digestion is about 9 kb for SpeI and 3 kb with XbaI.
14. The BAC can be stored at this step (before transformation) as a dry pellet at −20 °C for use in transformation up to several months.
15. Handle BACs with care as they contain large inserts and are low-copy number plasmids. Do not vortex the BAC or use harsh pipetting. Gently handle the BAC throughout the protocol.
16. When water is replaced with 10 % glycerol, electrocompetent cells can be stored in −80 °C, but this practice is not recommended for BAC transformation, as frozen electrocompetent cells reduce both transformation efficiency and incorporation of large DNA.
17. Saturated bacteria culture enhances transformation efficiency of larger plasmids. For normal size plasmids (<20 kb), dilute saturated culture in 1/20 with LB and grow 3 h to around OD = 0.6.
18. Avoid generating air bubbles or any excessive forces, which would result in low efficiency of transformations.
19. If the size of the transgene and remaining vector are similar enough to hinder gel purification, it may be feasible to cut into the vector backbone with additional restriction enzyme digestion.
20. Time constants lower than 2 ms indicate there is excess salt or air bubbles trapped inside the chamber. Be sure to eliminate all bubbles and check the source of excess salt, which can come from the DNA sample or cell preparation. The electroporation cuvette can be reused after flushing with 2 ml ice-cold water.

21. Depending on the targeting cassette size and targeting site location, the *Red* operon can promote recombination with as little as 50 mer homologous arms. However, longer homologous arms provide better efficiency and specificity. The distance of two homologous arms also has an effect during targeting. The closer the arms are together, the better the targeting efficiency.
22. Single arm homologous recombination used in GENSAT projects is possible [3]. However, two homologous arms are preferred to achieve "clean" targeting without including the vector backbone, to avoid including duplicate targeting arms that need to be resolved, and to silence the targeted gene. Furthermore, only with two arms can one move along the target locus to generate a deletion or remove unwanted sequences.
23. When using two homologous arms, it is always necessary to make sure both arms are properly flanking the transgene. When using a high-fidelity enzyme during PCR, the overhanging primer (50 mer) may be slightly chewed-in by the action of its proofreading activity. In this case, PCR confirmation of the arms is needed. When using longer homologous arms, overlapping PCR reactions can be adapted to stitch the arms together and provide a multiple cloning site for the ligation of transgene. Arm-extension, by inserting the short-arms-target cassette to a larger DNA fragment flanking the same target site as the short arms, is another way to keep and extend homologous arms when shorter homologous arms are insufficient for targeting. Targeting efficiency, when recombination machinery is kept constant, is dependent upon the number of targeting sites (copy of plasmid) and the length of homologous arms. Higher copy number and longer homology are known to increase the recombination efficiency [29].
24. To insert a transgene into a specific locus, traditional cloning procedures have limitations. PCR reactions, in part or in whole, are required to generate a targeting cassette especially when the insertion site is crucial. Thus, the benefits and limitations of a PCR reaction will need to be considered. For instance, large transgenes over 4 kb are more difficult to be amplified, and PCR amplification increases the chance of mutations. For high-throughput approaches, excellent primer binding sites that flank transgenes facilitate the generation of a targeting cassette by PCR.
25. The insertion of a transgene via homologous recombination is high fidelity, so any sequence in between homologous arms will be inserted. A well-characterized transgene is essential in making a BAC transgene. Any mutation or unwanted feature in a transgene will be carried to the BAC and it is very difficult to correct any error in a BAC-tg construct. When a transgene such as toxin or recombinase is expressed in bacteria, it might

cause an unwanted consequence and require the change in the transgene or insertion site. Sequential targeting can be used to increase the complexity of a transgene where there is a shared 5′ and 3′ DNA sequence.

26. If there are no PCR products, the wrong BAC was used or the BAC may be truncated or rearranged. Isolate new BAC and use genomic DNA (1 μg) as a positive control to check primer design and the PCR reaction conditions. If there are GC-rich areas, DMSO may be added to reduced secondary structures; note that this may increase mutations.
27. If no correct DNA fragments are observed, check that the required DNA concentration of 0.5–1 μg is used for the safe DNA staining, which use DNA dye SYBr Green and blue light boxes to visualize DNA on agarose gel to avoid UV light exposure. Check each individual enzyme digestion (including those in Subheading 3.2.2). The DNA fragments may be the wrong size if the arms or the transgene are missing any restriction enzyme sites. Usually raising the DNA quality, the reaction volume, and also using appropriate enzyme reaction buffers help to avoid partial digestion.
28. The clarity, integrity, and quantity of a targeting cassette are vital for successful targeting and can be monitored with gel electrophoresis or the OD 260/280 ratio.
29. DNA concentration of the targeting cassette should be 100–300 ng for short homologous arms and 50–200 ng for longer arms with a volume less than 3 μl. Large volume will dilute cell and introduce salt concentration. When targeting cassette concentration is too low, DNA precipitation can be used to increase DNA concentration.
30. EL250/350 cannot be selected by Tet. The colonies appear opaque, pale yellow, and grow slowly (overnight culture at 32 °C appear pinhead-sized). Contamination is observed when the colonies are the wrong color or grow very quickly. Thus, use disposable materials and rigorous aseptic technique to avoid contamination.
31. If there are no colonies, repeat the transformation with fresh BAC and competent cells (cells should be at a high OD > 1.0). Prepare the cells gently with careful pipetting, avoiding air bubbles and keep cells on ice. Confirm BAC DNA recovery through enzyme digestion and gel electrophoresis (1 % gel).
32. Based on experience, multiple targeting events can destabilize the BAC, which might be caused by leftover components of λ-red after each targeting. Transform BAC away from a recombinogenic environment if necessary (e.g., into DH10B cells).
33. Using a pool of BAC-tg can reduce the effort of characterizing individual colonies, especially when BAC-tg constructs are

used in cell culture study. It also dilutes the random mutation during construct fabrication. For making transgene animals or plants, it is better to use individual clones because individual colonies behave differently sometimes in terms of growth and it may be difficult to weed out the undesired product in a pool.

34. It is desirable to use simple and straightforward PCR reactions without resorting PCR reaction optimization and special PCR protocol (such as touch down PCR) to identify the proper integration of a transgene. These same PCR reactions may be used for the screening of the founder generation of transgene animals and plants.

35. When using homologous recombination with genomic DNA that contains repeats, be aware that the BAC insert may recombine after manipulation. Quality control is needed for each step before using BAC to expression a transgene in vivo (*see* **Notes 32** and **33**). Common recommendations are listed and additional requirements may arise from the facility that performs the BAC-tg injection.
    (a) Sequence the transgene while in the BAC-tg or HF-PCR product.
    (b) Always include the original unmodified BAC as control.
    (c) Follow BAC preparation with genotyping PCR that uses the bacteria culture as template.
    (d) Check BAC ends (junctions of the BAC vector and insert) with primers against BAC vector and insert ends (which determine BAC coverage).
    (e) BAC integrity can be checked with pulse-field gel electrophoresis (PFGE).
    (f) If PFGE is unavailable, PCR scanning can be performed with 10–20 kb interval along BAC coverage to rule out truncation.
    (g) DNA purity is important; either gel purification or CsCl banding is recommended.
    (h) Linearize BAC DNA with unique sites such as homing enzyme (PI-SceI) that is available in many BAC vectors.

Acknowledgements

The authors thank Dr. B.L. Wanner of Purdue University for the pKD46 plasmid, which was distributed by the E. Coli Genetic Stock Center at Yale. The authors also thank Dr. N. Copeland for providing DY380 and EL250/350. The protocol described here has been used in the UNC Neuroscience Center's BAC Engineering Core and is supported by an NINDS Center Grant.

References

1. Gong S, Zheng C, Doughty ML et al (2003) A gene expression atlas of the central nervous system based on bacterial artificial chromosomes. Nature 425(6961):917–925
2. Poser I, Sarov M, Hutchins JRA et al (2008) BAC TransgeneOmics: a high-throughput method for exploration of protein function in mammals. Nat Meth 5(5):409–415
3. Yang XW, Model P, Heintz N (1997) Homologous recombination based modification in Escherichia coli and germline transmission in transgenic mice of a bacterial artificial chromosome. Nat Biotech 15(9):859–865
4. Copeland NG, Jenkins NA, Court DL (2001) Recombineering: a powerful new tool for mouse functional genomics. Nat Rev Genet 2(10):769–779
5. Suster M, Sumiyama K, Kawakami K (2009) Transposon-mediated BAC transgenesis in zebrafish and mice. BMC Genomics 10(1):477
6. Muyrers JPP, Zhang Y, Benes V et al (2000) Point mutation of bacterial artificial chromosomes by ET recombination. EMBO Rep 1(3):239–243
7. Datsenko KA, Wanner BL (2000) One-step inactivation of chromosomal genes in Escherichia coli K-12 using PCR products. Proc Natl Acad Sci 97(12):6640–6645
8. Zhang Y, Buchholz F, Muyrers JPP et al (1998) A new logic for DNA engineering using recombination in Escherichia coli. Nat Genet 20(2):123–128
9. Lee EC, Yu D, Martinez de Velasco J et al (2001) A highly efficient Escherichia coli-based chromosome engineering system adapted for recombinogenic targeting and subcloning of BAC DNA. Genomics 73(1):56–65
10. Bouvier J, Cheng J-G (2001) Recombineering-based procedure for creating Cre/loxP conditional knockouts in the mouse. In: Ausubel FM, Brent R, Kingston RE, Moore DD, Seidman JG, Smith JA, Struhl K (eds) Current protocols in molecular biology. Wiley, New York, NY
11. Hollenback SM, Lyman S, Cheng J (2001) Recombineering-based procedure for creating BAC transgene constructs for animals and cell lines. In: Ausubel FM, Brent R, Kingston RE, Moore DD, Seidman JG, Smith JA, Struhl K (eds) Current protocols in molecular biology. Wiley, New York, NY
12. Baron U, Gossen M, Bujard H (1997) Tetracycline-controlled transcription in eukaryotes: novel transactivators with graded transactivation potential. Nucleic Acids Res 25(14):2723–2729
13. Hayashi S, McMahon AP (2002) Efficient recombination in diverse tissues by a tamoxifen-inducible form of Cre: a tool for temporally regulated gene activation/inactivation in the mouse. Dev Biol 244(2):305–318
14. Sauer B (1998) Inducible gene targeting in mice using the Cre/lox system. Methods 14(4):381–392
15. Fritze CE, Anderson TR (2000) Epitope tagging: general method for tracking recombinant proteins. Methods Enzymol 327:3–16
16. Martinez-Salas E (1999) Internal ribosome entry site biology and its use in expression vectors. Curr Opin Biotechnol 10(5): 458–464
17. Nott A, Meislin SH, Moore MJ (2003) A quantitative analysis of intron effects on mammalian gene expression. RNA 9(5): 607–617
18. Zufferey R, Donello JE, Trono D et al (1999) Woodchuck hepatitis virus posttranscriptional regulatory element enhances expression of transgenes delivered by retroviral vectors. J Virol 73(4):2886–2892
19. Cockerill PN, Garrard WT (1986) Chromosomal loop anchorage of the kappa immunoglobulin gene occurs next to the enhancer in a region containing topoisomerase II sites. Cell 44(2):273–282
20. Ebert MS, Neilson JR, Sharp PA (2007) MicroRNA sponges: competitive inhibitors of small RNAs in mammalian cells. Nat Meth 4(9):721–726
21. Bode J, Schlake T, Iber M et al (2000) The transgeneticist's toolbox: novel methods for the targeted modification of eukaryotic genomes. Biol Chem 381(9–10):801–813
22. Bochkov YA, Palmenberg AC (2006) Translational efficiency of EMCV IRES in bicistronic vectors is dependent upon IRES sequence and gene location. Biotechniques 41(3):283–284, 286, 288
23. Provost E, Rhee J, Leach SD (2007) Viral 2A peptides allow expression of multiple proteins from a single ORF in transgenic zebrafish embryos. Genesis 45(10):625–629
24. Sheng Y, Mancino V, Birren B (1995) Transformation of *Escherichia coli* with large DNA molecules by electroporation. Nucleic Acids Res 23(11):1990–1996
25. Higuchi R, Krummel B, Saiki R (1988) A general method of in vitro preparation and specific mutagenesis of DNA fragments: study of protein and DNA interactions. Nucleic Acids Res 16(15):7351–7367

26. Osoegawa K, Tateno M, Woon PY et al (2000) Bacterial artificial chromosome libraries for mouse sequencing and functional analysis. Genome Res 10(1): 116–128
27. Hill F, Benes V, Thomasova D et al (2000) BAC trimming: minimizing clone overlaps. Genomics 64(1):111–113
28. Jong PD (2013) CHORI: Children's Hospital Oakland Research Institute. Available from: https://bacpac.chori.org/about.htm. Accessed on August 24, 2013
29. Shen P, Huang HV (1986) Homologous recombination in *Escherichia coli*: dependence on substrate length and homology. Genetics 112(3):441–457

Chapter 5

Directing Enhancer-Traps and iTol2 End-Sequences to Deleted BAC Ends with loxP- and lox511-Tn10 Transposons

Pradeep K. Chatterjee

Abstract

A step-by-step detailed procedure is presented to progressively truncate genomic DNA inserts from either end in BACs. The bacterial transposon Tn10 carrying a loxP or a lox511 site is inserted at random into BAC DNA inside the bacterial cell. The cells are then infected with bacteriophage P1. The Cre protein expressed by phage P1 generates end-deletions by specifically recombining the inserted loxP (or lox511) with the loxP (or lox511) endogenous to and flanking insert DNA in BACs from the respective end. The Cre protein also helps phage P1 transduce the BAC DNA by packaging it in P1 heads. This packaging by P1 not only recovers the rare BAC clones containing Tn10 insertions efficiently but also selects end-truncated BACs from those containing inversions of portions of their DNA caused by transposition of Tn10 in the opposite orientation. The libraries of end-deleted BACs generated by this procedure are suitable for numerous mapping studies. Because DNA in front of the loxP (or lox511) arrowheads in the Tn10 transposon is retained at the newly created BAC end, exogenous DNA cassettes such as enhancer-traps and iTol2 ends can be efficiently introduced into BAC ends for germline expression in zebrafish or mice. The methodology should facilitate functional mapping studies of long-range *cis*-acting gene regulatory sequences in these organisms.

Key words Cre-lox recombination, End-deletions, P1 transduction, Tn10 transposon retrofitting, Enhancer-trapping, iTol2 BACs, Enhancer-trap BACs, Physical mapping of markers, Identifying regulatory sequences

1 Introduction

Locating genetic markers on a physical map of the chromosome was a challenge prior to the availability of entire genome sequences, but was facilitated by easy access to large collections of BACs truncated from an end of insert DNA [1, 2]. Because high-quality DNA can be easily prepared from BACs, they have become valuable reagents for direct use in biologic experiments that seek to functionally evaluate the sequences contained in them [reviewed in [3]].

Kumaran Narayanan (ed.), *Bacterial Artificial Chromosomes*, Methods in Molecular Biology, vol. 1227,
DOI 10.1007/978-1-4939-1652-8_5, © Springer Science+Business Media New York 2015

Thus, both physical mapping of genetic markers and functional mapping of long-range gene-regulatory sequences should benefit from efficient procedures to make progressive truncations from either end of insert DNA in BACs [4, 5].

Procedures for deleting pieces sequentially from one or both ends of cloned genomic DNA have been developed using a variety of transposon systems but are limited because these methods necessitated additional steps of modifications and sub-cloning of the BAC [6, 7]. While the approach described in this chapter also uses transposons, it is more convenient because sub-cloning of the genomic DNA is not required and it can be used directly on all BACs from the public domain. And in contrast to procedures that use sequence homology-based recombination to engineer end-deletions, which can only produce one alteration at a time, our approach with loxP or lox511 transposons can generate large collections of end-deleted BACs in a single experiment [3]. It is significant that the recombinases involved in our approach, Tn10-transposase and Cre protein, respectively, do not act upon repetitive sequences and/or other recombinogenic sites in the genomic DNA insert, therefore minimizing the risk of generating unwanted rearrangement products that are commonly associated with recombination-based methods. Thus BACs spanning loci with highly repetitive DNA content such as those commonly found in the human and other genomes are also amenable to our methodology [2].

The first step of our approach is the introduction of a small (6–7 kb) Tn10 transposon plasmid into the *E. coli* host carrying the BAC clone by calcium chloride transformation. Depending on which end of BAC insert DNA is chosen for progressive end-truncations, a loxP or lox511 site is incorporated into the Tn10 transposon. It should be noted that most BACs in the public domain are constructed in the pBACe3.6 and related vectors, and each already carry a loxP and a lox511 site flanking the genomic DNA insert [3, 8]. Contrary to several earlier reports, the problem of cross-recombination between loxP and lox511 does not occur in our approach where Cre protein is synthesized in the bacterial host by phage P1 infection. It is unclear why a much higher stringency of recombination is observed with phage P1-produced Cre protein as opposed to when Cre is made constitutively in an uninfected cell. It might be due to possible modification of Cre by phage-encoded proteins (*see* ref. 5 for detailed discussions).

Once the loxP or lox511 Tn10 transposon plasmid is in the same host cell as the BAC plasmid, transposition of lox sites is induced by adding isopropyl-β-D-thiogalactopyranoside (IPTG) to the cultures. The BAC DNA with transposon insertions is then transduced with P1 phage. Infection with P1 phage supplies Cre protein *in trans*, enabling the transposed lox site to recombine specifically with the endogenous lox site of the same type. Thus although transposition of the lox site-containing Tn10 into BAC

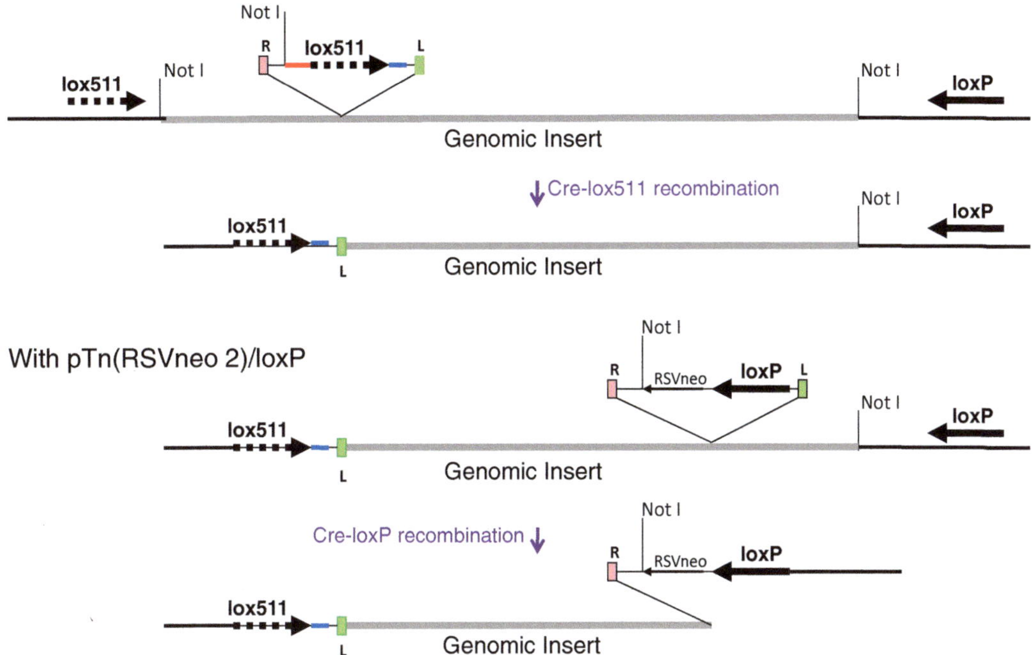

Fig. 1 Schematic representation of deletions generated from both ends of DNA insert in a BAC. The *upper panel* shows insertion of a lox511 transposon (*inverted triangle*) and then truncation from the lox511 end of genomic DNA insert in a BAC using the lox511 transposon pTnLox511(B)markerless 1 [5]. Note that the sequence (shown in *blue*) in front of the lox511 *arrowhead* (shown as *broken thick arrow* on *inverted triangle*) in the Tn10 transposon is retained in the BAC after the lox511–lox511 recombination by Cre protein that generates the truncation. However, the sequence marked in *red* behind the lox511 *arrowhead* in the transposon is lost after the lox511–lox511 recombination that creates the deletion. The Not I sites in transposon and BAC, as well as the 70 bp inverted repeat end of the Tn10 (marked in *pink*), are also lost. The largest clone from that library was then truncated from the opposite loxP end with a loxP transposon pTn(RSVneo 2)/loxP [5]. The loxP–loxP recombination by Cre in this case eliminates the 70 bp inverted repeat end of the Tn10 (marked in *green*) and the original Not I site in the BAC DNA. Note however that a Not I site and the RSVneo gene from the transposon is now juxtaposed in front of the loxP arrowhead after the BAC end-deletion, which ultimately produces a Not I–Not I BAC vector DNA band of a different size (because of RSVneo segment addition) compared to the starting BAC (illustrated in refs. 3, 5, 9, 12). Specific changes in size to the BAC vector DNA band upon Not I digestion is used as a routine diagnostic test of authentic lox-Cre deletions [3, 9]. Figure adapted from ref. 5

DNA is equally probable in one of two orientations, only one can lead to a deletion. The limited packaging capacity of the P1 phage head ensures that only deletions are recovered when the starting BAC is >110 kb. Using these procedures, represented schematically in Fig. 1, one can generate libraries of BACs deleted from either the loxP end, the lox511 end or both ends of insert DNA [*see* refs. 3, 9 for recent overview of the procedures].

The BAC end-deletion procedure has been used in a variety of ways to address the following needs:

(a) Locate genetic markers on a physical map of the chromosome [1, 2].
(b) Identify distal gene regulatory elements functionally by expressing end-deleted BACs with reporter genes in mice [10], or enhancer-trap BACs in zebrafish [11, 12].
(c) Prepare libraries of end-deleted integration-ready BACs with iTol2 ends for germline expression in zebrafish [13].
(d) Replace the loxP site with a lox66 site at one end of genomic DNA insert in BACs [14].

Applications such as enhancer-trapping, or generating libraries of iTol2 end-sequence attached BACs, require introducing DNA sequence cassettes exogenous to the BAC. These can be placed, together with the lox site, within the 70 bp inverted repeat ends of the Tn10 transposon that gets inserted. However, not all sequences transposed into the BAC DNA by Tn10 survive the Cre-lox recombination. As shown in Fig. 1, the directionality of the Cre-lox recombination reaction is such that only sequence in front of the loxP or lox511 arrowheads in the Tn10 transposon are retained at the newly created end of BAC DNA, while those behind the arrowheads are lost [3, 9]. This is a key feature of the BAC end-deletion procedure utilized in the diverse applications noted above. Here I describe a general methodology for transposing either a loxP or a lox511 Tn10 transposon into BAC DNA and generating deletions from either end of the genomic DNA insert in BACs through lox-Cre recombination. Because simultaneous delivery of DNA sequence exogenous to the BAC at the newly created ends of genomic insert is integral to the method, the step-by-step procedure described here applies to all lox-transposons regardless of the end purpose, i.e., enhancer-trapping, introducing iTol2 end-sequences, replacing loxP with lox66 or merely end-deletions for mapping studies.

2 Materials

All chemicals, unless specified, need to be at least Molecular Biology grade. Solutions should be made with deionized water, autoclaved, and/or sterile filtered through a 0.2 μm Nylon Membrane Disposable Sterile Syringe Filter.

1. Antibiotic stock solutions: make all solutions in ultrapure distilled water. This water can be purchased from companies such as ACROS, USA. Following are the concentrations of the antibiotics used in LB agar plates: 150 μg/mL Ampicillin (Amp, stock: 10 mg/mL in water, penicillin-G potassium); 10 μg/mL kanamycin (Kan, stock: 10 mg/mL kanamycin

monosulfate in water); and 10 μg/mL chloramphenicol (Cm, stock: 2.5 mg/mL in water). Use 60 μg/mL Amp for cultures in liquid LB, but those on LB agar plates require 150 μg/mL Amp to suppress satellite colonies. Filter-sterilize all antibiotics solutions. Store stock solutions frozen at −20 °C.

2. 100 mM IPTG stock solution in water. Store in 1-mL aliquots and freeze at −40 °C.
3. Bacto agar, Bacto tryptone, yeast extract.
4. Glass capillary tubes (microhematocrit capillary tubes), for streaking bacteria and recovering phage plaques from agar plates, and disposable plastic bacterial cell spreaders (sterile). Use of capillaries and disposable spreaders obviates the need for a Bunsen flame and ethanol bath at the work bench.
5. Polypropylene 14–15 mL snap-cap round-bottomed disposable tubes (sterile).
6. Phage and bacterial strains: NS3529 and NS3516 are *lac* Iq host strains that do or do not express Cre protein constitutively, respectively. YMC (rec+, supE) is the bacterial host used to grow P1 phage stocks. The P1^{vir} used is a virulent form of P1 phage. BACs from the public domain exist in the DH10B host. All phage and bacterial strains listed here have been described previously [15, 16] and are available on request.
7. Vectors and Transposon plasmids: The Tn10 transposon plasmid pTn(RSVneo2)/loxP is a representative of the set of loxP transposon plasmids used for truncating from the loxP-end of genomic insert DNA in a BAC. Within the 70 bp inverted repeat ends of the transposon resides it contains a RSVneo cassette in front of the wild type loxP sequence arrowhead [5]. After the Cre-loxP deletion it leaves behind the RSVneo cassette at the new end, which can confer resistance to kanamycin in bacteria and neomycin in mammalian cells. An ampicillin resistance gene is located outside the 70 bp inverted repeat ends of all of the transposon plasmids and is not inserted into BAC DNA [5, 17]. BAC deletion clones are selected in LB plates containing both kanamycin and chloramphenicol. The corresponding transposon plasmid used to truncate from the lox511 end of BAC DNA is pTnlox511(B) RSVneo2 [5].

   If a markerless transposon is desired for the first round of deletions, truncations from the loxP end can be made with pTnMarkerless2 [17], while truncations from the lox511 end made with pTnlox511(B)markerless 1 [5]. The vector backbones of these are identical to the corresponding RSVneo-containing Tn-plasmids above, except they are devoid of the RSVneo cassette. BACs deleted using the markerless transposons are selected on plates containing only chloramphenicol. All other steps remain unchanged (*see* **Note 1**).

8. LB medium: 10 g bacto-tryptone, 5 g bacto yeast extract, 10 g sodium chloride, in 1 L of deionized water and autoclave. LB-agar plates: LB medium, 15 g/L bacto-agar, and autoclave. Store LB growth medium and LB-agar plates at 4 °C. Mix all antibiotic stock solutions with autoclaved LB-agar medium after they are cooled to 50 °C.
9. Enriched LB medium: 10 g bacto-tryptone, 10 g bacto yeast extract, 10 g sodium chloride in 1 L of deionized water and autoclave. "Enriched LB medium" is used for growing BAC deletion clones to a higher cell density.
10. 0.5 M $CaCl_2$ stock solution: 5.55 g in 100 mL deionized distilled water, stored at room temperature in a tightly capped 50 mL sterile polypropylene orange/blue-cap tube.
11. S.O.C. medium: 20 g bacto-tryptone, 5 g bacto-yeast extract and 0.5 g NaCl in 1 L of deionized water. After dissolving nutrients, add 10 mL of 0.25 M KCl and adjust pH to 7.0 with 5 M NaOH and autoclave. After cooling to 40 °C, supplement the sterile 1 L solution with 5 mL sterile 2 M $MgCl_2$ and 20 mL of a sterile 1 M solution of glucose.
12. LCa plates: 15 g of Bacto agar in 1 L of LB medium and autoclave. After cooling to 60 °C, add calcium chloride (from 250 mM stock, sterile filtered) to a final concentration of 2 mM (*see* **Note 2**).
13. Top Agar: 0.7 g of Bacto agar to 100 mL distilled water and autoclave. After cooling to 55 °C, add 1 M $MgSO_4$ stock solution to a final concentration of 10 mM.
14. $P1^{vir}$ phage stock (available upon request): Bacteriophage P1 is used in the BAC end-deletion procedure for two purposes. P1 infection provides Cre protein in *trans*, that allows the lox-Cre recombinations to occur. Newly synthesized P1 phage heads also package the BAC DNA efficiently and allows easy recovery of end-deleted BACs. A virulent form of phage P1 ($P1^{vir}$) is used in the procedure.
15. Solution I resuspension buffer: 25 mM Tris-Cl, pH 8.0, 50 mM glucose, 10 mM EDTA.
16. Solution II lysis buffer: 0.2 N NaOH, 1 % SDS.
17. Solution III neutralizing buffer: 60 mL 5 M potassium acetate (KOAc), 11.5 mL glacial acetic acid (HOAc) and 28.5 mL water.
18. 50:50 mixture of phenol/chloroform (buffered and saturated with 25 mM Tris-Cl pH 8.0 + 0.1 mM EDTA). Stored in dark colored bottle at 4 °C.
19. TE + RNase A solution: 10 mM Tris-Cl, pH 8.0, 0.1 mM EDTA, 5 μg/mL RNase A.

20. Make up 2 mL molten top agar to 10 mM in $MgSO_4$.
21. Cotton gauze for crude filtration of floculant precipitate.
22. Paraffin tape for sealing holes in the caps of tubes after chloroform treatment.
23. Oak Ridge tubes: Clear polycarbonate tubes from NALGENE Nunc International, Rochester, NY. USA.
24. Special Elution buffer: Take 3×5 mL Qiagen elution buffer QF for each column (i.e., 12×15 = 180 mL for eluting 12 tips) and add 1 mL of 5 M NaCl for each 35 mL of elution buffer (i.e., add 5 and 1/7 mL of 5 M NaCl to the 180 mL of QF buffer) in a 250 mL Pyrex orange-capped bottle. Mix well, and microwave for 1 min at highest setting. Check temperature by inserting a clean thermometer into the solution, and place/float the Pyrex bottle with its contents in a large plastic beaker 2/3 full of water at 85 °C.

3 Methods

3.1 End-Deletion Procedure

3.1.1 Transforming BAC Clones with Tn10 Transposon Plasmid DNA

1. Streak out a small aliquot (2 μL) from a glycerol stock of a BAC clone on an agar plate containing 10 μg/mL of Cm and incubate overnight at 37 °C (*see* **Note 3**).
2. Pick a single BAC colony and grow it as a suspension culture in LB containing 10 μg/mL Cm overnight at 37 °C.
3. Take 20 μL of the overnight culture and inoculate into 5 mL of LB containing 10 μg/mL of Cm in a 50 mL sterile polypropylene orange/blue-cap (VWR) with a couple of small holes punched in the cap with a 22-gauge needle to facilitate aeration (*see* **Note 4**).
4. Incubate the tube with vigorous shaking at 37 °C for 2 h (early log phase). It is not necessary to check OD_{600} (*see* **Note 5**).
5. Remove 1.6 mL of culture into an Eppendorf tube (filled to the top), and centrifuge the cells for 1 min gently at approximately 2,000×*g* in a table-top microfuge at room temperature. Pelleting cells at higher speeds prevents easy resuspension and results in reduced viability.
6. Discard the supernatant (remove as much of the supernatant as possible by shaking the tube inverted), and place the tube in an ice-water slurry.
7. Add 1 mL of prechilled (4 °C) 0.1 M calcium chloride solution in water (sterile filtered), and resuspend the pellet by gentle vortexing (setting of "low" to "medium"). Try to prevent frothing (vortex the holding tube vertically), and keep tube cold at all times by frequently putting it back in ice-water slurry. Pipetting with a chilled tip also works well.

8. Pellet the cells at 2,000 × *g* for 0.5 min at 4 °C.
9. Remove all the supernatants as in **step 6**, and resuspend the cells in 0.5 mL of prechilled (4 °C) 0.1 M calcium chloride solution in water.
10. Repeat **steps 8** and **9**.
11. Resuspend the pellet in 150 μL chilled (4 °C) 0.1 M calcium chloride solution.
12. Incubate on ice for 3 min. The cells are now competent.
13. Dilute a crude preparation (miniprep) of the DNA from any one of the transposon plasmids pTn(RSVneo2)/loxP, pTnlox511(B) RSVneo2, or the markerless ones pTnMarkerless2 and pTnlox511(B)markerless 1 in distilled water, such that approximately 10 ng of DNA is in a 4 μL volume in a prechilled Eppendorf tube. Add 50 μL of the competent cells to the DNA (*see* **Note 6**).
14. Mix well by gentle tapping (keeping it chilled), and then leave on ice for 3 min.
15. Heat shock the suspension by placing each tube in a water bath at 37 °C (a heat block containing water in the wells serves fine) for exactly 5 min.
16. Immediately transfer the tubes to an ice bath, and leave for 2 min.
17. Place the tubes at room temperature for 2 min with the caps open.
18. Add 500 μL of room temperature S.O.C. (available from Cellgro-Corning), and incubate at 37 °C for 1 h with gentle shaking of the tubes in an air shaker.
19. Prewarm the LB agar plates containing Kan + Cm + Amp, or Cm + Amp if the markerless transposons are used for the transformation, to room temperature so that the plates are dry (*see* **Note 7**).
20. Add approximately 100 μL of transformed cell suspension (S.O.C. culture) to the LB agar plates containing Kan + Cm + Amp, or Cm + Amp only, and spread well. Allow the plates to dry before inverting and placing in an incubator at 37 °C overnight (16–20 h). Multiple BAC clones can be transformed in parallel with any of the transposon plasmids listed in Subheading 2 using this procedure.

3.1.2 Transposition of loxP and lox511 Cassettes into BACs and Phage P1 Transduction of Retrofitted Clones

1. Pool the several hundred colonies from a plate by scraping them with a cell-spreader and resuspend into 2 mL of LB. Take 5 μL of this culture to inoculate 2 mL of LB + Kan + Cm + Amp and grow overnight at 37 °C. Using a plate full of colonies instead of a single one is recommended [18]. Alternatively, an

aliquot of 2 μL from the suspension of pooled colonies can be diluted and grown for IPTG induction (*see* **Note 8**).

2. Take 40 μL of the overnight culture (or 2 μL from the suspension of pooled colonies) and dilute it into 5 mL of LB +10 μg/mL of each antibiotic Kan+Cm and 60 μg/mL Amp, contained in a 50-mL polypropylene orange/blue-cap tube with several holes punched in the cap with a 1 cc, 25-gauge needle for aeration.
3. After 2 h of vigorous shaking at 37 °C, add 95 μL of 100 mM IPTG and continue shaking for further 3 h (*see* **Note 9**).
4. Centrifuge cells down at 1,600 × *g* for 5 min at room temperature in a swinging-bucket Sorval tabletop centrifuge. Let centrifuge come to a stop without the brakes on.
5. Pour off the supernatant and resuspend the cell pellet in 500 μL of fresh prewarmed LB + 5 mM calcium chloride (no antibiotics here).
6. Infect each culture with 250 μL of $P1^{vir}$ phage stock that has been de-chloroformed (*see* **Note 10**).
7. Incubate without shaking for 30 min at 37 °C, to adsorb the phage.
8. Add 4 mL of prewarmed LB containing 5 mM calcium chloride to each tube and shake vigorously for 2 h (*see* **Note 11**).
9. Add 200 μL of chloroform (8–10 drops with a glass Pasteur pipet) to each culture, continue shaking for an additional 3 min at 37 °C, and then store overnight at 4 °C after sealing the holes in the cap with paraffin tape (*see* **Note 12**).
10. Transfer the contents into 15-mL polypropylene tubes (for tighter packing of the cell debris and chloroform), and centrifuge the lysates at 1,600 × *g* for 15 min at room temperature in a swinging-bucket rotor in the Sorval tabletop centrifuge. Carefully remove the clear supernatant with a 1-mL plastic tip attached to a pipet into a new 50-mL orange/blue-cap polypropylene tube, leaving the chloroform and cell debris behind.
11. De-chloroform the clear lysate by shaking the tubes very gently (50 rpm) for 1 h at 37 °C. The caps need to have several holes in them.
12. Recover the end-deleted BAC plasmids that are now packaged within the P1 heads by infecting freshly grown NS3516 Cre^- cells (or NS3529 Cre^+ cells if lox511 Tn10 transposon is used to delete BAC ends from the lox511 side of insert) with the de-chloroformed lysate from **step 11** (*see* **Note 13**).
13. Inoculate 10 mL of LB (this volume of cells is good for each BAC clone) with NS3516 Cre^- cells, and grow to mid-log in a 50-mL orange/blue-cap polypropylene tube with vigorous agitation at 37 °C.

14. Concentrate the NS3516 Cre$^-$ cells by centrifuging at 1,600 × *g* for 8 min at room temperature in a swinging-bucket rotor in the Sorval tabletop centrifuge. Resuspend the cell pellet in 500 μL of LB + 5 mM calcium chloride.

15. Add concentrated cell suspension to the 5 mL of dechloroformed lysate from **step 11**, and leave for 1 h at 37 °C without shaking (phage adsorption).

16. Add 1 mL of fresh prewarmed LB + 5 mM calcium chloride and shake at 37 °C for an additional 90 min.

17. Plate the entire 6 mL of culture in eight plates of LB + Cm + Kan (10 μg/mL of each antibiotic), i.e., approximately 700 μL of culture/plate. To save time, plates should be predried by leaving at room temperature for 1 h with lids partially open prior to adding culture. Spread well and leave at room temperature with lids partially open for an hour on bench top to dry. When dry, incubate plates at 37 °C for about 36 h. The plates should be LB + Cm, only, if one of the "markerless" transponsons such as pTnMarkerless2 and pTnlox511(B)markerless 1, OR the enhancer-trap transposons (which are also markerless because of limited "cargo capacity" of Tn10 transposons, *see* ref. 12) are used [5, 17].

18. There should be about 100–400 colonies per plate. Pick 60 clones and miniprep them for BAC DNA as described in Subheading 3.1.4. Analyze the BAC DNA after Not I digestion by Field Inversion Gel Electrophoresis (FIGE). If the analyses indicate that more than 30 % of the deletion clones contain co-transduced transposon plasmid (co-transduction of transposon plasmid can occur in deletion clones and their frequency can vary between <1 % to almost 80 % [1, 19], then go for Ampicillin Sensitivity Screen described in Subheading 3.1.3 below, *see* **Note 14**). Streak the BAC deletion clones in parallel in LB agar plates containing either Cm or Amp, and follow the steps described in Subheading 3.1.3. However, if the transduced transposon plasmid appears in less than 30 % of clones in the BAC deletion library, then they can be easily identified and weeded out after miniprepping the BAC DNA and analyzing on FIGE. Deletion libraries usually comprise of several thousand clones and cover the entire ~110 kb length of BAC with high density [2]. With most BACs, the transduced transposon plasmid occurs in 2–10 % of the clones in the deletion library and does not pose a serious enough problem to go through the Ampicillin Sensitivity Screen.

3.1.3 Ampicillin Sensitivity Screen

1. Two 96-well plates are prepared with 150 μL/well of either liquid LB medium or solid LB agar. Wells in one plate contain Cm (10 μg/mL), while the other has Amp (60 μg/mL for liquid or 150 μg/mL for solid agar). This can be done by the

sequential use of a 12-channel pipettor. Inoculate a single colony into each well in parallel in the two 96-well plates.

2. After overnight growth at 37 °C, pick the colonies from the Cm plate that failed to grow in Amp plate.

3.1.4 Isolating BAC DNA from End-Deletion Clones

This section describes a rapid and simple miniprep procedure that has been adapted from the standard alkali/sodium dodecyl sulfate (SDS) lysis method used for isolating small multi-copy plasmids [16, 20, 21].

1. Inoculate 2 mL cultures of "enriched LB medium" + 10 μg/mL chloramphenicol with a single colony from the BAC deletion library in 15 mL snap cap tubes and grow overnight (16 h) at 37 °C with caps lose for aeration and vigorous shaking (250 rpm in temperature controlled air shaker bath) as described earlier [1, 21] (*see* **Notes 15**).
2. Decant 1.7 mL of the culture into a Safelock Eppendorf tube. Leave remaining 0.2 mL of culture in snap-cap tube (sealed shut) at 4 °C. Make sure that each is numbered, along with the eppendorf tubes (*see* **Note 16**).
3. Spin the eppendorf tubes in a microfuge at room temperature for 1 min at 3,600 × *g*.
4. Decant the supernatant immediately, leaving behind the last 50 μL of medium.
5. Add 180 μL of Solution I and vortex gently (medium setting) to resuspend the cell pellet thoroughly, and leave at room temperature. This should be done as quickly after pelleting as possible.
6. Set up an ice-water tray. Make up fresh Solution II (*see* **Note 17**).
7. Add 310 μL of Solution II quickly to a set of ten tubes of cell suspension at room temperature. Close caps, and very gently mix the contents of the tube. Once suspension becomes clear (cell lysis is complete in about 30 s) place tubes immediately in ice-water tray (*see* **Note 18**).
8. After 5 min on ice-water tray, open caps of tubes and add 250 μL of an ice-cold Solution III. Close caps securely, check that all tubes have equal volumes, and mix the contents of the tube thoroughly, by inverting the tubes back and forth gently 10–12 times, before putting them back in ice-water tray. There can be a break here for overnight storage in cold room, if needed.
9. Spin tubes (after wiping away moisture on the outside of tube) in the microfuge for 3 min at 3,600 × *g*.
10. Carefully decant supernatants into fresh numbered tubes. The supernatants contain the BAC plasmid DNA along with cellular RNA, small quantities of protein and other soluble small molecules in the cell.

11. Add 200–300 μL of a 50:50 mixture of phenol/chloroform (the bottom layer from the bottle containing buffer saturated phenol/chloroform) to each tube. Use a fume hood for the procedure. Cap tightly and shake by inverting the rack of tubes back and forth gently (place an empty rack on top) for 1–2 min to mix the two immiscible layers.
12. Spin tubes in a microfuge at room temperature for 2 min at 3,600 × *g*.
13. Recover the top layer carefully (avoid any phenol/chloroform) and transfer into a fresh labeled tube.
14. Add isopropanol to fill tubes. If recovery is good at each step, there should be about 750–800 μL of aqueous layer in each tube after the phenol/chloroform extraction, and thus an equal volume (800–900 μL) of isopropanol should be ideal to precipitate all nucleic acids.
15. Cap tubes tightly and mix gently by inverting the tubes back and forth several times. Let stand at room temperature for 1–2 h. There can be a break here for the day, if needed (*see* **Note 19**).
16. Spin tubes for 5 min at 6,000 × *g* in a microfuge at room temperature. Discard the supernatant.
17. Add 1 mL of a solution at room temperature of 70 % ethanol in water to each tube. Wash pellet by inverting securely capped tube several times. The pellet may get dislodged.
18. Spin tubes at 6,000 × *g* for 2 min. Discard supernatant carefully, ensuring the pellet remains in the tube!
19. Remove as much of the wash solution as possible using a yellow tip, then tap pellet to the bottom of tube. This allows the dried pellet to be dissolved in a small volume of buffer.
20. Leave pellet to air dry overnight with caps open. Cover with tissue paper.
21. Resuspend each pellet in 40 μL of TE + RNaseA solution. Gently tap to mix after 15 min incubation at room temperature. Spin lightly to get rid of trapped air pockets in liquid at bottom. The BAC DNA should be resuspended and ready to use in 30 min. Aliquot 4–5 μL out of each for Not I digest and FIGE analysis. Store the remaining DNA at −20 °C (*see* **Note 20**).

3.1.5 QIAGEN Tip Purification of BAC DNA from Deletion Clones

This method produces the highest quality and quantity of BAC DNA from clones from the deletion library, and can be used for both FIGE analyses and direct BAC sequencing with Big Dye terminators [1, 21]. The DNA is also suitable for injecting into zebrafish embryos [11–13]. The major drawback is its low throughput.

1. Streak out BAC deletion clone from a glycerol stock onto LB + Cm plates.

2. Inoculate single colony in 85 mL of "enriched LB" containing 10 μg/mL of Cm, OR inoculate with 1 mL of an overnight culture from a single colony. Use 250 mL Kimax (Kimble USA) culture flasks with indentations on bottom for superior agitation.
3. Shake vigorously in air-shaker bath at 37 °C for 16 h (overnight).
4. Fill two Oak Ridge tubes (each can hold 40 mL) up to the mark on the neck. Cap the tubes tightly. Save the remaining 3–5 mL of culture in the flask in the cold room to make glycerol stock, on the same day.
5. Spin tubes in SA-600 rotor in a Sorvall centrifuge at 4 °C for 10 min at 3,600 × *g* with the brakes set in the "ON" position.
6. Pour off clear supernatant (bleach the pooled supernatants later to kill bacteria) and add 6 mL of Qiagen buffer P1 (make sure RNase A is included) to each tube.
7. Cap tubes tightly and vortex (at medium setting) till all of the cell pellet is thoroughly resuspended. Leave tubes at room temperature.
8. Add 6 ½ mL of Qiagen buffer P2 into a separate 15 mL white plastic tube (keep P2 buffer on foam lid to avoid precipitation of SDS/NaOH over cool bench surface!).
9. Swirl the tube of cell suspension and quickly add buffer P2 all at once. Roll the tube after capping tightly. Be gentle, so as not to shear chromosomal DNA! Let sit at room temperature for 3–4 min (*see* **Note 21**).
10. Once the contents of the tube become clear (cell lysis is complete), place in ice bath.
11. Add 6 mL of cold Qiagen buffer P3 and mix gently by inverting several times. Leave on ice for 20 min, mixing well by inverting every 5 min.
12. Spin in SA-600 rotor at 4 °C for 30–45 min at 7,000 × *g*, with the brakes set in the "OFF" position.
13. Get cotton gauze ready. Cut each into two.
14. The precipitate is light and some tend to float. Therefore, pour off and collect the supernatant through the cotton gauze into a clean Oak Ridge tube. If floculant precipitate spills over, make a note of it and avoid loading the precipitate later into the Qiagen columns (tips). This precipitate does not dissolve in the Qiagen buffer QBT used to dissolve the DNA pellet in a subsequent step (*see* **Note 22**).
15. Add 1/20 volume of sterile-filtered 3 M sodium acetate (approximately 2 mL) into each tube, mix by swirling. Add equal volume of room temperature isopropanol. Cap tightly and mix well at room temperature by inverting.

16. Spin tubes in SA-600 rotor at 8,000 × *g* for 15 min at 4 °C with the brakes set in the "ON" position. Pour off supernatant, and invert tubes on paper towel to drain liquid completely. It is NOT necessary to dry pellets.
17. Dissolve wet pellet in each tube immediately in 5 mL QBT (column equilibrating) buffer.
18. Set up Qiagen Tip-500 columns in 250 mL empty erlenmeyer flasks with blue plastic collar provided by the vendor. Mark each column clearly with the clone number. Once the DNA pellets are dissolved, equilibrate each column with 6 mL buffer QBT.
19. Once the buffer has completely drained out of the tip, add 2 × 5 mL of the DNA solution from the two tubes (filled initially with same clone culture) into each column.
20. Wait for the liquid to drain through completely. Wash column with 3 × 14 mL of Qiagen wash buffer QC. Each time let the liquid drain through completely.
21. Make up fresh "Special Elution buffer."
22. After the third 14 mL wash in QC buffer, remove the Qiagen tips from the flasks and mount on clean Oak Ridge tubes held upright on a foam rack. Apply three 5 mL aliquots of the hot (~85 °C) Special Elution buffer directly to the sintered glass disk within each Qiagen column (*see* **Note 23**). Collect the flowthrough in the tubes and precipitate the DNA by adding 12 mL isopropanol at room temperature to each tube. Mix well by inverting capped tubes several times.
23. Spin the Oak Ridge tubes after 1 h at room temperature in a SA-600 rotor at 8,000 × *g* for 30 min at 4 °C. Mark the position on tube where the DNA pellet is expected.
24. Decant supernatant and invert the tubes on tissue paper. After half hour, wipe off the lips with a Kimwipe tissue and add 750 μL of a 0.3 M sodium acetate solution in TE buffer (the DNA pellet need not be completely dry).
25. Dissolve the DNA by swirling, spin the tubes at 2,000 × *g* for 2 min, and recover the solution into labeled eppendorf tubes. Wash Oak Ridge tubes with another 100 μL of the same buffer, and add that to the eppendorf tube. The tube should now have about 850 μL of aqueous phase containing the DNA.
26. Extract with 300 μL of a 25 mM Tris-Cl pH 8.0 + 0.1 mM EDTA buffer-saturated 50:50 phenol/chloroform mixture. Collect aqueous phase in new tube.
27. Extract with 300 μL of pure chloroform.
28. Repeat the chloroform extraction from **step 27**.

29. The volume of aqueous phase should be between 800 and 850 μL. Fill the tube with isopropanol at room temperature. Cap the tube and mix. If the preparation is good, you should see a good fibrous precipitate of pure BAC deletion clone DNA!
30. Spin tubes at room temperature in eppendorf centrifuge at 7,000 × *g* for 5 min. Decant supernatant. Invert tube on tissue paper towel.
31. Add 1.2 mL of 70 % ethanol at room temperature. Cap and invert the tube several times to wash the pellet off excess salt. Spin for 2 min at 7,000 × *g*. Decant supernatant (Careful! Pellet may be loose).
32. Let pellets dry thoroughly (a couple of hours at least). Dissolve pellet in 100–200 μL of 1/5× TE buffer for Big Dye Terminator fluorescent sequencing and also for injection into zebrafish embryos (*see* **Note 24**). DNA pellet should be incubated at room temperature for at least half hour in buffer to dissolve. Gently tap and mix contents of tube several times during this period. The DNA concentration will be approximately 0.5–1 μg/μL and should be accurately determined by measuring absorbance at 260 nm.

3.2 Preparation of P1^{vir} Lysate

3.2.1 Preparation of Magnesium Chloride Stock of YMC Strain

1. Streak out YMC strain onto an LB plate. Grow overnight at 37 °C.
2. Pick a single colony and inoculate a 10-mL culture in LB with shaking at 37 °C overnight.
3. Centrifuge cells for 5 min at 1,600 × *g* at room temperature in a Sorvall tabletop centrifuge.
4. Resuspend cells in 10 mL of 10 mM magnesium chloride solution (sterile filtered).
5. Centrifuge cells again for 5 min at 1,600 × *g*.
6. Resuspend cells in 2 mL of 10 mM magnesium chloride solution.
7. Store at 4 °C. Keep on ice or cold at all times (*see* **Note 25**).

3.2.2 Creation of a Single Phage Plaque

1. Prewarm one LCa plate at 43 °C for at least 30 min.
2. Prepare bacterial overlay in a 5-mL polypropylene round-bottomed tube by quickly mixing the following in a heat block set at 37 °C: 2 mL of LB made up to 10 mM $MgCl_2$, 25 μL of YMC cell stock in magnesium chloride, 2 mL of molten top agar made up to 10 mM in $MgSO_4$ (*see* **Note 26**).
3. Cover top of tube with a piece of Parafilm, invert the tube three times, pour the contents onto the LCa plate, and let solidify for 10 min at room temperature.

4. Very gently streak P1^{vir} lysate (10^{-6} dilution of a 6-month-old stock) with a glass capillary across the surface once (Remove chloroform from the lysate before streaking on plate by shaking at 37 °C for 30 min).
5. Incubate the plate un-inverted overnight at 37 °C (*see* **Note 27**). Plaques of P1^{vir} phage are clear and approximately 0.7 mm in diameter (much smaller than those of λ phage).

3.2.3 Creation of a High-Titer Phage Lysate

1. Prewarm LCa plates to 43 °C for at least 30 min.
2. Put 50 μL of YMC stock in a 5 mL polypropylene tube and add $CaCl_2$ to final concentration of 5 mM. Add two similar-sized plaques of phage P1^{vir} and mix (*see* **Note 28**).
3. Incubate at 37 °C for 5 min in a heat block.
4. Add 2 mL of LB (prewarmed to 37 °C) containing 10 mM $MgCl_2$ + 5 mM $CaCl_2$ to the tube of phage plaque + cells.
5. Quickly add 2 mL molten top agar (kept at 50 °C in a heat block) containing 10 mM $MgSO_4$.
6. Cover the mouth of the tube with a piece of Parafilm, quickly invert three times, and pour onto a warm LCa plate (*see* **Note 29**).
7. Immediately place in a 37 °C incubator un-inverted, and incubate for 5–6 h until a clear lysate covers the entire surface of the plate.

3.2.4 Harvesting Phage Lysate

1. Add 2 mL of LB + 10 mM $MgCl_2$ to each plate containing the P1^{vir} lysate, scrape the sloppy top agar layer with a plastic cell spreader, and pour into a polypropylene centrifuge tube (Caution: polycarbonate tubes corroded by chloroform). Repeat to recover all the lysate.
2. Add 100 μL of chloroform to the polypropylene tube and place in 37 °C shaker for 5 min.
3. Centrifuge 5 min at 4,000 × g in a Sorvall SS-34 or SA-600 rotor at 4 °C (with the brake turned to the "OFF" position) to pellet the debris, agarose, and chloroform.
4. Carefully pipette the lysate (supernatant) into a 5-mL polypropylene snap-cap tube (keep cold and out of light).
5. Add 2 mL of additional LB + 10 mM MgCl2 to the pellet in the centrifuge tube, mix well by shaking, and centrifuge again.
6. Pipette supernatant into the previous tube of lysate.
7. Add a drop of chloroform, and store at 4 °C protected from light (wrap tube in aluminum foil).

3.2.5 Titering Phage Lysate

1. Prewarm LCa plates to 43 °C for at least 30 min.
2. Prepare a 10^{-5} and 10^{-6} dilution of phage lysate in LB + 10 mM $MgCl_2$.

3. Mix the following in a 5-mL tube in a heat block at 37 °C:
   (a) 10-μL dilution of phage to 50 μL of YMC magnesium chloride stock.
   (b) 0.6 μL of 500 mM $CaCl_2$.
   Incubate at 37 °C for 5 min.
   (c) 2 mL of LB at 37 °C, 100 μL of 100 mM $MgCl_2$, and then 2 mL of top agar (containing 10 mM $MgSO_4$, pre-warmed in heat block at 50 °C).
4. Cover the top of the tube with Parafilm and invert three times quickly to mix well.
5. Pour onto LCa plate. Let solidify for 5 min at room temperature. Incubate un-inverted overnight at 37 °C.
6. Count the plaques and calculate the titer. They should be in the high 10^9–10^{10} pfu/mL range, or better. Homogeneous plaque size should be observed.
7. Check the lysate for sterility by plating 50 μL on an LB plate; if colonies of bacteria appear, repeat chloroform treatment and recheck.

4 Conclusion

The lox-Tn10 mini-transposons can insert into genomic DNA in either orientation, and thus potentially generate both deletions and inversions with equal efficiency after recombining specifically with the respective lox sites endogenous to the BAC. Deletions reduce BAC size, while inversions do not. Because the P1 head cannot package both lox sites from an inversion event if the starting BAC is larger than the P1 headful packaging capacity of ~110 kb, the DNA from an inversion is not recovered. The procedure generates deletions exclusively when the starting BAC clone is larger than ~110 kb [3, 9].

Most BAC libraries available today comprise clones with genomic inserts that average 150 kb and flanked by loxP and lox511 sites [22]. While all end-deletions from a clone of this size are not recoverable, deletions ending at any specific location can be obtained by using a different suitably overlapping BAC clone from the contig [11, 12]. Because most available libraries have over 20-fold coverage, the method is ideally suited for generating closely spaced deletions of genomic insert DNA. BAC clones that are <110 kb also produce end-deletions with high efficiency. However, approximately half of all clones in such pools are inversions, and these can be readily identified by field inversion gel electrophoresis (FIGE) after being linearized with *Not I* enzyme. Because the loxP transposons contain a Not I site, the BAC DNA with a lox-Cre inversion has an additional Not I–Not I band in the FIGE, and its length varies depending on where the initial transposition occurred. The total length of DNA bands on a FIGE is also larger than the starting BAC, unlike the situation in a lox-Cre deletion.

The procedure described here can be scaled up to generate insert-end deletions in multiple BACs: as many as 30 clones have been processed simultaneously. The real hindrance in scaling appears to be clone-to-clone variation in growth characteristics. Some clones appear to grow very slowly, and require individual attention. Thus, induction with IPTG and the P1 transduction steps can get out of phase with the rest and the process becomes tedious.

A several thousand-member deletion library is readily generated from a BAC clone using the procedure described in this chapter. Typically, size analysis of DNA from 60 deletion clones by FIGE after Not I digestion indicates that approximately 72 % are of unique length (unpublished observations). An additional 3–8 % is found to be unique after end sequencing, suggesting transposon insertions at different sites on genomic DNA as their origin [2]. More closely spaced deletions can be isolated if one is willing to screen a larger number of clones [2]. The deletions occur through transposition of lox sites at random, and largely without a sequence preference [2, 18].

5 Notes

1. Over the years I have designed and used a variety of Tn10 transposons to do specific tasks including progressively truncating from either the loxP or lox511 end of genomic DNA inserts in a BAC [1, 5], enhancer-trapping using BACs [11, 12], replacing loxP with lox66 at BAC ends [14], and inserting iTol2 end-sequences at BAC DNA ends [13]. Two Tn10 transposon plasmids served as the basic backbone for constructing the additional features required for each of these tasks. They are the transposon plasmids pTnMarkerless2 [17] and pTnLox511(B)markerless 1 [5], and carry the loxP site and the lox511 site respectively. The loxP site-containing transposon series is specific for making truncations from the loxP end of genomic insert DNA in BACs, while the lox511 series of transposons specifically delete from the lox511 end of BAC DNA. DNA sequence cassettes located in front of either the loxP or lox511 sites are retained at the newly created deleted end of genomic insert DNA in BACs by each of these transposons. Note that deletions with a markerless transposon can only be performed in the first round of deletions when the objective is to delete from both ends of DNA insert as discussed in [5].

2. LCa plates = LB + 2 mM $CaCl_2$ plates (Make sure plates are wet and moist. Freshness of plates is not as important as moisture content). Phage titers tend to be low when relatively dry plates are used.
3. It is not necessary to thaw the BAC clone glycerol stock stored at −70 °C, since the viability falters with repeated freeze thaw. Scraping off a few specs with a glass capillary from the surface of the frozen stock is sufficient to produce a few hundred colonies. Make sure the agar plates are free of liquid condensate and relatively dry before streaking.
4. Making competent cells from BAC clones require that they be grown with good aeration. Using 5 mL culture in a 50 mL tube provides a larger air–liquid interface for the purpose.
5. Transformation of cells containing the BAC DNA with the transposon plasmid is very efficient because the plasmid is topologically intact and supercoiled. Thus the BAC-containing cells need not be highly competent.
6. Minipreps of transposon plasmid DNA are prepared by alkali/SDS lysis and phenol/chloroform extraction, and typically produce 2–3 μg of plasmid DNA/mL of culture (unpublished observations). The final pellet of plasmid DNA containing excess of cellular RNA from a 1.6 mL culture is dissolved in 50 μL of 10 mM Tris-Cl, pH 8.0, plus 0.1 mM EDTA buffer containing 5 μg/mL RNase A. This crude preparation of supercoiled transposon plasmid DNA is diluted tenfold with distilled water, and 4 μL of this diluted stock used for transforming competent cells made from a BAC clone.
7. The loxP or lox511 transposon plasmids introduced into a BAC clone carry either a kanamycin resistance (as in pTn(RSVneo2)/loxP, and pTnlox511(B) RSVneo2) gene or no selectable marker (as in pTnMarkerless2, and pTnlox511(B) markerless 1) within the 70 bp inverted repeat ends of the transposon. All transposon plasmids however carry an ampicillin resistance gene outside the 70 bp inverted repeat sequences that does not get inserted into the BAC DNA. Therefore, the selection for BAC clones transformed with a transposon plasmid requires LB plates containing either kanamycin plus chloramphenicol plus ampicillin (for the RSVneo containing transposons), or only chloramphenicol and ampicillin (for the markerless transposons).
8. Picking a single colony of the BAC transformed with the transposon plasmid to expand the culture can be risky. This is because certain BACs appear to de-repress the transposase gene in the Tn-plasmid during the transformation process [18]. It leads to the transposase protein being expressed and insertion of the transposon into BAC DNA in a cell that expands into a colony. Such undesirable expansion of a single

cell with a transposon-inserted BAC eventually produces a distribution of insertions that is highly skewed, rather than random, incorrectly indicating insertion of the transposon as site-specific [18].

9. Induction of the *tac* promoter with IPTG to express transposase needs to be done during early log phase of cell growth. Measuring absorbance at 600 nm becomes tedious when large numbers of clones are being processed. Look out for a faint silky white appearance when swirling the tube of cells as an indicator of early log phase. The period of cell growth (2 h) to achieve this may vary a little depending on the BAC clone. Appearance of faint silky white turbidity in culture is best indicator that growth is optimal. Some large BACs are notoriously slow growing and may require longer than 2 h to attain that silky white turbidity. Therefore end-deletions in these should be done separately.
10. $P1^{vir}$ phage stocks are stored at 4 °C in the presence of a drop of chloroform to prevent anything from growing in them. Prior to infecting cells, the phage needs to be thoroughly de-chloroformed to ensure good viability of the cells they infect. To de-chloroform, place approximately 8 × 250 μL of phage suspension in a 50 mL polypropylene tube with several holes in its orange cap, and shake gently for ~45 min at 37 °C.
11. The medium for phage adsorption is made 5 mM in calcium chloride by adding an appropriate amount of a filter-sterilized stock solution of 0.5 M calcium chloride.
12. Cultures of BACs infected with $P1^{vir}$ phage do not undergo cell-lysis well because the host is deficient in certain recombination functions (the deficiency is otherwise beneficial for faithful propagation of genomic inserts) and is unable to support efficient growth of the phage. Therefore, the chloroform treatment at the end of infection is required to attain lysis. Even then, the suspension does not clear as in lysis with phage lambda. Some cell membrane debris is visible when held against the light after chloroform treatment only if the cell density is low. The debris is somewhat stringy in appearance.
13. Although comparable, Cre-recombination of lox511 sites is somewhat less efficient than loxP sites. Thus recovering BACs deleted from the lox511 end in the Cre^- host NS3516 often leads to some of the BAC plasmid DNA appearing unresolved topologically. The problem can be avoided using the Cre^+ host NS3529 to recover the P1 packaged BAC DNA in such cases.
14. Ampicillin sensitivity screen: Recovery of the loxP or lox511 transposon plasmids in clones carrying the end-deleted BAC varies widely, from less than 1 % to as high as 80 % [1]. Although the reason for this variance is unclear, the high transduction frequency of the transposon plasmid along with

the BAC is attributed to the presence of the loxP or lox511 sites in them [5, 19]. Nevertheless, many BAC deletion clones can be rendered free of the transposon plasmid if the clone is isolated after streaking on plates containing only the antibiotic whose resistance gene is carried by the BAC (i.e., chloramphenicol) but lacking the antibiotic specific to the transposon plasmid [1]. Note that antibiotic resistance genes within the 70 bp repeat ends of the transposon are collinear with the BAC DNA after transposition, and no longer require selection separately. Thus a lack of selection against the resistance genes in the transposon plasmid (in this case ampicillin, which is outside the 70 bp repeat ends, as well as any within them) allows the Tn-plasmid to be diluted in a few generations.

15. Grow BAC deletion clone cultures with good aeration. Having more than 2 mL of culture per 15 mL tube is therefore not desirable.
16. Aliquots of BAC cultures saved at 4 °C are viable for about 10 days. This time interval is sufficient to prepare the DNA, and determine its size, from several hundred BAC deletion clones by FIGE so as to decide which ones to save.
17. Addition of 10 % SDS to 10 M sodium hydroxide will lead to precipitation. Therefore, add SDS to larger volume of water before adding alkali.
18. Vigorous shaking must be avoided after cell lysis with Solution II (alkali-SDS) so that shearing of chromosomal DNA does not occur. Contaminating chromosomal DNA pieces are a hindrance to obtaining good sequence reads.
19. The isopropanol precipitation of nucleic acids (BAC + cellular RNA) needs to be done at room temperature and not in the cold so as to avoid precipitating excess salt and proteins.
20. It is important to save BAC DNA obtained from the miniprep procedure at −20 °C. The preparation is relatively crude, and prolonged storage at 4 °C leads to degradation of the large BAC DNA as judged by FIGE analysis. DNA isolated by the Qiagen tip method is much purer, and can be stored at 4 °C for longer periods of time (several months) without breakage.
21. Alkali/SDS lysis of cells at high density is most efficient when all cells are exposed to reagent simultaneously. Thus the measured aliquot of alkali/SDS (Qiagen buffer P2) solution is taken in a separate tube and added all at once to the cell suspension that is kept swirling gently. Once added, the tube containing cells is capped tightly and rocked very gently while rolling it slowly for thorough mixing of contents without shearing of chromosomal DNA. Pipetting the alkali/SDS solution gradually lyses cells on the periphery, and DNA from the lysed cells makes the remaining cells in the suspension impervious to fresh reagent.

22. A convenient way to filter the supernatant is to tightly wrap the cotton gauze over the mouth of the Oak Ridge tube containing solution and precipitate, and slowly pouring out the liquid at a low angle. Care should be taken not to block the entire mouth of tube with liquid while pouring. Otherwise, air will bubble into the tube to occupy displaced liquid resulting in stirring up the light precipitate.
23. The flow of warm elution buffer through the Qiagen tip stops occasionally due to moisture condensing underneath the tip on the outside. The liquid appears to form an air tight seal with the mouth of the Oak Ridge tube. Wipe the bottom of tip if necessary.
24. The final BAC DNA pellet is dissolved in a very low concentration of Tris-Cl, pH 8.0, plus EDTA buffer, 1/5× TE buffer. Higher concentrations of Tris-Cl are detrimental for both injection into zebrafish embryos and Big Dye Terminator sequencing (unpublished observations).
25. The magnesium chloride stock of YMC cells is good for up to 2 weeks if kept cold.
26. Top agar is conveniently made in a 100 mL Gibco serum glass bottle, and can be used multiple times. Microwave the bottle of solidified Top agar each time and shake its contents well to homogenize. Then place bottle in a heat block set at 50 °C for duration of experiment.
27. The plate containing YMC cells in top agar overlay should be incubated un-inverted, otherwise the entire overlay will slide out. This precaution is also necessary when growing the P1 phage stocks on plates.
28. Phage plaques are picked using VWR brand micro-hematocrit glass capillary tubes. After piercing the top and bottom agar layers of a plaque, the agar-containing plaque is extruded by pulling out the capillary at an angle with the index finger sealing the top of the capillary. The agar plug containing the plaque is then ejected out of the capillary using a plastic tip attached to a pipetteman/eppendorf pipettor. The 100 μL of air pumped into the capillary by the pipette tip is sufficient to squirt the phage plaque into the receiving tube containing YMC cells.
29. LCa plates need to be prewarmed to 43 °C. To prevent them from cooling, place the plates on a foam surface (lid of a dry-ice shipping container is ideal) and not on the cool stone bench-top. Do this until the top agar layer is poured. Then place the plate back in the incubator.

Acknowledgments

Development of the methodology over a number of years was supported by MBRS-SCORE grant #SO 608049 from NIGMS, grant # 1U56 CA92077-01 from NCI, EXPORT grant # 1P20 MD00175-01 from the NIH and funds from the North Carolina Biotechnology Center. I am truly grateful to Dr. Ken Harewood for his unwavering support and encouragement for developing this technology and its applications. Funding support in part has also been obtained from Award Number P20MD000175 from the National Center on Minority Health and Health Disparities (NCMHD), for which I thank Dr. Sean Kimbro. The content is solely the responsibility of the author and does not necessarily represent the official views of the NCMHD or the National Institutes of Health. I thank Drs. Nancy Shepherd, Derek Norford, Delores Grant, Leighcraft Shakes and other colleagues at JLC-BBRI for encouragement and support, and to Dr. Pieter de Jong for encouragement and kind gifts of numerous BAC clones. I thank Ms. Rosalind Grays, Camilla Felton, Connie Keys, Crystal McMichael, Jody Lewis and Darlene Laws for support and encouragement at JLC-BBRI.

References

1. Chatterjee PK, Yarnall DP, Haneline SA, Godlevski MM, Thornber SJ, Robinson PS, Davies HE, White NJ, Riley JH, Shepherd NS (1999) Direct sequencing of bacterial and P1 artificial chromosome nested-deletions for identifying position-specific single nucleotide polymorphisms. Proc Natl Acad Sci U S A 96:13276–13281
2. Gilmore RC, Baker J Jr, Dempsey S, Marchan R, Corprew RNL Jr, Byrd G, Maeda N, Smithies O, Bukoski RD, Harewood KR, Chatterjee PK (2001) Using PAC nested-deletions to order contigs and microsatellite markers at the high repetitive sequence containing Npr3 gene locus. Gene 275:65–72
3. Chatterjee PK, Shakes LA, Wolf HM, Mujalled MA, Zhou C, Hatcher C, Norford DC (2013) Identifying distal cis-acting gene-regulatory sequences by expressing BACs functionalized with loxPTn10 transposons in zebrafish. Roy Soc Chem Adv 3:8604–8617
4. Chatterjee PK, Coren JS (1997) Isolating large nested deletions in bacterial and P1 artificial chromosomes by in vivo P1 packaging of products of Cre-catalyzed recombination between the endogenous and a transposed loxP site. Nucl Acids Res 25:2205–2212
5. Shakes LA, Garland DM, Srivastava DK, Harewood KR, Chatterjee PK (2005) Minimal cross-recombination between wild type and loxP511 sites in vivo facilitates truncating both ends of large DNA inserts in pBACe3.6 and related vectors. Nucl Acids Res 33:e118
6. Ahmed A (1987) Use of transposon-promoted deletions in DNA sequence analysis. Methods Enzymol 155:177–204
7. Wang G, Blakesley RW, Berg DE, Berg CL (1993) pDUAL: a transposon-based cosmid cloning vector for generating nested deletions and DNA sequencing templates *in vivo*. Proc Natl Acad Sci U S A 90:7874–7878
8. Frengen E, Weichenhan D, Zhao B, Osoegawa K, van Geel M, de Jong PJ (1999) A modular, positive selection bacterial artificial chromosome vector with multiple cloning sites. Genomics 58:250–253
9. Wolf HM, Iranloye O, Norford DC, Chatterjee PK (2011) Functionalizing bacterial artificial chromosomes with transposons to explore gene regulation. In: Chatterjee P (ed) Bacterial artificial chromosomes. Rijeka, Croatia, InTech, pp 45–62 (www.intechopen.com). ISBN 978-953-307-725-3
10. Chi X, Chatterjee PK, Wilson W III, Zhang S-X, DeMayo F, Schwartz RJ (2005) Complex cardiac Nkx2-5 gene expression activated by noggin sensitive enhancers followed by chamber specific modules. Proc Natl Acad Sci U S A 102:13490–13495
11. Shakes LA, Malcolm TL, Allen KL, De S, Harewood KR, Chatterjee PK (2008) Context

dependent function of APPb enhancer identified using enhancer trap-containing BACs as transgenes in zebrafish. Nucl Acids Res 36:6237–6248

12. Shakes LA, Du H, Wolf HM, Hatcher C, Norford DC, Precht P, Sen R, Chatterjee PK (2012) Using BAC transgenesis in zebrafish to identify regulatory sequences of the amyloid precursor protein gene in humans. BMC Genomics 13:451
13. Shakes LA, Abe G, Eltayeb MA, Wolf HM, Kawakami K, Chatterjee PK (2011) Generating libraries of iTol2-end insertions at BAC ends using loxP and lox511 Tn10 transposons. BMC Genomics 12:351
14. Chatterjee PK, Shakes LA, Stennett N, Richardson VL, Malcolm TL, Harewood KR (2010) Replacing the wild type loxP site in BACs from the public domain with lox66 using a lox66 transposon. BMC Res Notes 3:38
15. Sternberg NL, Shepherd NS (1996) Construction of bacteriophage P1 libraries with large inserts. In: Dracopoli NC, Haines JL, Korf BR, Moir DT, Morton CC, Seidman CE, Seidman JG, Smith DR (eds) Current protocols in human genetics. Wiley, New York, NY, pp 5.3.1–5.3.26
16. Sambrook J, Fritsch EF, Maniatis T (1989) Molecular cloning: a laboratory manual, 2nd edn. Cold Spring Harbor Laboratory Press, Cold Spring Harbor, NY
17. Chatterjee PK, Mukherjee S, Shakes LA, Wilson W III, Harewood KR, Byrd G (2004) Selecting transpositions of a markerless transposon using phage P1 headful packaging: new transposons for functionally mapping long range regulatory sequences in BACs. Anal Biochem 335:305–315
18. Chatterjee PK, Briley LP (2000) Analysis of a clonal selection event during transposon-mediated nested-deletion formation in rare BAC and PAC clones. Anal Biochem 285:121–126
19. Chatterjee PK, Shakes LA, Srivastava DK, Garland DM, Harewood KR, Moore KJ, Coren JS (2004) Mutually exclusive recombination of wild type and mutant loxP sites in vivo facilitates transposon-mediated deletions from both ends of genomic DNA in PACs. Nucl Acids Res 32:5668–5676
20. Birnboim HC, Doly J (1979) A rapid alkaline extraction procedure for screening recombinant plasmid DNA. Nucl Acids Res 7:1513–1523
21. Chatterjee PK, Baker JC Jr (2004) Preparing nested-deletion template DNA for field inversion gel electrophoresis analyses and position-specific end-sequencing with transposon primers. In: Zhao S, Stodolsky M (eds) Bacterial artificial chromosomes, vol 1, Methods in molecular biology series. Humana Press Inc, Totowa, NJ, pp 243–254
22. Osoegawa K, Mammoser AG, Wu C, Frengen E, Zeng C, Catanese JJ, de Jong PJ (2001) A bacterial artificial chromosome library for sequencing the complete human genome. Genome Res 11:483–496

Chapter 6

Recombining Overlapping BACs into Single Large BACs

George Kotzamanis and Athanassios Kotsinas

Abstract

BAC clones containing the entire genomic region of a gene including the long-range regulatory elements are very useful for gene functional analysis. However, large genes often span more than the insert of a BAC clone, and single BACs covering the entire region of interest are not available. Here, we describe a general system for linking two or more overlapping BACs into a single clone. Two rounds of homologous recombination are used. In the first, the BAC inserts are subcloned into the pBACLink vectors. In the second, the two BACs are combined together. Multiple BACs in a contig can be combined by alternating use of the pBACLInk vectors, resulting in several BAC clones containing as much of the genomic region of a gene as required. Such BACs can then be used in gene expression studies and/or gene therapy applications.

Key words Bacterial artificial chromosomes, Linking, Recombineering

1 Introduction

Bacterial artificial chromosomes (BACs), originating from the human genome project, contain genomic loci inserts of approximately 180 kb on average and cover the entire human genome [1]. These sequenced BACs can accommodate most genes along with their long-range regulatory elements and have proven extremely useful in the analysis of gene function and in transgene expression.

However, there are many large human genes which are larger than the average insert size of the BAC libraries and for these it is often difficult to find a single BAC spanning the entire gene locus with all its controlling elements. The genomic region of such genes is usually covered by two or more contiguous overlapping BACs. In cases like these it will be useful to be able to link together available BAC clones into a single clone spanning the desired region.

Highly efficient homologous recombination systems have been developed in *E. coli* that allow modifications of large BACs without the use of restriction enzymes and ligases, a process called recombineering [2, 3]. A very efficient recombineering method is based on the Red recombination system of bacteriophage λ [4].

Kumaran Narayanan (ed.), *Bacterial Artificial Chromosomes*, Methods in Molecular Biology, vol. 1227, DOI 10.1007/978-1-4939-1652-8_6, © Springer Science+Business Media New York 2015

Modified DH10B strains have also been produced with the Red recombination genes (exo and bet), and the gam gene, integrated into the *E. coli* chromosome under the control of the temperature-sensitive λcI857 repressor [5]. The repressor is inactivated at 42 °C allowing the genes to be expressed so that recombination can take place.

Here we describe a general method, which allows linking of the inserts of two, or more, overlapping BACs into a single BAC clone using the Red homologous recombination system.

2 Materials

2.1 E. coli Strains

1. DH10B: F-*mcr*A Δ(mrr-hsdRMS-mcrBC) φ80dlacZΔM15 ΔlacX74 deoR recA1 endA1 araD139 Δ(*ara, leu*)7697 *gal*U *gal*K λ^- *rps*L *nup*G (Invitrogen). Used for BAC electroporations.
2. EL350: DH10B[*lcI*857(*cro-bio*A)<>*ara*C-pBAD*cre*] [5]. Used for recombination mediated by the Red system.

2.2 Antibiotics

1. Spectinomycin: 50 mg/ml in H_2O (stock), working concentration of 50 μg/ml (store at −20 °C).
2. Chloramphenicol: 25 mg/ml in 100 % ethanol (stock), working concentration of 12.5 μg/ml (store at −20 °C).
3. Gentamicin: 50 mg/ml in H_2O (stock), working concentration of 5 μg/ml (store at Room Temperature).
4. Zeocin: 100 mg/ml in H_2O (stock), working concentration of 25 μg/ml (store at −20 °C).

2.3 Media

1. Luria-Bertani (LB) medium: 1 % bacto tryptone, 0.5 % yeast extract, 1 % NaCl.
2. Solid LB plates: 1 % bacto tryptone, 0.5 % yeast extract and 1 % NaCl, 1.5 % bacto agar.
3. SOB medium: 2 % bacto tryptone, 0.5 % yeast extract, 0.05 % NaCl, 2.5 mM KCl.

2.4 Buffers/Solutions

1. P1: 50 mM Tris–HCl pH 8.0, 10 mM EDTA, 100 μg/ml RNase A.
2. P2: 200 mM NaOH, 1 % SDS.
3. P3: 3 M potassium acetate pH 5.5.
4. QC: 1 M NaCl, 50 mM MOPS pH 7.0, 15 % isopropanol.
5. QF: 1.25 M NaCl, 50 mM Tris–HCl, pH 8.5, 15 % isopropanol.
6. TBE: 45 mM Tris–borate, 1 mM EDTA.
7. TE: 10 mM Tris–HCl pH 7.4, 1 mM EDTA pH 8.

2.5 Plasmids

1. pBACLinkSp [6]. The plasmid contains the single-copy BAC origin of replication, a spectinomycin resistance gene and convenient sites for the cloning of two PCR-amplified homology regions located at the ends of a BAC insert. It is used for subcloning of the BAC insert. Following subcloning (selected for spectinomycin resistance), a new single-copy BAC is generated, which contains the portion of the initial BAC insert sequence found between the two homology regions.
2. pBACLinkGm [6]. The plasmid contains the single-copy BAC origin of replication, a gentamicin resistance gene and convenient sites for the cloning of two PCR-amplified homology regions located at the ends of a BAC insert. It is used for subcloning of the BAC insert. Following subcloning (selected for gentamicin resistance), a new single-copy BAC is generated, which contains the portion of the initial BAC insert sequence found between the two homology regions.
3. pBlSHomABeloHomOloxZeo [7]. The plasmid contains the single-copy BAC origin of replication, a zeocin resistance gene and two PCR-amplified homology regions located at the ends of a BAC insert. It is used for subcloning of the BAC insert. Following subcloning (selected for zeocin resistance), a new single-copy BAC is generated, which contains the portion of the initial BAC insert sequence found between the two homology regions.

2.6 Primer Sequences

Sp6: ATTTAGGTGACACTATAG, used for sequencing the ends of each BAC insert in order to determine its orientation.
T7: TAATACGACTCACTATAGGG, used for sequencing the ends of each BAC insert in order to determine its orientation.

1. TTTGTCACAGGGTTAAGGGC, used in PCR to confirm correct recombination events.
2. TTTGCAACTGCGGGTCAAGG, used in PCR to confirm correct recombination events.
3. GGTTATGTGGACAAAATACCTGG, used in PCR to confirm correct recombination events.

3 Methods

3.1 Growth and Maintenance of E. coli

1. Grow DH10B bacteria at 37 °C and EL350 at 30 °C in LB medium or on solid LB plates.
2. Select bacteria that contain either the original or the modified BACs, using the appropriate antibiotics.
3. Store bacterial cultures at −70 °C in the presence of 15 % (v/v) glycerol for long-term storage or at 4 °C on agar plates for up to 3 weeks.

3.2 Preparation of Electrocompetent Bacteria

1. Inoculate 1 l of SOB medium with 1 ml of a saturated overnight culture and incubate at 37 °C for DH10B and 30 °C for EL350, with agitation at 200 rpm until an OD600nm = 0.6–0.8 (7–8 h for EL350).
2. Transfer the culture to ice and leave for 30 min.
3. Harvest the bacteria in four 250-ml bottles and centrifuge at 4,000 × *g* at 4 °C for 10 min.
4. Wash the cell pellets twice by resuspension and recentrifugation with 200 ml of ice-cold 10 % glycerol.
5. Resuspend in a small volume of ice-cold 10 % glycerol and transfer into a single 50-ml tube for one more centrifugation at 6,000 × *g* at 4 °C for 10 min.
6. Resuspend in 2 ml of ice-cold 10 % glycerol, divide in 40 μl aliquots, freeze in a dry-ice/ethanol bath for 5 min and store at −80 °C.
7. Use each 40 μl aliquot for a single electroporation.

3.3 Preparation of Electrocompetent EL350 Bacteria Induced for Red Recombination

Follow the same protocol described in Subheading 3.2 with all incubations at 30 °C except the following:

1. After an OD_{600} of 0.6–0.8 is reached, incubate the culture at 42 °C for 15 min.
2. Exactly after 15 min, transfer the culture to ice and complete the procedure.

3.4 Electroporation of E. coli with BAC DNA

1. Thaw 40 μl of electrocompetent bacteria on ice.
2. Mix cells with 2 μg of BAC DNA and leave on ice for 10 min.
3. Transfer into cold electroporation cuvettes (0.1 cm, Bio-Rad) and electroporate at 1.8 kV/cm, 200 Ω, 25 μF.
4. Add 1 ml of S.O.C. medium (Invitrogen) to the cuvette and then transfer into new 15 ml round bottom sterile tubes for growth at 30 °C for 1.5 h for EL350 and at 37 °C for 1 h for DH10B with vigorous shaking.
5. Dilute electroporated bacteria and spread onto LB plates containing the appropriate antibiotic.

3.5 Large-scale Preparation of BAC DNA

Large-scale preparation of BAC DNA was performed using a commercial kit (Qiagen Plasmid Maxi Kit) as follows:

1. Grow 400 ml of LB cultures overnight at 37 °C for DH10B and at 30 °C for EL350 with vigorous shaking.
2. Centrifuge the bacteria at 6,000 × *g* for 15 min at 4 °C and resuspend the pellet in 10 ml buffer P1.
3. Add 10 ml of buffer P2 and leave the mixture at room temperature for 5 min.
4. Add 10 ml of buffer P3, mix the contents by inversion and incubate on ice for 20 min.

5. Centrifuge at maximum speed in a microcentrifuge for 30 min at 4 °C.
6. Transfer the supernatant quickly to clean tubes and recentrifuge for another 15 min.
7. Apply the cleared lysate to a Qiagen tip 500 anion exchange resin column.
8. Wash the column twice with buffer QC.
9. Elute the DNA by addition of high salt buffer QF (1.25 M NaCl, 50 mM Tris–HCl pH 8.5, 15 % isopropanol).
10. Precipitate with 0.7 volumes of isopropanol at room temperature.
11. Pellet the DNA by centrifugation at maximum speed in a microcentrifuge at 4 °C for 30 min.
12. Wash the pellet with 15 ml of cold 70 % ethanol and centrifuge for 10 min under the same conditions.
13. Air-dry the pellet and suspend in 30 μl of TE pH 8.0.

3.6 Purification of Linear DNA Fragments from Agarose Gels

For the purification of DNA fragments smaller than ~12 kb in size, use the QiaQuick gel extraction kit according to the manufacturer's instructions.

1. Following electrophoresis of the DNA in an agarose gel, excise the required band using a clean scalpel under low-intensity long-wave ultraviolet light and put in a tube.
2. Add three volumes of buffer QG to 1 volume of gel and incubate the tube at 50 °C until the gel dissolves.
3. Add one gel volume of isopropanol to the dissolved gel and place the mix in a QiaQuick column and centrifuge for 1 min at full speed in a microcentrifuge.
4. Discarding the flow-through, add 0.5 ml of buffer QG to the column and centrifuge for 1 min at full speed in a microcentrifuge.
5. Add 750 μl of the wash buffer PE to the column and centrifuge for 1 min at full speed in a microcentrifuge.
6. Discard the flow-through and recentrifuge for 1 min at full speed in a microcentrifuge.
7. Add 30 μl of the elution buffer EB to the column, place the column in a clean tube and centrifuge for 1 min at full speed in a microcentrifuge to collect the purified DNA.

3.7 Pulsed Field Gel Electrophoresis

Use pulsed field electrophoresis to analyze DNA over 15 kb in size.

1. Cast 1 % agarose gels in 0.5× TBE.
2. Conduct electrophoresis in 0.5× TBE at 4 °C at 5.9 V/cm for 18 h with switching times from 5 to 15 s using a Chef-DRIII (BioRad).

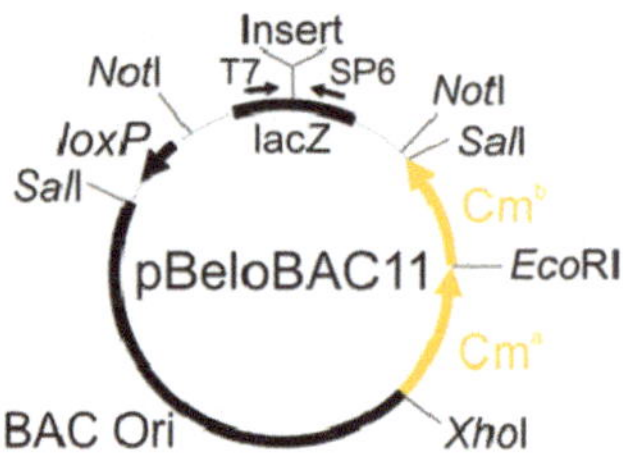

Fig. 1 Diagram of vector pBeloBAC11 showing the primers T7 and Sp6 flanking the insert. Cm^a and Cm^b are two parts of the chloramphenicol resistance gene (Cm^R)

3. Following electrophoresis, stain the gels in ethidium bromide for 30 min and visualize under UV light.

Strategy part: The method applies to the linking of the inserts of overlapping BAC clones originating from a library that is based on the pBeloBAC11 vector [8]. Usually, the aim is to identify the BAC clones covering a specific region of an already sequenced genome and to link their inserts into one single BAC.

3.8 Identification and Characterization of the BAC Clones Covering a Genomic Region

1. Identify the BAC clones representing the sequence of the desired region in the genome by sequence alignment.
2. Sequence the ends of the inserts of each of the identified BAC clones, using the T7 and Sp6 promoters as primers. These promoters flank the cloning site in the middle of the *lacZ* gene of the pBeloBAC11 vector (Fig. 1).
3. Align the obtained sequences of the ends of each BAC insert against the sequence of the genomic region, to determine the orientation of the inserts on the BACs.

3.9 Linking the Inserts of Overlapping BACs

Depending on the orientation of the inserts of each BAC clone, different strategies need to be developed for their linking. From all possible combinations of the orientation of the inserts, the one presented in Fig. 2 will be used as an example. In this example, three BAC clones cover the desired region and need to be linked together. The inserts of BAC2 and BAC3 have the same orientation and the insert of BAC1 has the opposite orientation. The first step is to link BAC1 and BAC2 together to generate BAC1-2. The second step is to link BAC3 to BAC1-2 to generate BAC1-2-3. However, before the inserts of the BACs are linked together, it is necessary to "subclone" them, by homologous recombination, into new BAC vectors. These new BAC vectors contain unique restriction sites for their linearization, as linearized DNA is required in the subsequent linking reactions, and also an appropriate selectable marker gene, which is required to select for correctly retrofitted clones after the linking.

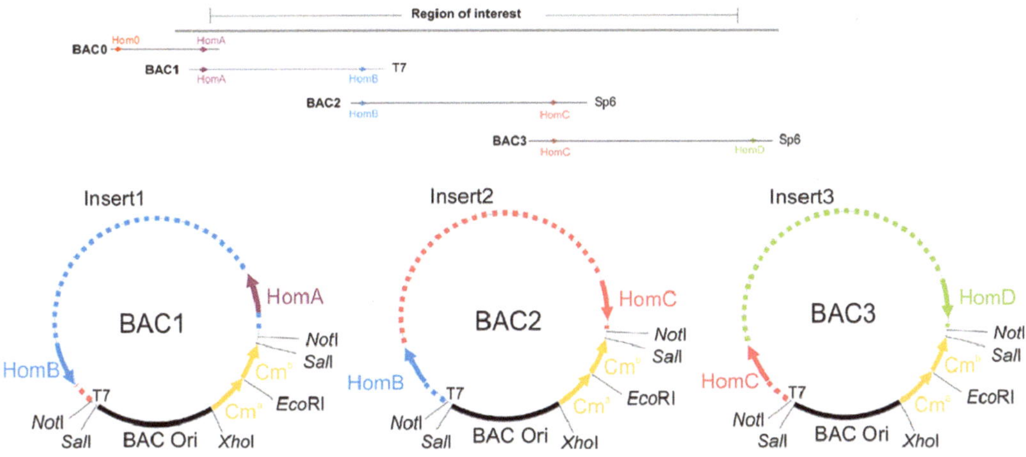

Fig. 2 Diagram of the BAC inserts, which cover the region of interest and need to be linked together and maps of BAC1, BAC2 and BAC3, showing the orientation of their inserts with respect to primer T7 and the regions of homology HomA, HomB, HomC, HomD, and Hom0 that are used in homology recombination reactions to subclone the inserts of the BACs before linking. The insert of BAC1, BAC2, BAC3 and BAC0 is subcloned to a modified BAC vector using HomA and HomB, HomB and HomC, Hom C and HomD, and Hom0 and HomA regions, respectively. The subcloned insert 1 is then linked to the subcloned insert 2 by homologous recombination within the overlapping region of homology between insert1 and insert2, to generate BAC1-2. Then, the subcloned insert 3 is linked to BAC1-2 by homologous recombination within the overlapping region of homology between insert 2 and insert 3, to generate BAC1-2-3. Finally, the subcloned insert 0 is linked to BAC1-2-3 by homologous recombination within the overlapping region of homology between insert0 and insert1, to generate BAC0-1-2-3. Details are provided in the text and the following figures

*3.9.1 Subcloning of BAC1 Insert (See **Note 1**)*

1. Select two regions of homology HomA located at the beginning of the BAC insert and HomB located in the overlapping region between BAC1 and BAC2 (Fig. 2) (*see* **Notes 2–4**).
2. Amplify these homology regions from BAC1 by PCR (*see* **Note 5**) and clone into the vector pBACLinkSp to create the subcloning vector pBACLinkSpAB (Fig. 3).
3. Introduce BAC1 into EL350 bacteria by electroporation (*see* **Notes 6** and **7**).
4. Linearize pBACLinkSpAB with *Not*I, purify the fragment containing HomA and HomB at its ends and electroporate 500 ng of it into induced EL350 bacteria containing BAC1.
5. Select for spectinomycin resistance and also for chloramphenicol sensitivity as the vector backbone on BAC1 is replaced by the vector fragment from pBACLinkSpAB by homologous recombination to give BAC1LinkSp (Fig. 3).
6. Confirm correct subcloning by pulsed field gel electrophoresis analysis after digestion with I-*Ppo*I and by PCR using primers that amplify across HomA (primers 1 and 5) and across HomB (primers 2 and 4) only after correct homologous recombination as shown in Fig. 3. Primers 1, 2 and 3, shown in Fig. 3,

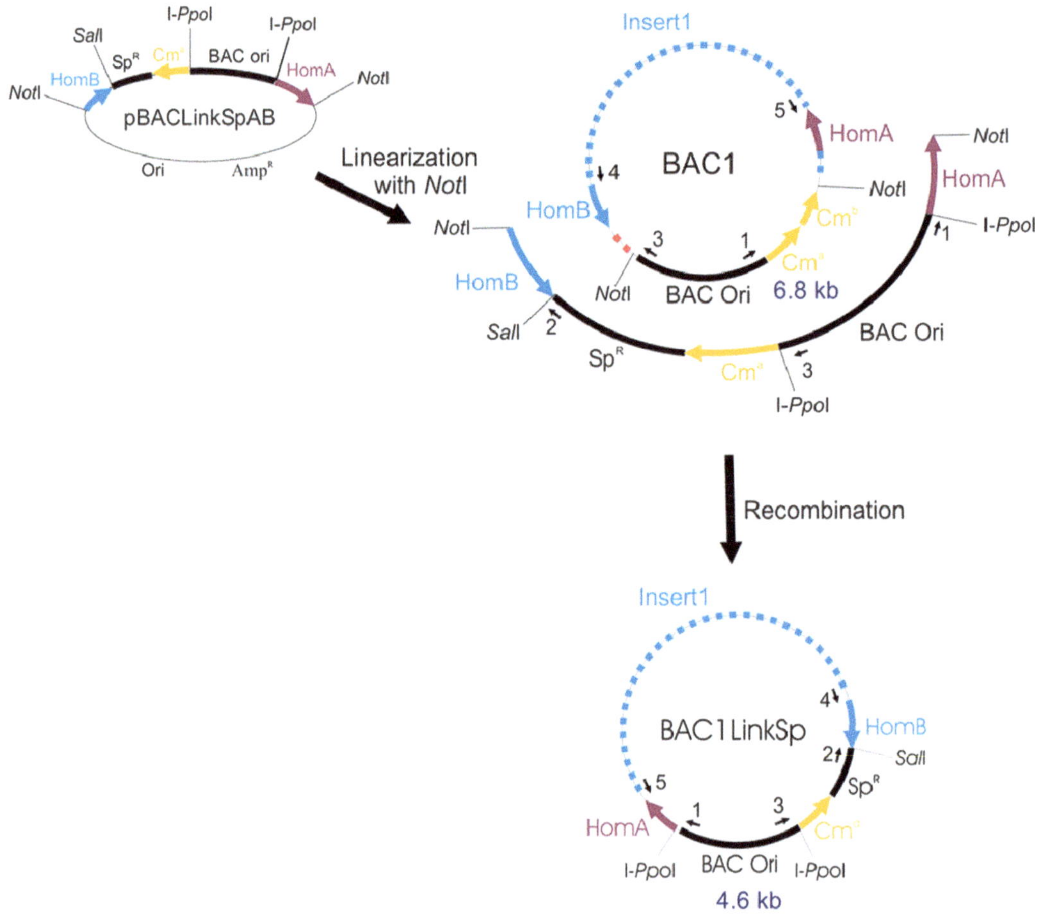

Fig. 3 Subcloning of BAC1 insert. Two PCR-amplified regions of homology HomA and HomB are cloned into plasmid pBACLinkSp to generate pBACLinkSpAB. pBACLinkSpAB is linearized and transfected into induced for Red system EL350 bacteria containing BAC1. Following homologous recombination at HomA and HomB, insert1 between these homology regions on BAC1 is subcloned to vector BAC1LinkSp. BAC1LinkSp contains insert1, a spectinomycin selectable marker and two unique I-*Ppo*I sites used to generate a linear fragment with HomA and Cm^a at its ends. These HomA and Cm^a regions are used in subsequent homology recombination reactions. After digestion of BAC1LinkSp with I-*Ppo*I, the characteristic 4.6 kb fragment (shown in *blue*) and a large fragment, the size of which is determined by the size of insert1, are expected and are indicative of correct subcloning. The primers 1–5 that are used for confirmation of the recombination event are shown

are standard for any subcloning using pBACLinkSp, regardless of the sequence of BAC1 insert to be subcloned, and their sequence is provided in Subheading 2.6. Primers 4 and 5, which are also used for screening, are designed according to the sequence of BAC1.

3.9.2 Subcloning of BAC2 Insert (See ***Note 1****)*

1. Select two regions of homology HomB located in the overlapping region between BAC1 and BAC2 and HomC located at the end of the BAC insert (Fig. 2) (*see* **Notes 2–4**).

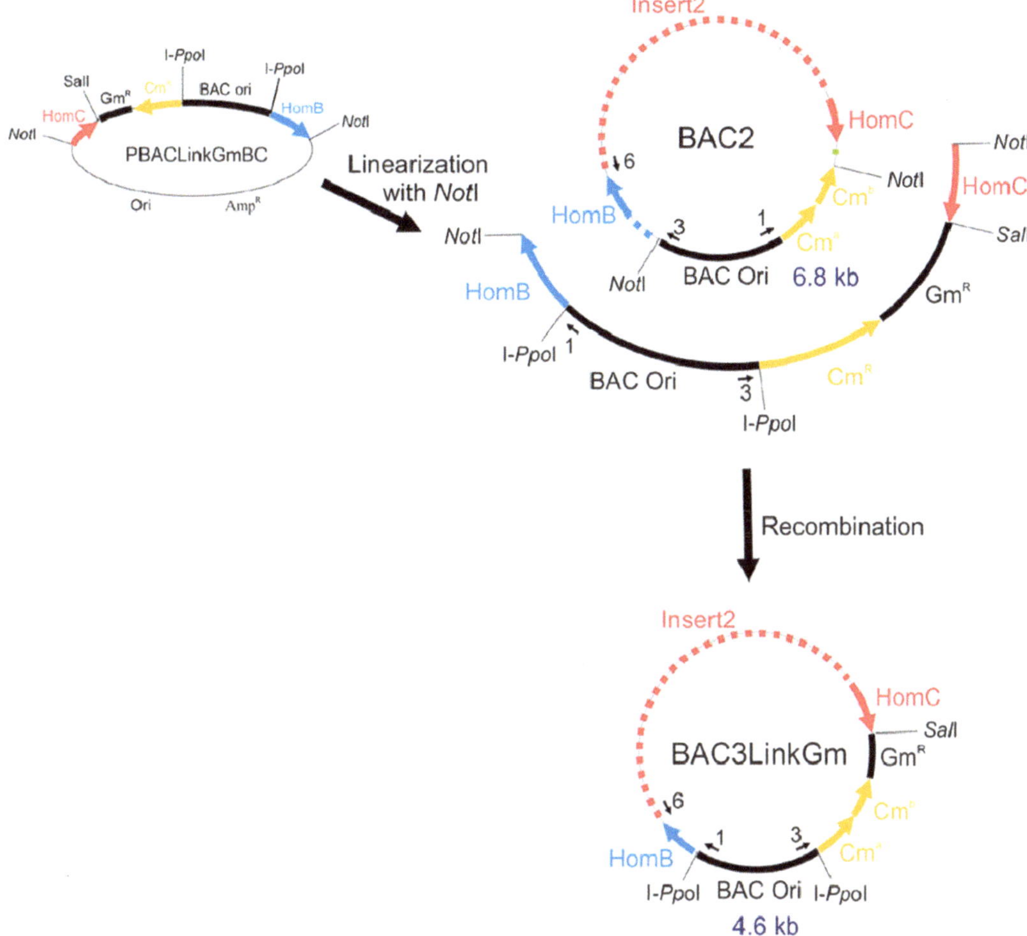

Fig. 4 Subcloning of BAC2 insert. Two PCR-amplified regions of homology HomB and HomC are cloned into plasmid pBACLinkGm to generate pBACLinkGmBC. pBACLinkGmBC is linearized and transfected into induced for Red system EL350 bacteria containing BAC2. Following homologous recombination at HomB and HomC, insert2 between these homology regions on BAC2, is subcloned to vector BAC2LinkGm. BAC2LinkGm contains insert2, a gentamicin selectable marker and two unique I-*Ppo*I sites. After digestion of BAC2LinkGm with I-*Ppo*I, the characteristic 4.6 kb fragment (shown in *blue*) and a large fragment, the size of which is determined by the size of insert2, are expected and are indicative of correct subcloning. The primers 1–6 that are used for confirmation of the recombination event are shown

2. Amplify these homology regions from BAC2 by PCR (*see* **Note 5**) and clone into the vector pBACLinkGm to create the subcloning vector pBACLinkGmBC (Fig. 4).
3. Introduce BAC2 into EL350 bacteria by electroporation (*see* **Notes 6** and 7).
4. Linearize pBACLinkGmBC with *Not*I, purify the fragment with HomB and HomC at its ends and electroporate 500 ng of it into induced EL350 bacteria containing BAC2.

5. Select for gentamicin resistance as the vector backbone on BAC2 is replaced by the vector fragment from pBACLink-GmBC by homologous recombination to give BAC2LinkGm (Fig. 4).
6. Confirm correct subcloning by pulsed field gel electrophoresis analysis after digestion with I-*Ppo*I and by PCR using primers that amplify across HomB (primers 1 and 6) only after correct homologous recombination as shown in Fig. 4. Primer 6 is designed according to the sequence of BAC2.

*3.9.3 Linking of the Inserts of BAC1 and BAC2 (See **Notes 2** and **8**)*

1. Digest BAC2LinkGm with I-*Ppo*I and electroporate 2 μg of the digested DNA into induced EL350 bacteria that already contain BAC1LinkSp (*see* **Notes 6** and 7).
2. Select for gentamicin resistance and also for spectinomycin sensitivity as following recombination in the HomB and Cm^a regions, the Sp^R gene on BAC1LinkSp should be replaced by the Gm^R and Cm^R genes from BAC2LinkGm (Fig. 5).
3. Confirm correct subcloning by pulsed field gel electrophoresis analysis after digestion with I-*Ppo*I and by PCR using primers located in the inserts of the two BACs on either side of HomB (primers 4 and 6 in Fig. 5).

3.9.4 Subcloning of BAC3 Insert

1. Select two regions of homology: HomC located in the overlapping region between BAC2 and BAC3 (as shown in Fig. 2) and HomD located at the end of the BAC3 insert (as shown in Figs. 2 and 6) (*see* **Notes 2–4**).
2. Amplify these homology regions HomC and HomD from BAC3 by PCR (*see* **Note 5**) and clone into the vector pBA-CLinkSp to create the subcloning vector pBACLinkSpCD (Fig. 6).
3. Introduce BAC3 into EL350 bacteria by electroporation (*see* **Notes 6** and 7).
4. Linearize pBACLinkSpCD with *Not*I, purify the fragment containing HomC and HomD at its ends and electroporate 500 ng of it into induced EL350 bacteria containing BAC3.
5. Select for spectinomycin resistance and also for chloramphenicol sensitivity as the vector on BAC3 is replaced by the vector from pBACLinkSpCD by homologous recombination to give BAC3LinkSp (Fig. 6).
6. Confirm correct subcloning by pulsed field gel electrophoresis analysis after digestion with I-*Ppo*I and by PCR using primers that amplify across HomC (primers 1 and 7) and across HomD (primers 2 and 8) only after correct homologous recombination as shown in Fig. 6. Primers 7 and 8 are designed according to the sequence of BAC3.

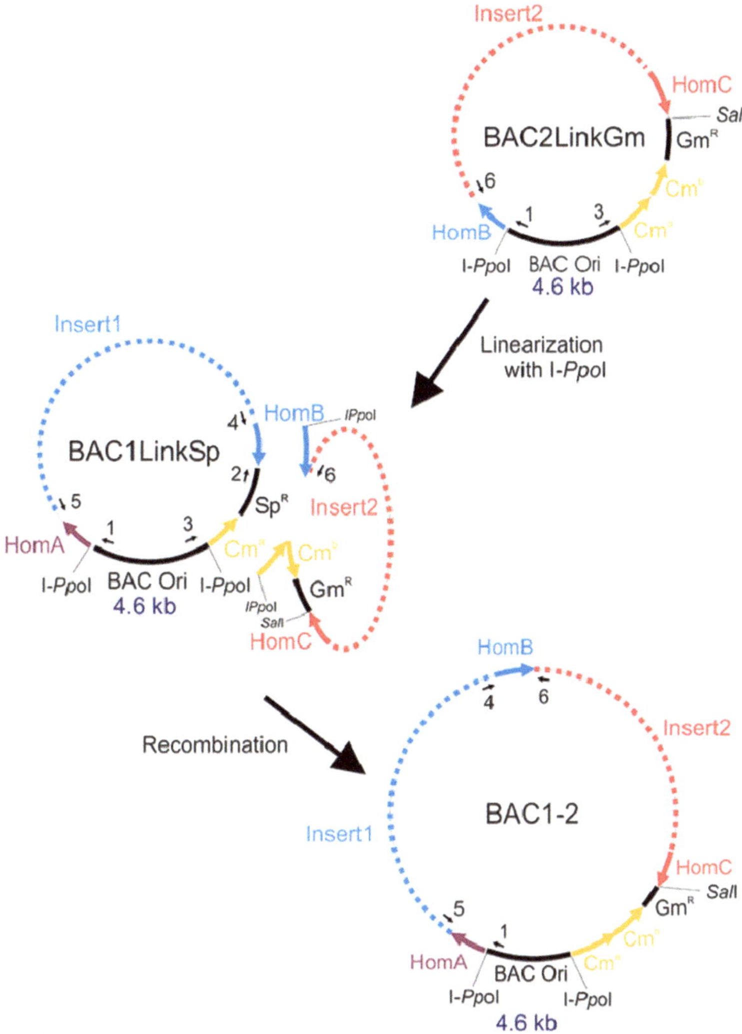

Fig. 5 Linking BAC1 and BAC2, BAC2LinkGm is linearized with I-*Ppo*I and the linear fragment with HomB and Cm[a] at its ends is transfected into induced for the Red system EL350 bacteria containing BAC1LinkSp. Following recombination within the HomB and Cm[a] regions of homology, insert1 and insert2 are linked together. After digestion of BAC1-2 with I-*Ppo*I, the characteristic 4.6 kb fragment (shown in *blue*) and a large fragment, the size of which is determined by the size of insert1 and insert2, are expected and are indicative of correct subcloning. The primers 4–6 that are used for confirmation of the recombination event are shown

3.9.5 Linking of the Inserts of BAC1-2 and BAC3

1. Digest BAC3LinkSp with I-*Ppo*I and electroporate 2 μg of the digested DNA into induced EL350 bacteria containing BAC1-2.
2. Select for spectinomycin resistance and also for gentamicin sensitivity as following recombination in the HomC and Cm[a]

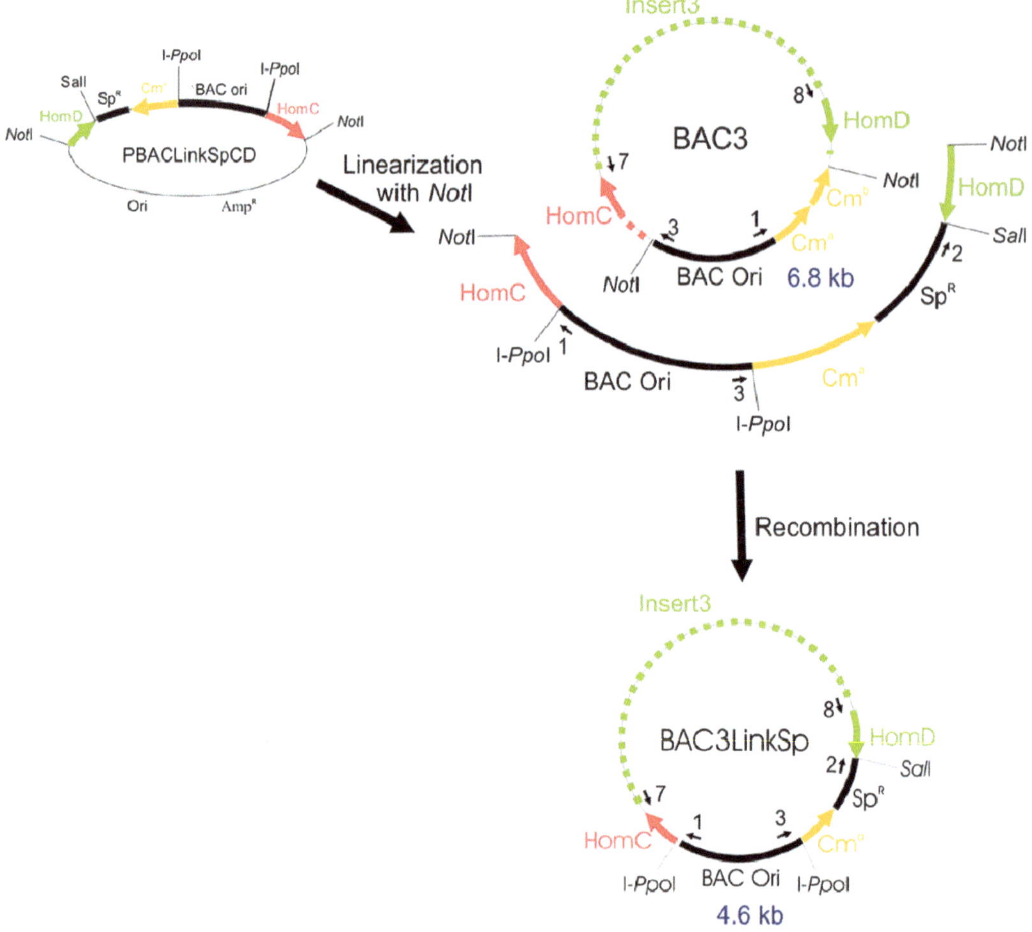

Fig. 6 Subcloning of BAC3 insert. Two PCR-amplified regions of homology HomC and HomD are cloned into plasmid pBACLinkSp to generate pBACLinkSpCD. pBACLinkSpCD is linearized and transfected into induced for Red system EL350 bacteria containing BAC3. Following homologous recombination at HomC and HomD, insert3 between these homology regions on BAC3 is subcloned to vector BAC3LinkSp. BAC3LinkSp contains insert3, a spectinomycin selectable marker and two unique I-*Ppo*I sites used to generate a linear fragment with HomC and Cm^a at its ends. These HomC and Cm^a regions are used in subsequent homology recombination reactions. After digestion of BAC3LinkSp with I-*Ppo*I, the characteristic 4.6 kb fragment (shown in *blue*) and a large fragment, the size of which is determined by the size of insert3, are expected and are indicative of correct subcloning. The primers 1–7 and 2–8 that are used for confirmation of the recombination event are shown

regions, the Gm^R gene on BAC1-2 should be replaced by the Sp^R gene from BAC3LinkSp (Fig. 7).

3. Confirm correct subcloning by pulsed field gel electrophoresis analysis after digestion with I-*Ppo*I and by PCR using primers located in the inserts of the two BACs on either side of HomC (primers 7 and 9 in Fig. 7). Primer 9 is designed according to the sequence of BAC2.

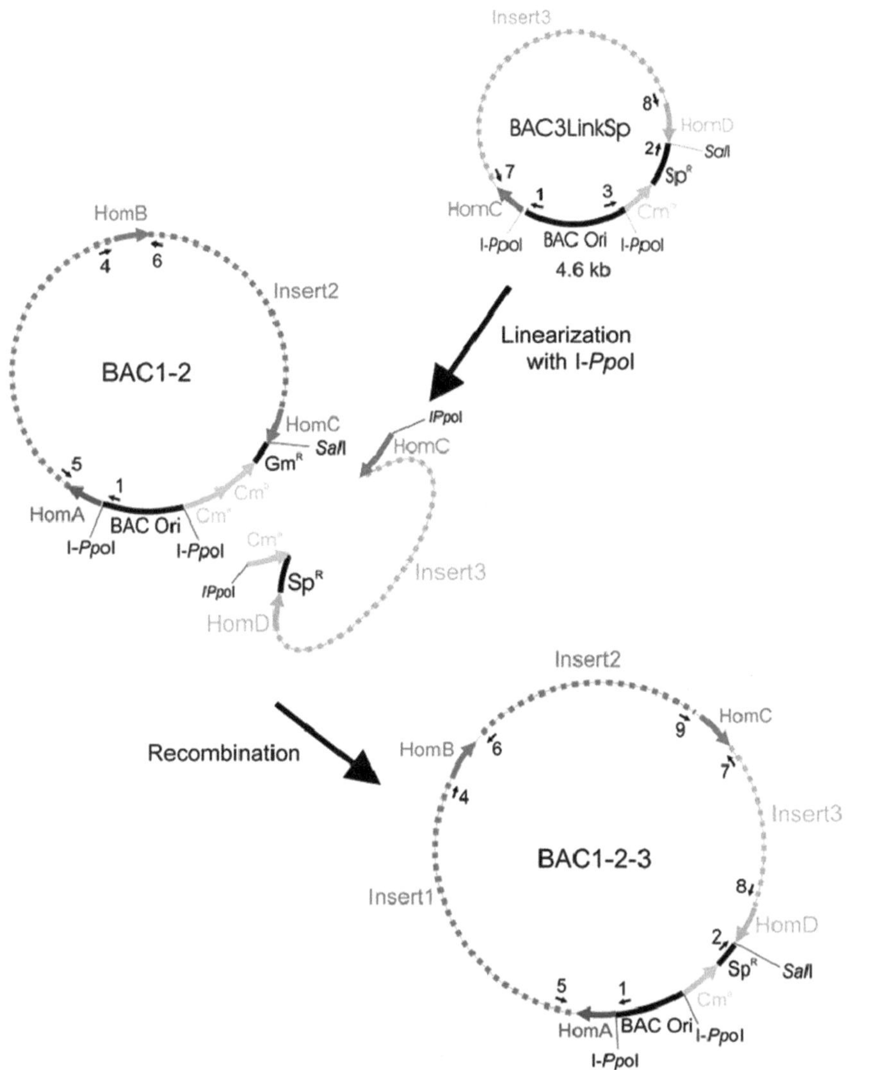

Fig. 7 Linking BAC1-3 and BAC3. BAC3LinkSp is linearized with I-*Ppo*I and the linear fragment with HomC and Cm[a] at its ends is transfected into induced for the Red system EL350 bacteria containing BAC1-2. Following recombination within the HomC and Cm[a] regions of homology, insert3 and insert1-2 are linked together. After digestion of BAC1-2-3 with I-*Ppo*I, the characteristic 4.6 kb fragment (shown in *blue*) and a large fragment, the size of which is determined by the size of insert1, insert2 and insert3, are expected and are indicative of correct linking. The primers 7–9 that are used for confirmation of the recombination event are shown

3.9.6 Linking Additional BACs to BAC1-2-3 (See Notes 2 and 8)

Additional BACs can be added after insert3 of BAC3 by the alternating use of vectors pBACLinkSp and pBACLinkGm. An additional BAC can also be added before insert1 of BAC1 by using a slightly different recombineering strategy:

1. Identify the BAC representing the region before insert1 of BAC1 (BAC0 in Fig. 2).

2. Determine the orientation of the insert of BAC0 as in Subheading 3.2, **step 1**.
3. Select two regions of homology Hom0 located at the beginning of the BAC insert and HomA located in the overlapping region between BAC0 and BAC1 (Fig. 2) (*see* **Notes 2–4**).
4. Amplify these homology regions from BAC0 by PCR (*see* **Note 5**) and clone into the subcloning vector pBlSHomABeloHom0loxZeo (Fig. 8).
5. Introduce BAC0 into EL350 bacteria by electroporation (*see* **Notes 6** and 7).
6. Linearize pBlSHomABeloHomOloxZeo with *Not*I, purify the fragment with HomA and HomB at its ends and electroporate 500 ng of it into induced EL350 bacteria containing BAC0.
7. Select for zeocin resistance and also for chloramphenicol sensitivity as the vector on BAC0 is replaced by the vector from pBlSHomABeloHomOloxZeo by homologous recombination to give BAC0LinkZeo (Fig. 8).
8. Confirm correct subcloning by pulsed field gel electrophoresis analysis after digestion with I-*Ppo*I and by PCR using primers that amplify across HomA (primers 3 and 10) only after correct homologous recombination as shown in Fig. 8. Primer 10 is designed according to the sequence of BAC0.
9. Digest BAC0LinkZeo with *Sal*I and electroporate 2 μg of the digested DNA into induced EL350 bacteria containing BAC1-2-3.
10. Select for zeocin and gentamicin resistance.
11. Confirm correct subcloning by pulsed field gel electrophoresis analysis and by PCR using primers located at either side of HomA (primers 5 and 10 in Fig. 8).

4 Notes

1. We find that subcloning the BAC insert(s) into pBACLinkSp and pBACLinkGm works with fairly high efficiency and this step has not been a problem.
2. We find that recombining one BAC onto the other is much harder as you are transfecting a much larger fragment of DNA. So, make the BAC that you are adding as small as possible.
3. We have had no success using small oligo-sized regions of homology. We have only been successful with regions of about 1 kb amplified by PCR.
4. Put the region of homology right at the end of the longer BAC so that you are adding the smallest amount possible and so that you can design a pair of primers spanning this region

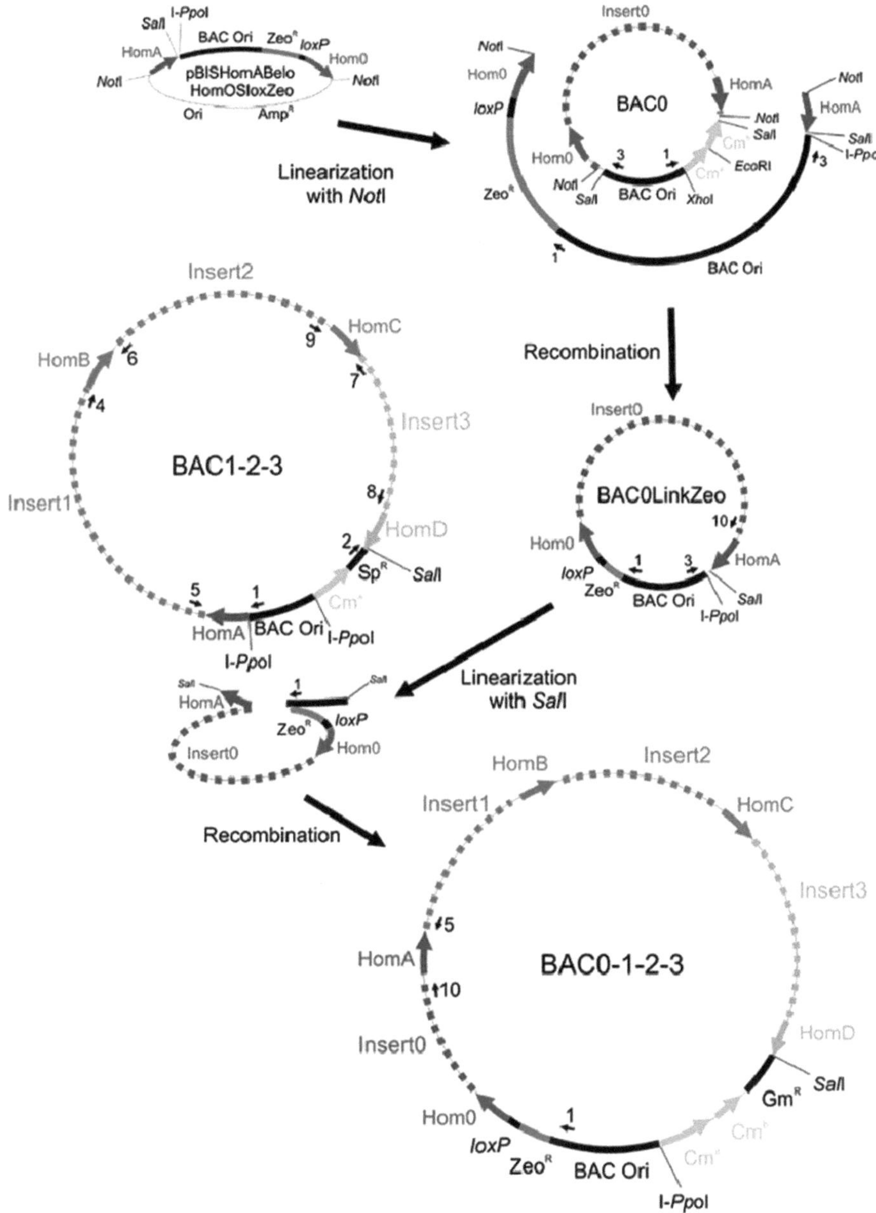

Fig. 8 Subcloning of BAC0 insert and linking of BAC1-2-3 and BAC0. Two PCR-amplified regions of homology Hom0 and HomA are cloned into plasmid pBISHomABeloHom0loxZeo. pBISHomABeloHom0loxZeo is linearized with *NotI* and transfected into induced for Red system EL350 bacteria, containing BAC0. Following homologous recombination at Hom0 and HomA, insert0 between these homology regions on BAC0 is subcloned to vector BAC0LinkZeo. BAC0LinkZeo contains insert0, a zeocin selectable marker and one unique I-*PpoI* site used to generate a linear fragment with HomA and BAC Ori at its ends. After digestion of BAC0LinkZeo with I-*PpoI*, a large fragment, the size of which is determined by the size of insert0, is expected and is indicative of correct subcloning. The primers 3–10 that are used for confirmation of the recombination event are shown. Linearized with I-*PpoI* BAC0LinkZeo is transfected into induced for the Red system EL350 bacteria containing BAC1-2-3. Following recombination within the HomA and BAC Ori regions of homology, insert0 and insert1-2-3 are linked together. After digestion of BAC0-1-2-3 with I-*PpoI*, a large fragment, the size of which is determined by the size of insert0, insert1, insert2 and insert3 is expected and is indicative of correct linking. The primers 5–10 that are used for confirmation of the recombination event are shown

of homology which will allow you to screen for homologous recombination events.

5. For the amplification of the regions of homology, use a proofreading polymerase in order to minimize introduction of mutations. Follow the instructions of the manufacturer. Include in the PCR primers the sequence of the restriction sites to be used in cloning of the homology regions into the pBACLink plasmids. Use the following conditions: 4 min at 94 °C, 30 cycles of [30 s at 94 °C, 30 s at 55 °C, 1.5 min at 68 °C], and 10 min at 68 °C.
6. Test the transformation efficiency of your electrocompetent EL350 bacteria with an Amp^R plasmid with a colE1 origin, e.g., pUC19, and ensure it is higher than 1×10^7 CFU per μg of DNA. We have found that transformation efficiency will drop by 1,000-fold when electroporating large BAC DNA and therefore it is crucial to use highly electrocompetent cells. If the efficiency is low, prepare fresh electrocompetent bacteria and retest.
7. Make a good maxi-prep of the BAC DNA that is to be transfected, about 1 μg per μl. Divide in 10 μl aliquots and store the BAC DNA at −20 °C. Thaw on ice only the amount of BAC DNA required before use. Digest at least 10 μg and then extract with phenol/chloroform, precipitate with ethanol, and resuspend in 5 μl TE. Check DNA by pulsed filed gel electrophoresis. Electroporate with 2 μg of the digested DNA. No purification of the linear fragment is necessary.
8. Check that your BAC that will be added does not contain an I-*Ppo*I site. Do this by digestion with I-*Ppo*I, do not rely on the sequence of the BAC.

Acknowledgement

Financial support was from the European Commission FP7 project INsPiRE.

References

1. Osoegawa K, Mammoser AG, Wu C et al (2001) A bacterial artificial chromosome library for sequencing the complete human genome. Genome Res 11(3):483–496
2. Copeland NG, Jenkins NA, Court DL (2001) Recombineering: a powerful new tool for mouse functional genomics. Nat Rev Genet 2:769–779
3. Muyrers JPP, Zhang Y, Stewart AF (2001) Techniques: recombinogenic engineering – new options for cloning and manipulating DNA. Trends Biochem Sci 26:327–331
4. Yu D, Ellis HM, Lee EC et al (2000) An efficient recombination system for chromosome engineering in *Escherichia coli*. Proc Natl Acad Sci U S A 97:5978–5983
5. Lee EC, Yu D, de Velasco JM et al (2001) A highly efficient *Escherichia coli*-based chromosome engineering system adapted for recom-

binogenic targeting and subcloning of BAC DNA. Genomics 73:56–65

6. Kotzamanis G, Huxley C (2004) Recombining overlapping BACs into a single larger BAC. BMC Biotechnol 4:1
7. Kotzamanis G, Abdulrazzak H, Gifford-Garner J et al (2009) CFTR expression from a BAC carrying the complete human gene and associated regulatory elements. J Cell Mol Med 13(9A):2938–2948
8. Kim UJ, Birren BW, Slepak T et al (1996) Construction and characterization of a human bacterial artificial chromosome library. Genomics 34:213–218

Part III

Gene Expression Using BACs as Vectors

Chapter 7

Generation of Knockout Alleles by RFLP Based BAC Targeting of Polymorphic Embryonic Stem Cells

Tahsin Stefan Barakat and Joost Gribnau

Abstract

The isolation of germ line competent mouse Embryonic Stem (ES) cells and the ability to modify the genome by homologous recombination has revolutionized life science research. Since its initial discovery, several approaches have been introduced to increase the efficiency of homologous recombination, including the use of isogenic DNA for the generation of targeting constructs, and the use of Bacterial Artificial Chromosomes (BACs). BACs have the advantage of combining long stretches of homologous DNA, thereby increasing targeting efficiencies, with the possibilities delivered by BAC recombineering approaches, which provide the researcher with almost unlimited possibilities to efficiently edit the genome in a controlled fashion. Despite these advantages of BAC targeting approaches, a widespread use has been hampered, mainly because of the difficulties in identifying BAC-targeted knockout alleles by conventional methods like Southern Blotting. Recently, we introduced a novel BAC targeting strategy, in which Restriction Fragment Length Polymorphisms (RFLPs) are targeted in polymorphic mouse ES cells, enabling an efficient and easy PCR-based readout to identify properly targeted alleles. Here we provide a detailed protocol for the generation of targeting constructs, targeting of ES cells, and convenient PCR-based analysis of targeted clones, which enable the user to generate knockout ES cells of almost every gene in the mouse genome within a 2-month period.

Key words Bacterial artificial chromosome (BAC), Restriction fragment length polymorphism (RFLP), Embryonic stem (ES) cells, Targeting strategy, Knockout alleles, PCR analysis, Homologous recombination, BAC recombineering

1 Introduction

Mouse ES cells are characterized by their ability to self-renew and differentiate into all cell types of the three germ layers, both in vitro and in vivo [1–3]. One crucial finding was that mouse ES cells are permissive for homologous recombination, allowing gene targeting approaches to alter the mouse genome, allowing the creation of loss-of-function (knockout) and gain-of-function (knock-in) mouse models to study gene function [4–9]. More than two decades after its initial description, gene targeting approaches have resulted in thousands of knockout mouse models, and large scale,

Kumaran Narayanan (ed.), *Bacterial Artificial Chromosomes*, Methods in Molecular Biology, vol. 1227,
DOI 10.1007/978-1-4939-1652-8_7, © Springer Science+Business Media New York 2015

high-throughput projects aiming to disrupt every single gene in the mouse are underway [10–14]. In addition to generating knockout alleles, gene targeting approaches can also be used to generate fusions between the gene of interest and a reporter gene (e.g., EGFP, mCherry, LacZ, and others), to facilitate in vivo tracing of gene expression, or to generate tagged versions of genes (Flag-tag, His-tag, FTAB-tag, and others), which can be used in affinity purification and proteomic studies to characterize protein interaction partners [15, 16]. Besides the advantages for generating animal models, ES cell and homologous recombination techniques also provide valuable tools for in vitro studies, in which modified ES cells can be used to study the outcome of gene ablation or modification on a wide range of processes, including DNA damage repair, signaling pathways, epigenetics, and X chromosome inactivation amongst many others.

Conventional studies on homologous recombination in ES cells make use of plasmid constructs containing a positive selection cassette, rendering targeted cells resistant to a particular antibiotic, flanked by homology arms, together with a negative selection cassette eliminating cells with a randomly integrated transgene. The size of the homologous DNA sequence was found to be crucial for the efficiency of a successful targeting event, with longer stretches of homologous arms facilitating more recombination events [17, 18]. The degree of homology between the targeting construct and the genome of the ES cell is also important, as the presence of SNPs differing between the sequence of the targeting vector and the endogenous DNA sequence can significantly affect the targeting efficiency [17]. Therefore, it is advisable to use isogenic DNA for the construction of the targeting vector [19]. The need for long, isogenic homology arms to construct successful targeting vectors can be challenging and time-consuming as conventional methods like restriction enzyme-based cloning will be difficult when large vectors (>20 kb) are being generated. To circumvent these problems, BACs have been introduced as targeting constructs for homologous recombination [20–26].

BAC constructs were originally developed to allow cloning of large pieces of genomic DNA (up to >200 kb) which could be used for physical mapping and sequencing approaches, and are characterized by a high-clonal stability and low level chimerism, compared to YAC based vectors, which are prone to spontaneous rearrangements [27–36]. BAC plasmids are based on the *Escherichia coli* F-factor plasmid, which is maintained in low copy number in bacteria [27, 37, 38]. The large insert size of genomic fragments make BAC constructs popular in transgenesis approaches, as complete genes including all regulatory sequences can be included [39–41]. Several well annotated BAC libraries exist from different species, including a large variety of mouse strains, which can be easily accessed through public available genome databases like the

UCSF (http://genome.ucsc.edu/) or Ensemble genome browsers (http://www.ensembl.org/), and the NCBI Clone Registry (http://www.ncbi.nlm.nih.gov/clone/). After choosing the BAC clone of interest, these can be ordered through noncommercial sources and commercial companies, including the BACPAC Resources Center at Children's Hospital Oakland Research Institute (CHORI), Oakland, California (https://bacpac.chori.org), the RIKEN BioResource Center (http://www.brc.riken.jp/inf/en/index.shtml), or Invitrogen (http://www.invitrogen.com).

A main advantage of the use of BACs is that they can be easily modified through BAC recombineering techniques, facilitating insertion of selectable markers into the BAC, deletions of unwanted genomic areas, and substitutions of complete regions, thereby enabling the creation of isogenic targeting constructs for subsequent homologous recombination in ES cells, since BACs from the desired mouse strain can be used [42–45, 21–24, 46–53]. Most of the BAC recombineering techniques rely on the RecA protein, and inducible induction of the recombination machinery in bacteria, which enables recombination of the BAC with a recombination substrate transformed into the bacteria (Fig. 1). Excellent reviews and protocols on all the details of BAC recombination have been published, and we refer the reader to those sources for a more in depth overview of these techniques [53, 52, 54–57] (http://recombineering.ncifcrf.gov/). By combining the ease of BAC recombineering, and the use of isogenic, long homology arms, high targeting efficiencies for many different loci have been obtained in BAC targeting approaches, ranging from 7 to 28 % of clones being properly targeted [23].

Beside the above mentioned advantages of BACs as targeting constructs, their widespread use in the scientific community is currently hampered by the difficulties in identifying correct targeted

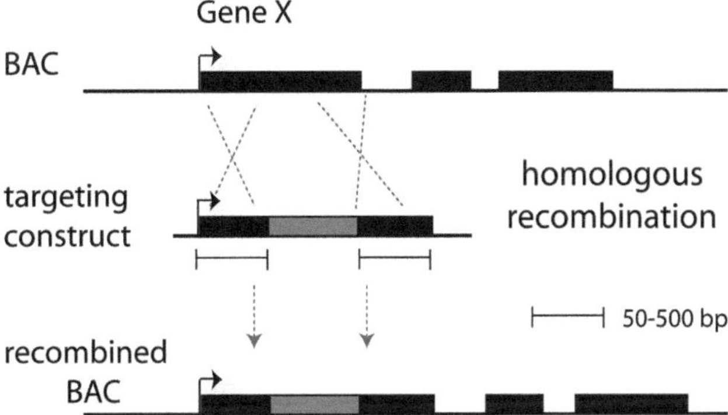

Fig. 1 Principle of BAC recombination. Using homologous recombination in bacteria, a selection cassette (depicted in *blue*), cloned between homologous arms, is integrated in the coding sequence of gene X (exons in *black*) located on a schematic BAC (*black line*) (color figure online)

clones after BAC based targeting approaches. Conventional Southern Blotting techniques are difficult to employ, since the presence of large homology arms impedes the use of external probes [21]. Therefore, to identify correct targeted clones, quantitative real time PCR (qPCR) and DNA-FISH techniques have been applied [21, 23]. In the qPCR approach, primers are chosen which amplify a fragment spanning the projected deletion/insertion, together with a qPCR amplifying a fragment located in one of the arms. A change in copy number detected by the first and not the second primer set is indicative for a correct targeting event. In the DNA-FISH approach, two probes are applied, one spanning the targeting construct, the other spanning the genomic location flanking the gene of interest. When both probe signals co-localize, it is likely that homologous recombination has occurred, and the targeting construct is not randomly integrated elsewhere in the genome.

However, all of these techniques are limited by a high percentage of false positive and negative clones. qPCR is prone to false positive results, and its use depends on the design of sensitive primer pairs, which might not always be available for every genomic location. DNA-FISH is a laborious technique, especially if many clones have to be screened, and may not detect small fragmented pieces of randomly integrated targeting constructs. To circumvent these problems, and fully benefit from the advantages of the BAC targeting approach, we have recently introduced a novel strategy which relies on the presence of RFLPs [58], and have applied this strategy to generate genetically modified ES cells used in different studies on X chromosome inactivation [59, 39].

For our strategy, we make use of highly polymorphic ES cells derived from crosses between C57Bl/6 females and male Cast/Ei mice. The classical inbred *Mus musculus domesticus* mouse strain C57Bl/6, is among the most widely used and best characterized mouse strains [60]. The C57Bl/6 genome has been completely sequenced, functions as the reference genome for the mouse species and several well characterized BAC libraries are available [34]. The Cast/Ei inbred strain has been derived from a wild population of the subspecies *Mus musculus castaneus*, has been sequenced, and a BAC library is available. Cast/Ei mice are more difficult to breed, but offer the advantage of a distinct genetic background. Therefore, Cast/Ei mice have been extensively used in crosses to other strains in studies which required distinction of the maternal and paternal genome by means of SNPs, as for example to study genomic imprinting and X inactivation. The Cast/Ei and C57Bl/6 strains are highly polymorphic, with one estimated SNP per 311 bp [61]. Therefore, RFLPs that can distinguish between both strains will be present in virtually every gene. In our BAC targeting strategy, we exploit this fact by designing BAC targeting vectors in such a way that after proper targeting, an RFLP is destroyed, or transcription through an RFLP is abrogated (Fig. 2). By combining PCR analysis with restriction digestion of the PCR product, an easy and straightforward readout is obtained, which permits the reliable identification of properly targeted clones.

Our strategy involves the design of different variants to generate knockout alleles by insertional mutagenesis, introduction of a premature transcriptional stop, and even the generation of inducible knockout and rescue alleles using a single targeting event. In the first approach (Fig. 2a), the BAC targeting vector disrupts an RFLP located within the coding region of a gene, and inserts a selection cassette into the open reading frame of the gene of interest, thereby ablating gene function. This approach is especially helpful when an RFLP located in the proximal region of the gene can be found, and when important domains for the encoding protein are located downstream of the RFLP. In another approach, a splice acceptor polyadenylation stop cassette (SA-tpA) together with a selection cassette is introduced into a proximal intron of the gene of interest, thereby prematurely stopping gene transcription (Fig. 2b). The SA-tpA selection cassette can be either introduced into an RFLP located in the intron, thereby allowing efficient screening for properly targeted clones as in the first approach (Fig. 2b I), or an RFLP downstream of the insertion site can be used to screen for absence of transcription across the RFLP in candidate clones (Fig. 2b II). The introduction of an intronic SA-tpA cassette also allows the generation of inducible knockout or rescue alleles (Fig. 2c). The SA-tpA cassette can be flanked by asymmetrical Lox sites resulting in a locked inversion of the cassette upon expression of the Cre recombinase protein [62–65]. To generate an inducible knockout allele, the SA-tpA cassette, flanked by the Lox sites, is introduced in antisense orientation in an RFLP located in an intron in the proximal region of the gene of interest. In this orientation, the SA is not recognized, and gene transcription is not inhibited. Upon expression of Cre, the SA-tpA cassette is inverted, and gene transcription will be abrogated (Fig. 2c I). To generate an allele in which expression of the gene can be inducible rescued, the SA-tpA cassette is first introduced in the sense orientation, and Cre expression is used to disable the premature stop of transcription (Fig. 2c II).

The generation of targeted ES cells using our strategy consists of a three step process (Fig. 3). In the first step, a targeting plasmid for recombining a BAC containing the gene of interest is developed. In the second step, the BAC is recombined using BAC recombineering techniques in bacteria, and in the final step, the modified BAC is used for gene targeting in ES cells. As a first step, RFLPs present in the gene of interest are identified. Several SNP databases exist (http://www.informatics.jax.org/), in which SNPs and RFLPs between the C57Bl/6 and Cast/Ei mice can be identified. A decision on the approach must then be taken: (1) A conventional knockout allele based on insertional mutagenesis,

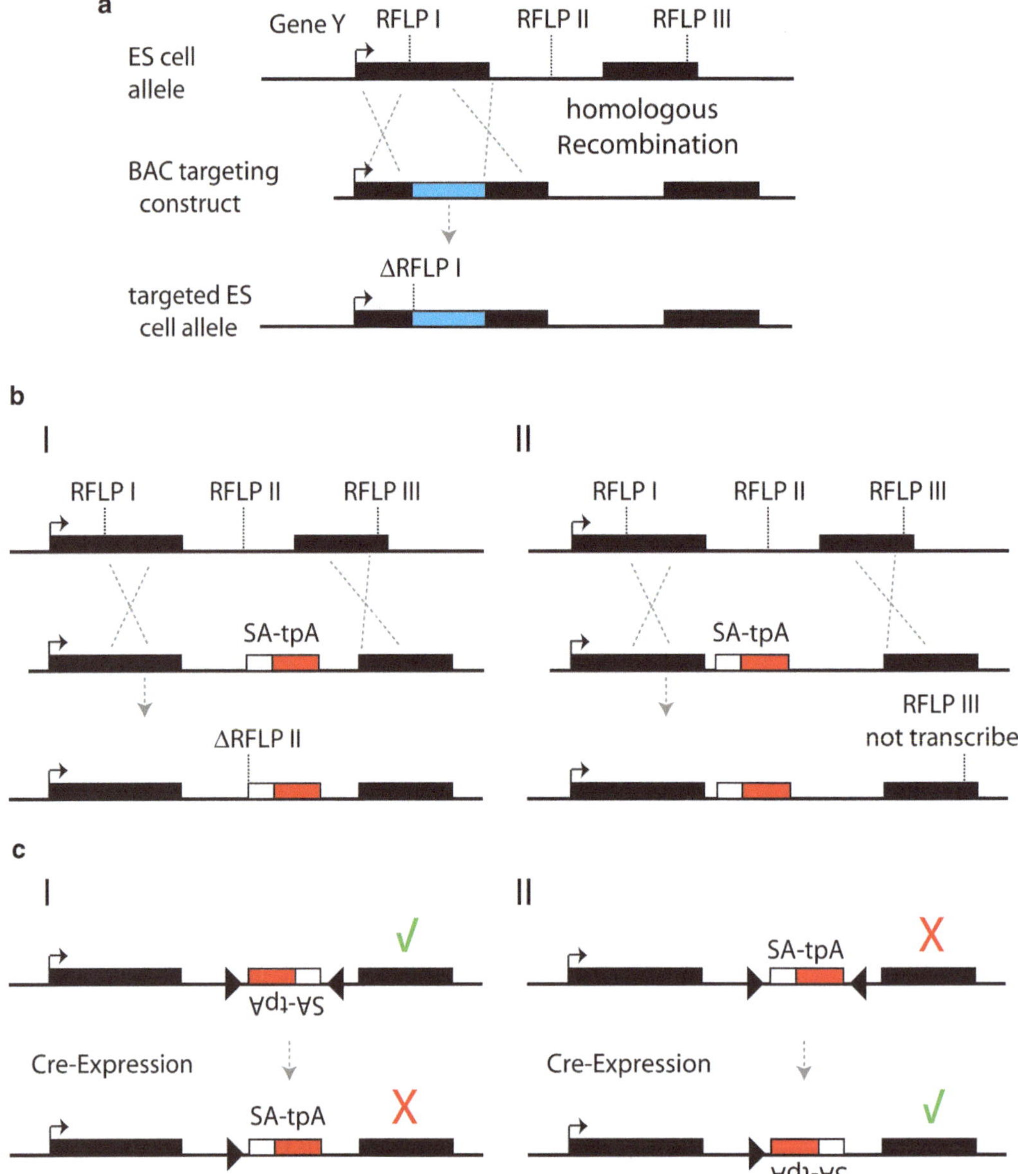

Fig. 2 RFLP based targeting strategy. (**a**) Insertional mutagenesis: a selection cassette (*blue*) is integrated in an exon (*black*) of gene Y, thereby disrupting RFLP I. Deletion of RFLP I can be detected by PCR. (**b**) Introduction of a SA-tpA cassette in an intron: the SA-tpA cassette consists of a splice acceptor (SA, *white*), a triple transcriptional stop sequence (tpA, *red*), and a resistance cassette flanked by either Lox or Frt sites (not shown). The SA-tpA cassette can be targeted to an intron, thereby disrupting the intronic RFLP II (option I), used to detect the correct targeting. Alternatively, proper targeting of the SA-tpA cassette can be detected by screening for expression of the downstream RFLP III (option II), which will not be expressed from the targeted allele in case of a correct targeting event. (**c**) When the SA-tpA construct is flanked by asymmetrical lox sites (*black arrow heads*), the construct can be used to generate inducible knockout (option I) or rescue alleles (option II). In the inducible knockout approach (option I), the SA-tpA construct is targeted to an intron as shown in (**b**), but in the antisense orientation. In this orientation, the SA-tpA does not inhibit transcription. After targeting, the resistance cassette, flanked by Frt sites (not shown) is looped out using transient expression of Flpase.

(2) introduction of a SA-tpA cassette or (3) generation of an inducible allele. After identification of a suitable RFLP and the desired approach, a BAC must be chosen which contains the RFLP and flanking regions. Location of BACs can be found in the UCSF or Ensemble genome browsers, or Mapviewer (http://www.ncbi.nlm.nih.gov/mapview/), from which the sequences can also be retrieved.

In general, it appears best to choose a BAC with a length between 150 and 200 kb, which is centrally located around the RFLP, allowing an equal length of homology arms on both flanking sites. To generate the targeting plasmid, an 800–1,000 bp fragment surrounding the selected RFLP is PCR amplified using a proofreading high fidelity polymerase, and cloned into a vector suitable for subcloning of PCR products (e.g., pCR-BluntII-Topo, Invitrogen). It is important that in the selected region the target RFLP represents an unique restriction site, which does not digest elsewhere in the amplified fragment, or in the vector backbone, enabling single step insertion of the selection cassette into the RFLP. To allow the fast generation of knockout BAC constructs, we have generated a couple of vectors (overview Table 1, all DNA available upon request), from which the selection cassette with or without the SA-tpA and Lox sites can be easily released by restriction digestion. These fragments, which contain all necessary elements to express resistance genes for both antibiotic selection in prokaryotes and eukaryotes, can then be cloned into the unique RFLP restriction site, resulting in the targeting plasmid, in which the selection cassette is now flanked by two homologous arms of around 500 bp. Although BAC recombination can be accomplished with homology arms of only 50 bp, increasing the length of these arms will result in a higher efficiency (our unpublished observations, and [50]). In the subsequent step, the generated plasmid is used for BAC recombineering. Proper modification of the BAC is verified by colony PCR and restriction digestion, and once established, the modified BAC can be used to target ES cells. After selection with the desired antibiotic, emerging colonies are picked, and screened for the correct recombination event using an RFLP-based allele-specific PCR.

The many RFLPs present in the hybrid C57Bl/6-Cast/Ei ES cells, combined with the different approaches presented above make it possible to modify virtually every gene using this BAC

Fig. 2 (continued) Upon Cre expression, the SA-tpA cassette is inverted, and transcription is disrupted. Asymmetrical Lox sites lock the inversion event. To generate an inducible rescue allele (option II), the SA-tpA is targeted in the sense orientation, thereby enforcing a preliminary stop of transcription. After FLPase-mediated loop out of the selection cassette (not shown), the SA-tpA can be inverted, thereby restoring transcription of gene Y

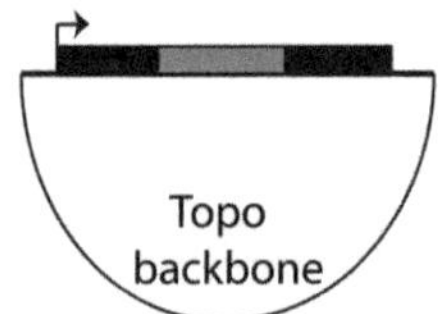

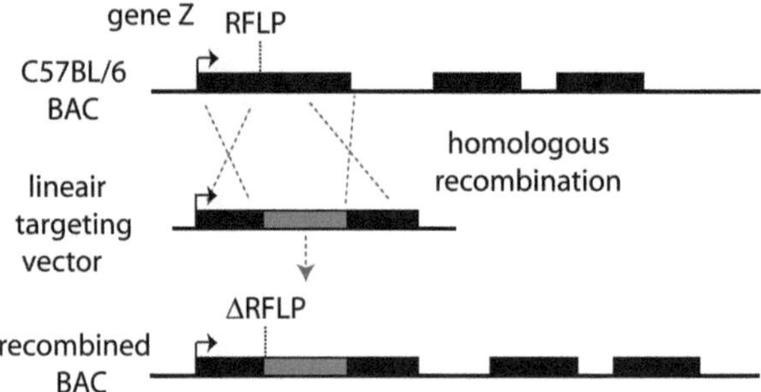

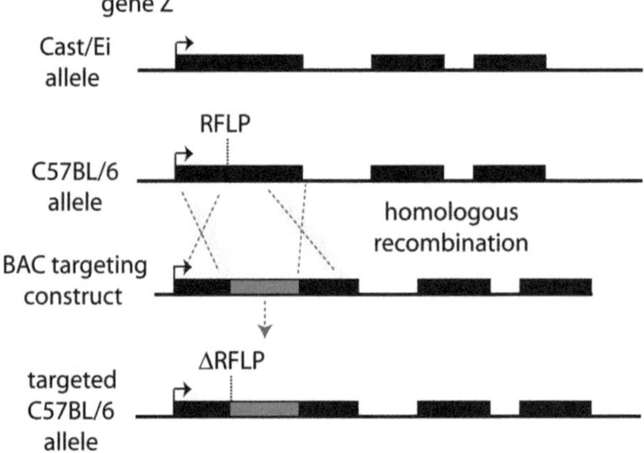

Fig. 3 The targeting strategy in three basic steps. In the first step (*I*), a targeting plasmid is developed, which will be used to modify a BAC. An 800–1,000 bp fragment surrounding the selected RFLP is PCR amplified, subcloned, and the RFLP is used to insert a selection cassette (*blue*). In the plasmid, the selection cassette is flanked by two homologous arms (*black boxes*). In the BAC recombination step (*II*), the linearized plasmid from step I is used for homologous recombination of, in this case, a BAC of C57Bl/6 origin. The selection cassette is integrated in the BAC, thereby disrupting the RFLP used to introduce the selection cassette. In the final step (*III*), the BAC targeting construct is used for homologous recombination in ES cells. Since a C57Bl/6 BAC was used to generate the targeting construct, targeting of the isogenic C57Bl/6 allele will preferentially occur, thereby disrupting the RFLP restriction site, which is only present at the C57Bl/6 allele, but not at the Cast/Ei allele. Loss of the RFLP at the C57Bl/6 allele can be detected by PCR analysis, thereby enabling a straightforward readout of proper targeting (color figure online)

Table 1
Overview of targeting cassettes

Cassette	Fragment size (bp)	Digest
Lox-Kana/Neo-Frt	1,995	EcoRl
Lox-Kana/Neo-Lox	1,978	EcoRl
Lox-Frt-Kana/Neo-Frt	2,369	Nhel-Aflll
Amp/puro	2,733	Sfil-Notl
Lox66-SA-tpA-Lox71	3,287	Dralll-Asel

Table summarizes the available selection cassettes, fragment sizes in base pairs, and the restriction digest used to release the cassette from its vector. Lox: Lox-site for Cre-recombination. Frt: Frt-site for Flpase-recombination. Kana/Neo: cassette with a dual prokaryotic/eukaryotic promoter driving a gene encoding a protein rendering bacteria resistant to kanamycin, and ES cells resistant to neomycin (Geneticin, G418). Amp/Puro: cassette with an ampicillin resistance gene allowing selection in bacteria, and a puromycin resistance gene for use in ES cells. Lox66 and Lox71: asymmetrical Lox sites. SA-tpA: cassette encoding a splice acceptor (SA) and a triple transcriptional stop signal (tpA)

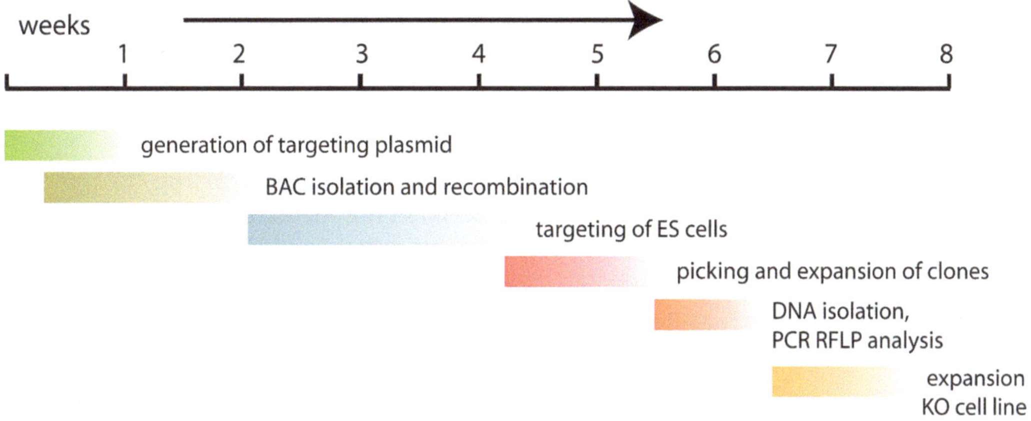

Fig. 4 Timeline. Predicted timeline for carrying out the RFLP-based targeting strategy following the different steps described in this protocol. See text for details

targeting strategy, benefitting from the high targeting efficiency of BAC targeting vectors. In this protocol, we provide detailed steps for the generation of a BAC targeting construct, the transfection of ES cells with the modified BAC constructs, and the identification of properly targeted clones. Several excellent protocols on BAC recombineering [52], BAC transgenesis [55], and design of targeting constructs in general [54] already exist. In this protocol we introduce crucial new steps related to BAC recombineering including time-saving alternatives with regard to previously published methods. Combined with our novel RFLP-based strategy, this protocol permits the generation of knockout ES cells in a very short time span (Fig. 4), for use in in vitro studies or for generation of mouse models through establishment of chimeric mice or through tetraploid embryo complementation [66–70].

2 Materials

2.1 Reagents for Cloning, BAC Recombination, and ES Cell Targeting

1. Plasmids containing selection cassettes and/or SA-tpA sequences (*see* Table 1) (all plasmids are available upon request).
2. Bacterial artificial chromosomes (BACs in DH10B host bacteria; BACPAC Resources at CHORI, Invitrogen, or Riken BioResource Center).
3. C57Bl/6-Cast/Ei ES cell lines (available upon request).
4. Recombination proficient bacteria, EL250 [44].
5. DH5α chemical competent bacteria (e.g., Max Efficiency, Invitrogen).
6. NucleoSpin gel extraction kit Extract II (Bioke).
7. Plasmid Maxi kit (Macherey-Nagel, or Qiagen).
8. PCR materials: PCR primers, Taq-polymerase, dNTP mixture (10 mM each, PCR grade), 50 mM $MgCl_2$, Phusion high fidelity polymerase (New England Biolabs).
9. Reagents for cloning: restriction enzymes, including PI-SceI, BSA (100×, purified), Calf intestinal alkaline phosphatase, T4 DNA ligase, Klenow enzyme, T4 DNA polymerase.
10. pCR-BluntII-Topo, Zero Blunt Topo PCR Cloning-Kit (Invitrogen).
11. Cell culture materials: 1 μg/ml Puromycin, 270 μg/ml Geneticin (G418), DPBS, Dulbecco's Phosphate Buffered Saline, Trypsin–EDTA, 0.25 % (v/v) trypsin/0.2 % (w/v) EDTA in PBS.
12. Mouse Embryonic Fibroblast (MEFs) commercial (for instance Millipore http://www.millipore.com) or self prepared from E13.5 embryos. *CAUTION* All experiments requiring animal handling should be conducted in accordance with international animal protection guidelines and local animal protection laws.
13. Reagents for PCR precipitation and digestion: 10 mg/ml yeast tRNA, 10 mg/ml RNase A.
14. Reagents for RNA isolation and cDNA synthesis: Trizol reagent, DNAseI, Super script III cDNA synthesis kit (all from Invitrogen).

2.2 Equipment

1. 96-well cell culture plates, both flat-bottom and U-shaped.
2. Bottle-top cell culture filter, 0.22 μm.
3. PCR disposable adhesive.
4. Eppendorf Refrigerated Microcentrifuge Model 5417R.
5. Eppendorf Refrigerated Centrifuge Model 5810R.

6. Eppendorf Thermomixer compact.
7. Bacterial incubator (New Brunswick Scientific, Innova43).
8. Beckman J6-MC centrifuge.
9. Electroporation system for bacteria (e.g., Gene Pulser System, BIO-RAD), including cuvettes.
10. Electroporation system for ES cells (e.g., Gene Pulser Xcell, BIO-RAD), including cuvettes.
11. Diaper for wrapping cell culture plates for freezing clones (e.g., TENA).

2.3 Solutions and Media

1. 1 M Tris–HCl: mix 121 g of Tris base and 1,000 ml of dH_2O. Adjust the pH to 7.4 with HCl. Store at room temperature.
2. 10 mM Tris–HCl, pH 7.4: dilute 1 M Tris–HCl, pH 7.4 100×, and store at room temperature.
3. 0.5 M EDTA, pH 8: mix 186.1 g of EDTA and 800 ml of dH_2O. Adjust the pH to 8 with ~20 g of NaOH pellets. Store at room temperature.
4. P1 buffer: 50 ml 1 M Tris–HCl, pH 8, 20 ml 0.5 M EDTA, pH 8, 10 ml RNase A (10 mg/ml), 920 ml dH_2O. Store at 4 °C.
5. P2 lysis buffer: 930 ml dH_2O, 50 ml 20 % SDS, 20 ml 10 M NaOH. Store at room temperature.
6. P3 buffer: 2.8 M KAc, pH 5.1: Add 274.8 g KAc to 400 ml dH_2O, set pH to 5.1 by adding HCl, and add dH_2O to 1 l. Store at 4 °C.
7. 50× TAE for gel electrophoresis: mix 1,210 g Tris base, in 500 ml 0.5 M EDTA pH 8, 285.5 ml Acetic Acid and add dH_2O to 5 l. Autoclave and store at room temperature.
8. 0.2 % gelatin solution: add 2 g gelatin to 1 l dH_2O. Autoclave and store at 4 °C.
9. 2× Freezing solution: add in following order 60 ml DMEM, 20 ml FCS, 20 ml DMSO. Mix and filter-sterilize. Store at 4 °C for up to a month.
10. Mouse Embryonic Stem Cell Medium: DMEM, 15 % fetal calf serum, 100 U/ml penicillin, 100 mg/ml streptomycin, nonessential amino acids, 0.1 mM β-mercaptoethanol, and 1,000 U/ml LIF. Filter-sterilize and store at 4 °C for maximal 2 weeks.
11. Low SDS lysis solution: 40 ml 0.5 M Tris–HCl pH 7.4, 2 ml 0.5 M EDTA, 2 ml 20 % SDS, 8 ml 5 M NaCl, add dH_2O to a total volume of 200 ml. Store at room temperature. Add 50 μl Prot.K (10 μg/ml)/ml lysis solution before use. *CAUTION* contains SDS. Protect eyes.

12. LB: mix 10 g of Bacto Tryptone, 5 g of yeast extract, 5 g NaCl, 1 l dH_2O, pH 7.2. Autoclave and store at room temperature.
13. LB-agar bacteria plates: add 10 g of LB-agar to a 500 ml LB medium bottle, and autoclave. Cool down, until bottle can be touched, then add appropriate antibiotic, and pour plates. Store plates at 4 °C for up to a month.
14. Ampicillin stock solution: 50 mg/ml in dH_2O, aliquot and store at −20 °C. Use at a final concentration of 25 μg/ml for plasmids, and 8 μg/ml for BACs.
15. Kanamycin stock solution: 50 mg/ml in dH_2O, aliquot and store at −20 °C. Use at a final concentration of 50 μg/ml for plasmids, and 15 μg/ml for BACs.
16. Chloramphenicol stock solution: 34 mg/ml in ethanol, aliquot and store at −20 °C. Use at a final concentration of 12.5 μg/ml for BACs.

3 Methods

3.1 Estimated Timing of the Various Steps Described in This Protocol

Subheading 3.2: In silico design of strategy: 1 day.

Subheading 3.3: Generation of targeting plasmid for BAC recombineering: 5–6 days.

Subheading 3.4: BAC isolation and recombination: 4 days.

Subheading 3.5: Transformation of BAC clone into recombination proficient bacteria: 5 days.

Subheading 3.6: BAC recombination: 1 day.

Subheading 3.7: Colony PCR to check proper recombination: 4 days.

Subheading 3.8: Targeting of mouse Embryonic Stem cells: 17 days.

Subheading 3.9: Picking of clones in 96-well plates for Freezing and DNA: 7 days.

Subheading 3.10: DNA purification in 96-well plates and PCR RFLP analysis to identify correct targeted clones: 2 days.

3.2 In Silico Design of Strategy (Timing: 1 Day)

1. Identify suitable RFLP in the gene of interest, and choose one or more BAC clones. Several computer-based databases exist, in which SNPs and RFLPs can be selected (for instance http://www.informatics.jax.org/). Retrieve genome sequences (http://www.ncbi.nlm.nih.gov/mapview/), and select the targeting cassette and approach. It is good practice to generate sequence files in which the modified, recombinant sequence is assembled. This can be done with several software programs (e.g., DNAMAN, Lynnon Biosoft, or Clone Manager, Scientific & Educational software), and will save valuable time later on, when primers or restriction digestions to verify cloning steps are chosen.

2. Order selected BAC clone. Take care that a BAC is chosen from an isogenic library, which is derived from the same mouse strain as the genotype of the allele being targeted.
3. Design primers to amplify an 800–1,000 bp fragment surrounding the selected RFLP. Primers can be designed using different software programs (e.g., DNAMAN, Clone Manager), or online applications such as Primer 3 (http://frodo.wi.mit.edu/). We usually design primers with an annealing temperature around 58 °C. *See* Subheading 4 (*see* **Note 1**).

3.3 Generation of Targeting Plasmid for BAC Recombineering (Timing: 5–6 Days)

1. PCR amplify homologous fragments surrounding the RFLP using the primers designed in Subheading 3.2, **step 3** above. Dilute primers to a 100 pM stock in 10 mM Tris–HCl, pH 7.4. Set up a PCR using the mix indicated below. For this PCR, use a high-fidelity Taq-polymerase with proofreading activity (e.g., Phusion, NEB) to minimize the introduction of mutations through PCR. Use either genomic DNA derived from the mouse strain of the allele being targeted, or use isogenic BAC DNA (50 ng of genomic DNA as PCR template). Isogenic DNA can also be ordered from the Jackson's laboratory (http://www.jax.org). It is advisable to verify the presence of the RFLP in PCR products amplified from genomic DNA of the ES cells which will be used for targeting, by restriction digestion or sequencing.

Component	Amount (μl)
5× High-Fidelity Buffer	10
dNTP (1 mM)	1
Forward primer (100 pM)	0.5
Reverse primer (100 pM)	0.5
Phusion polymerase	0.2
DNA template (50 ng/μl)	1
Sterile dH_2O	36.8
Total volume	50

2. Run the PCR on a thermocycler using the following parameters:

Cycle	Denature	Anneal	Extend
1	98 °C, 2 min		
2–31	98 °C, 15 s	58 °C, 30 s	72 °C, 1 min
32			72 °C, 10 min
33			Hold at 12 °C

3. Run the PCR product on a 1 % agarose gel using standard molecular cloning methods [71] together with a molecular weight marker. Visualize with a UV-transillumninator set at low UV intensity. Isolate the PCR band running at the expected size (*see* **Notes 2** and **3**).
4. Clean up the DNA fragment using a commercial gel extraction kit, following manufacturer's instructions. Elute DNA in a total of 20 μl elution buffer. Pause point: DNA can be stored at −20 °C for several years.
5. Clone the PCR product in pCR-BluntII-Topo using the pCR-BluntII-Topo-kit (Invitrogen) (or an alternative vector suitable for subcloning of blunt PCR products). Mix in an Eppendorf tube: 2 μl PCR product, 0.5 μl Salt solution, and 0.5 μl vector (Topo-vector). Vortex and spin down shortly. Incubate for 30 min at room temperature.
6. Transform the Topo-reaction product from Subheading 3.3, **step 5** into chemically competent DH5α bacteria by heat shock. Chemically competent bacteria are commercially available, or can be prepared following published protocols [71].
   (a) Thaw a vial of DH5α bacteria on ice, and mix with the Topo-reaction product. Incubate on ice for 15 min.
   (b) Heat-shock bacteria at 42 °C for 45 s, and put on ice immediately.
   (c) Add 1 ml LB-medium (without antibiotic selection) and incubate bacteria for 60 min at 37 °C, shaking at 700 rpm in a Thermomixer.
   (d) Pellet the bacteria by centrifuging for 1 min at 4,000 × *g* in a microfuge and remove LB medium by flicking tube upside down. Some fluid will remain in the tube.
   (e) Resuspend bacteria in the remaining LB present in the tube and plate bacteria on LB-agarose plates with kanamycin selection (10 μg/ml). Incubate at 37 °C overnight.
7. The next day, visible colonies should be present on the bacteria plates. Pick single colonies (between 12 and 24) with a pipette tip and inoculate in 3 ml of LB medium with kanamycin. Incubate these mini-cultures at 37 °C overnight in a bacterial shaker at 200 rpm (*see* **Note 4**).
8. Isolate the plasmid DNA from the mini-cultures ("mini-prep"):
   (a) Take 1.5 ml of bacterial culture, transfer to Eppendorf tube, and centrifuge at 16,200 × *g* for 1 min in a microcentrifuge.
   (b) Remove supernatant, and resuspend pellet in 300 μl buffer P1. Let shake at 1,400 rpm for approximately 5 min at 37 °C until all cells are in suspension.

(c) Add 300 μl lysis buffer P2, invert five times, and wait 20 s.

(d) Add 300 μl ice cold buffer P3. Invert tube five times and incubate on ice for 5 min to precipitate bacterial protein and DNA (*see* **Note 5**).

(e) Centrifuge at 16,200×*g* for 8 min in a microcentrifuge.

(f) Pour supernatant in a new Eppendorf tube, filled with 700 μl isopropanol, and shake vigorously to precipitate the plasmid DNA.

(g) Centrifuge at 16,200×*g* for 8 min in a microcentrifuge.

(h) Remove supernatant and wash DNA pellet with 500 μl 70 % ethanol. Centrifuge at 16,200×*g* for 5 min in a microcentrifuge.

(i) Remove supernatant, air-dry, and resuspend pellet in 100 μl 10 mM Tris–HCl, pH 7.4.

9. Check the isolated plasmid DNA by restriction digestion to verify the presence of a correct fragment inserted in Topo-vector. The choice of the restriction digest is determined by sequence of amplified fragment. In general, it is helpful to choose one enzyme located at the beginning or end of sequence, together with an enzyme which digests only once in the vector sequence. This will enable you to check both the presence of the proper fragment, and also determine the orientation of insertion in to the Topo-vector. The amplified RFLP can also be used for digestions. For digestion with multiple enzymes, use the NEB Double Digest finder (http://www.neb.com/nebecomm/DoubleDigestCalculator.asp), or other sources, to determine the appropriate buffer and incubation conditions. Digest 10 μl of mini-prep DNA, with 5 μl of 10× buffer, 0.5 μl 100× BSA (if required), 1 μl of each enzyme, in a total volume of 50 μl, and incubate at appropriate temperature (37 °C for most enzymes) for 2 h. For multiple digestions, this can be easily done in 96-well microtiter plates, sealed with tape, and placed in an incubator. Separate restriction products on a 1 % agarose gel, and continue with a clone showing the expected restriction fragments (*see* **Note 6**). Pause point: Mini-DNA can be stored at −20 °C for several years, bacteria can be stored at 4 °C for several weeks.

10. In the next step, the selection cassette of choice, with or without the SA-tpA cassette (*see* Table 1), or any other preferred selection cassette is cloned into the RFLP restriction site in the amplified fragment. First, release the cassette from its vector, by digestion with the appropriate enzyme (*see* Table 1) and generate blunt ends (except when generated overhangs are compatible with RFLP-site). Digest 2–5 μg of plasmid, together with 10 μl 10× Buffer, 1 μl 100× BSA (if required), 2 μl of each enzyme, in a total volume of 100 μl, at 37 °C for 2 h.

To generate blunt ends, use Klenow enzyme to fill in 5′ overhangs, or T4 polymerase to fill in 5′ overhangs or remove 3′ overhangs. For a Klenow reaction, add 2 μl of 10× Buffer, 3 μl dH_2O, 12 μl 330 μM dNTP, and 3 μl Klenow enzyme to the digestion mix after incubation, mix and incubate for 15 min at room temperature. Immediately run reaction product on a 1 % agarose gel. For T4 DNA polymerase, add 2 μl of 10× Buffer, 2.8 μl dH_2O, 0.2 μl 100× BSA, 12 μl 100 μM dNTP, and 3 μl T4 DNA polymerase, mix and incubate for 15 min at 12 °C. Immediately run reaction product on a 1 % agarose gel. Isolate the correct sized fragment, and clean up, as described in Subheading 3.3, **step 4** (*see* **Note** 7).

11. Open the amplified fragment (targeting arms) cloned in Topo from Subheading 3.3, **step 9** with the RFLP enzyme. Digest 20 μl of plasmid with 10 μl 10× Buffer, 1 μl 100× BSA (if required), 2 μl of enzyme, 1 μl of Calf intestinal phosphatase (CIP) in a total volume of 100 μl, at 37 °C for 2 h. Dephosphorylation of the 5′ termini will prevent self-closure of the plasmid upon ligation. Remove the 5′ or 3′ overhangs of the plasmid if necessary and clean up the DNA as described in Subheading 3.3, **step 10**.
12. Ligate the selection cassette from Subheading 3.3, **step 10** to the linearized plasmid from Subheading 3.3, **step 11**. Add in an Eppendorf tube the linearized plasmid, and selection cassette, in a 1:3 M ratio (60 ng end concentration), together with 1 μl 10× Ligation buffer and 0.5 μl T4 DNA Ligase in a total reaction volume of 10 μl, mix and incubate for 2 h at room temperature, or overnight at 16 °C.
13. Mix the ligation product with competent DH5α bacteria and heat shock transform the bacteria as described in Subheading 3.3, **step 6**. If an ampicillin resistance gene is included in the selection cassette, include ampicillin in the LB-agarose plates (25 μg/ml). If colonies are emerging the next day, start mini-cultures (*see* Subheading 3.3, **step 7**), isolate mini-DNA (*see* Subheading 3.3, **step 8**), and check final targeting plasmid by restriction digestions, to verify proper insertion in correct orientation (*see* Subheading 3.3, **step 9**) (*see* **Note 8**). Pause point: Plasmids can be stored at −20 °C for several years.
14. Digest the final targeting plasmid from Subheading 3.3, **step 13** to isolate the selection cassette flanked by homologous arms from the Topo-vector. This will increase the efficiency of BAC recombineering as the targeting cassette will not be able to replicate independently in bacteria. Digest 30 μl of plasmid, in a 50 μl reaction, run on an agarose gel, isolate the correct fragment, and clean up the DNA (*see* Subheading 3.3, **step 4**). Measure concentration by NanoDrop. Pause point: A linearized targeting fragment can be stored at −20 °C for at least 1 year.

3.4 BAC Isolation and Recombination (Timing: 4 Days)

1. Upon arrival of the BAC clone ordered in Subheading 3.2, **step 2**, plate the BAC clone (delivered in DH10B cells) on LB-agar plates with chloramphenicol selection (12.5 μg/ml), and incubate at 37 °C overnight.
2. Pick single colonies with a pipette tip and inoculate 3 ml LB (with chloramphenicol 12.5 μg/ml), and incubate at 37 °C overnight, shaking at 200 rpm. Although for the first steps of BAC recombineering only limited amounts of BAC DNA are required, which could be obtained from a mini-prep (*see* Subheading 3.3, **step 8**), we prefer to grow a maxi-culture from every new arriving BAC. This offers the advantage that cleaner DNA and larger quantities are obtained, resulting in more efficient transformation and is convenient for subsequent control experiments, and long-term storage (*see* **Note 9**).
3. Add complete content of the overnight cultures to a 2 l Erlenmeyer flask filled with 500 ml of LB with chloramphenicol (12.5 μg/ml). Incubate at 37 °C overnight, shaking at 200 rpm.
4. Prepare a glycerol freeze-stock of every new BAC clone: transfer 850 μl of BAC culture to a cryo-vial and add 150 μl 97 % Glycerol, mix, and immediately snap-freeze bacteria in liquid nitrogen. Store bacteria at −80 °C. Pause point: Glycerol stocks of bacteria can be maintained at −80 °C for several years.
5. Isolate Maxi-BAC-culture:
   (a) Transfer the 500 ml of culture to a bucket and centrifuge at 4,000 × *g* for 10 min at 4 °C in a Beckman J6-MC centrifuge.
   (b) Remove the supernatant and add 100 ml of buffer P1. Vortex the bucket for approximately 30 s, until all bacteria are resuspended.
   (c) Add 100 ml of lysis buffer P2. Invert bucket six times. Incubate for 1 min at room temperature (*see* **Note 10**).
   (d) Add 100 ml of ice cold buffer P3. Invert bucket six times and put on ice for 20 min. Proteins and bacterial genomic DNA will precipitate (*see* **Note 11**).
   (e) Centrifuge at 4,000 × *g* for 30 min at 4 °C.
   (f) Equilibrate a maxi-prep column (we use Macherey-Nagel Nucleo Bond columns, but Maxi kits from Qiagen work equally well), by adding 25 ml of Equilibration buffer, carefully wetting the complete filter insert.
   (g) Add the supernatant from **step e** to the column and let pass through by gravity. This will require several loading steps.
   (h) When all fluid has passed through, add 15 ml of Equilibration buffer to the column, carefully wetting the complete filter insert.

(i) Remove Filter and add 25 ml of Wash buffer. Allow to empty by gravity.

(j) Place a 50 ml Falcon tube under the column and add 15 ml of Elution buffer (preheated at 65 °C) to the column. The BAC DNA will now be eluted into the tube. Pause point: eluted BAC DNA can be stored overnight at 4 °C.

(k) Precipitate the BAC DNA by adding 10.5 ml Isopropanol to the tube, and shake vigorously. Centrifuge at 4,000 × *g* for 30 min at 4 °C.

(l) Remove supernatant, and wash DNA pellet with 70 % ethanol, centrifuge at 4,000 × *g* for 15 min at 4 °C.

(m) Remove supernatant, air-dry pellet, and dissolve in 200 μl 10 mM Tris–HCl, pH 7.4. To facilitate dissolving, incubate at 50 °C for 1 h, or 4 °C overnight. Collect content by short centrifugation. Pause point: BAC DNA can be stored at 4 °C for several months (*see* **Note 12**).

6. Check isolated BAC DNA by restriction digestion, or PCR analysis, to verify that correct clone was obtained. A standard restriction enzyme such as *Eco*RI or *Bam*HI, will result in a "finger-print" pattern of the respective BAC, which can easily be recognized. Use 2 μl of BAC DNA per digestion (approximately 1 μg/μl), 5 μl of 10× restriction buffer, 0.5 μl 100× BSA (if required) and 2 μl of enzyme, in a 50 μl digest. Incubate for 2 h at the appropriate temperature, and run on 1 % agarose gel. To check the BAC by PCR analysis, dilute an aliquot of BAC DNA 10×, and take 1 μl as template. Use any primer combination available, for example primers located in the gene of interest, or primers used to amplify the region surrounding the RFLP (*see* Subheading 3.3, **step 1**) (*see* **Note 13**).

3.5 Transformation of BAC Clone into Recombination Proficient Bacteria (Timing: 5 Days)

1. Streak recombination proficient bacteria [44] from a frozen glycerol stock onto LB-agar plates with no selection, and incubate at 30 °C overnight (*see* **Note 14**).
2. Pick a single colony, and inoculate a 5 ml LB-culture (no antibiotics), and grow overnight at 30 °C, shaking at 200 rpm.
3. Take 0.5 ml of the bacteria culture, and inoculate a 50 ml LB-culture (without antibiotics), and grow for 4–5 h at 30 °C, shaking at 200 rpm, until an OD_{600} of 0.6–0.8 is reached.
4. Collect bacteria in a 50 ml Eppendorf tube, and centrifuge at 2,000 × *g* for 10 min at 4 °C.
5. Recombination proficient bacteria will now be rendered electrocompetent, by washing steps in ice cold dH_2O. Remove the supernatant from Subheading 3.5, **step 4**, and resuspend the pellet in 10 ml ice cold dH_2O. Centrifuge at 2,000 × *g* for 10 min at 4 °C.

6. Remove the supernatant, and repeat the washing steps with ice cold dH_2O three times (4×10 min in total).
7. Remove the supernatant, and resuspend the pellet in 250 μl ice cold dH_2O, and store on ice. The bacteria are now electrocompetent, and the amount of bacteria is enough for five transformations.
8. Divide the bacteria over five Eppendorf tubes and add different concentrations of BAC DNA (1 μl is approximately 1 μg of BAC DNA): 0.2, 0.5, 1, 2, and 4 μl. Incubate for 15 min on ice. Add the bacteria–DNA mixture to a bacterial electroporation cuvette (Bio-Rad), prechilled on ice, and electroporate at 1.8 kV, 25 mF, 200 Ω, using a Bio-Rad Gene Pulser System. Add 1 ml LB medium, transfer to an Eppendorf tube and let recover for 60 min at 30 °C, shaking at 200 rpm (*see* **Notes 15** and **16**).
9. Spin down the bacteria at $4{,}000 \times g$ for 30 s and flick the tube upside down to remove the supernatant. Resuspend pellet in remaining fluid and plate on LB-agar plates containing chloramphenicol (12.5 μg/ml). Incubate overnight at 30 °C.
10. If colonies appear after 24 h, pick single colonies using a pipette tip and inoculate 3 ml LB cultures (12–24 colonies per BAC), with chloramphenicol selection (12.5 μg/ml). Incubate overnight at 30 °C.
11. Isolate BAC DNA the next morning, following the mini isolation protocol (*see* Subheading 3.3, **step 8**), with the exception of the last step, where DNA is dissolved in 50 μl 10 mM Tris–HCl, pH 7.4.

    Check the DNA by restriction digestion, using 25 μl of BAC DNA, using the same enzyme which was chosen to obtain a "finger-print" pattern of the BAC (*see* Subheading 3.4, **step 6**). If the same pattern is observed compared to the original BAC clone, it can be concluded that transformation was successful (*see* **Note 17**).
12. Select a clone from Subheading 3.5, **step 11**, which shows the expected restriction digestion pattern, and make a glycerol stock (*see* Subheading 3.4, **step 4**) since future applications of the BAC in recombination proficient bacteria might be required.

3.6 BAC Recombination (Timing: 1 Day)

1. Add 0.5 ml of the fresh bacteria culture from Subheading 3.5, **step 12** to a 200 ml Erlenmeyer flask containing 50 ml of LB-medium with chloramphenicol (12.5 μg/ml), and incubate at 30 °C for 4–5 h, shaking at 200 rpm, until $OD_{600} = 0.6–0.8$.
2. Place Erlenmeyer with bacteria culture in a 42 °C shaking water bath for 20 min. At 42 °C, the recombination machinery will be activated.

3. Collect the bacteria by centrifugation at 2,000 × *g* for 10 min at 4 °C, and make cells electrocompetent (*see* Subheading 3.5, **steps 5**–7). Take 50 μl of electrocompetent bacteria in an Eppendorf tube and add 100 ng of linearized targeting fragment (*see* Subheading 3.3, **step 14**). Electroporate as in Subheading 3.5, **step 8**, and let bacteria recover for 60 min at 30 °C, shaking at 200 rpm. Electroporation of the small targeting fragment is very efficient; therefore it is not necessary to try different concentrations.
4. Spin down the bacteria at 4,000 × *g* for 30 s and flick the tube upside down to remove the supernatant. Resuspend pellet in remaining fluid and plate onto LB-agar plates containing chloramphenicol (12.5 μg/ml) and, depending on the targeting cassette used, either kanamycin (3 μg/ml) or ampicillin (8 μg/ml). It might be helpful to plate 1/10th of the bacteria onto a separate plate, to prevent overgrowing of the plate in case recombination turns out to be highly efficient. Incubate overnight at 30 °C.

3.7 Colony PCR to Check Proper Recombination (Timing: 4 Days)

1. After 24 h, small colonies will appear. For some BACs, it can take more time until colonies emerge (up to 40 h). Check emerging colonies by colony PCR, to verify that the correct recombination event has occurred. Use primers which are located outside the homologous arms used for recombination, combined with primers located in the inserted selection cassette (Fig. 5) Primers used originally to amplify the homologous arms (*see* Subheading 3.3, **step 1**) can be used as a negative control, since after recombination PCR amplification with these primers should no longer give the expected product (either a longer band is obtained, including the selection cassette, or no band at all if fragment becomes too large to be amplified efficiently). In general, we use two primer combinations, and choose a clone that is positive both on the 5′ and 3′ site (*see* **Note 18**).
   (a) Pick between 20 and 48 colonies and mark and number them on the bacteria plate. Touch the colony with a pipette tip and drop the tip in an Eppendorf tube filled with 20 μl dH_2O. If the plate is too crowded to mark isolated colonies, touch complete colony with a pipette tip, streak bacteria to a new plate with proper antibiotics, on a marked location, and then drop the pipette tip in dH_2O. Incubate the replica plate at 30 °C for later use.
   (b) Prepare a master mix for PCR. Take 1 μl of template from the Eppendorf tubes from (I), and pipette this directly into the PCR plate. For this you can reuse the same pipette tip that was used to touch the colony.

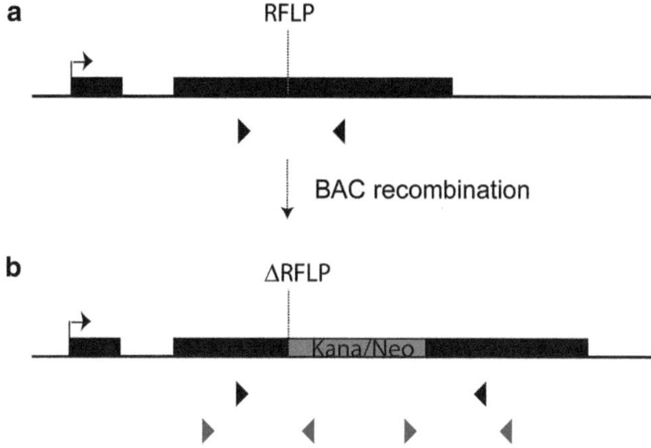

Fig. 5 Primer location for colony PCR screen. (**a**) Primers used to amplify the homologous arms surrounding the selected RFLP used for BAC recombination are indicated with the *black arrow heads*. (**b**) After BAC recombination, the Kana/Neo cassette disrupts the RFLP. The primers used in (**a**) will no longer amplify the genomic fragment, since insertion of the Kana/Neo cassette results in a fragment which is too long to be amplified using the same extension time. To check the colonies obtained after BAC recombination, the primer combinations shown in *red* are used. These consist of primers located within the Kana/Neo cassette, and primers located outside the homologous arms. Therefore these primers do not amplify the targeting plasmid, but will only result in a band when recombination has occurred (color figure online)

Component	Amount for 1 reaction (μl)	Amount for 50 reactions
10× PCR buffer	2.5	125 μl
$MgCl_2$ (50 mM)	0.9	45 μl
dNTP 1 mM	1	50 μl
Forward primer (100 pM)	1.25	62.5 μl
Reverse primer (100 pM)	1.25	62.5 μl
Taq-polymerase	0.4	20 μl
dH_2O	16.7	835 μl
Template	1	
Total volume	25	Divide 24 μl per well and add 1 μl template

2. Run the PCR in a thermocycler using the following parameters where annealing temperature and extension time are dependent on the primer and fragment size, respectively:

Cycle	Denature	Anneal	Extend
1	98 °C, 2 min		
2–31	98 °C, 15 s	58 °C, 30 s	72 °C, 1 min
32			72 °C, 10 min
33			Hold at 12 °C

3. Resolve the PCR amplified fragments on a 1 % agarose gel and choose several colonies that are positive for all primer combinations tested. Starting from the marked bacterial plate or the replica plate, grow a mini-culture of these colonies at 30 °C, shaking at 200 rpm, and isolate DNA using a BAC-mini protocol (*see* Subheading 3.5, **step 11**). Check by restriction digestion, whether BAC "finger-print" is still detectable. The "finger-print" might be modified by the recombination event, although the amount of DNA obtained with the BAC miniprep protocol might be too low to definitively judge this (*see* **Notes 19** and **20**).
4. Grow a maxi-culture (with the appropriate antibiotics) of two candidate clones containing the recombinant, modified BAC, and isolate BAC DNA as in Subheading 3.4, **step 5**. Also make glycerol stocks (*see* Subheading 3.4, **step 4**). Check the isolated BAC DNA by different restriction digestion reactions, to verify correct insertion of the targeting fragment. If all digests confirm the expected pattern based on the predicted, modified, sequence (*see* Subheading 3.2, **step 1**), the final BAC targeting vector for targeting of ES cells has been obtained. Pause Point: The BAC targeting vector can be stored for several months at 4 °C and reused, if required. A fresh preparation will however increase the targeting efficiency.
5. For targeting of ES cells, the final BAC targeting vector needs to be linearized, using the PI-SceI restriction enzyme (the PI-SceI restriction site is present in BAC-backbones derived from the vector pBACe3.6). For one targeting, we use between 60 and 80 μg of DNA, which will be approximately 60–80 μl of BAC DNA derived from the BAC-maxi isolation. The amount of DNA obtained following the BAC isolation protocol is more or less constant (we estimate the concentration by restriction analysis of 1 μl on an agarose gel), therefore digest BAC DNA as follows: 83 μl BAC DNA, 10 μl 10× PI-SceI buffer, 1 μl 100× BSA, 6 μl PI-SceI enzyme. Incubate at 37 °C overnight, and inactivate enzyme at 65 °C for 20 min. Store at 4 °C until use (*see* **Note 21**).

3.8 Targeting of Mouse Embryonic Stem Cells (Timing: 17 Days)

1. Detailed protocols on mouse ES cell culture have been described elsewhere [72–74]. Here we assume that the reader is familiar with the basics of ES cell culture, which are required to maintain ES cells in a pluripotent, undifferentiated state. Sterile culture techniques should always be applied. Both female and male hybrid C57Bl/6-Cast/Ei ES cell lines, have been described [58], are germ line competent, are available upon request (Table 2).
2. Coat a 25 cm^2 cell culture flask with 0.2 % Gelatin solution and incubate for minimal 5 min at room temperature.
3. Thaw a vial of Mouse Embryonic Fibroblasts (MEFs, $2 \times 10^4/cm^2$) to be used as feeder cells for ES cells. For long-term storage both MEFs and ES cells are stored at −180 °C. Also defreeze a vial of ES cells. Put both vials in a 37 °C water bath, until thawed. Clean vials with 70 % ethanol, and transfer to cell culture hood. Transfer the content of both vials drop-wise to a 15 ml Falcon tube filled with 5 ml pre-warmed (37 °C) ES cell medium and centrifuge for 5 min at $200 \times g$ to remove the DMSO used for freezing (*see* **Note 22**).
4. Remove the gelatin from the cell culture flask. Resuspend the cell pellet from Subheading 3.8, **step 3** in 5 ml ES medium, and transfer to cell culture flask. Rock plate to homogenously spread the cells and incubate in a cell culture incubator at 37 °C, 5 % CO_2.
5. Change the culture medium daily. Small, sharp edged, dome-shaped colonies should appear on top of the feeder layer (Fig. 6). Cells should become confluent after 3–4 days. If only a few colonies appear, cells must be split 1:1 (see below) (*see* **Note 23**).

Table 2
Overview of available ES cell lines

Line	Strain	Male/female	Germ line
E3	B6/Cast	♂	Yes
E5	B6/Cast	♂	N.D.
E8	B6/Cast	♀	N.D.
E14	B6/Cast	♂	Yes
E15	B6/Cast	♀	N.D.

Table summarizing the available C57Bl/6-Cast/Ei mouse ES cell lines. E3, E5, and E14 are male, whereas E8 and E15 are female. Several cell lines have been shown to be germ line competent. N.D.: not determined

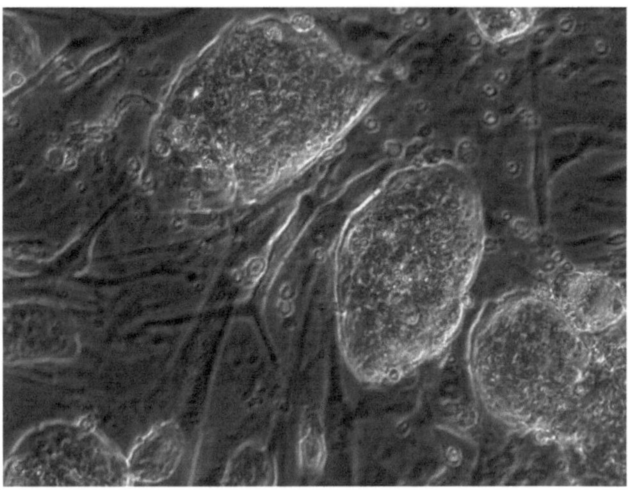

Fig. 6 ES cell morphology. Morphology of mouse ES cell colonies grown on MEF feeders. Note the shiny edge and dome-like colony shape of the ES cells

6. Split cells as follows:
   (a) Aspirate medium, and wash two times with cell culture PBS.
   (b) Remove PBS and add 1.5 ml Trypsin–EDTA (pre-warmed at 37 °C). Incubate at 37 °C for 7 min. After 3.5 min, gently shake the plate to break up colonies.
   (c) After incubation, add 4 ml of ES cell medium to inactivate the Trypsin–EDTA. Pipette cells up and down to obtain a single cell solution, and transfer cells to a 15 ml Falcon tube. Centrifuge 5 min at 200 × *g* (*see* **Note 24**).
   (d) If cells will be passaged, resuspend cells in ES medium, and transfer to a new culture flask that has been gelatinized and coated with MEFs. If cells are used for targeting, proceed with the next step. We normally grow our ES cells to 70 % confluency and split our cells 1 in 7.
   (e) Wash cells with pre-warmed (37 °C) DMEM and centrifuge for 5 min at 200 × *g*, to wash out the serum.
   (f) Resuspend cells in 320 μl DMEM and transfer to a 2 mm electroporation cuvette (BTX). Add 80 μl of linearized BAC DNA (*see* Subheading 3.7, **step 5**) using a cut pipette tip, and carefully pipette up and down a few times to mix cells and DNA (*see* **Note 25**).
   (g) Electroporate cells at 118 kV, 1,200 μF, ∞ Ω, using a BIO-RAD Xcell pulser (*see* **Note 26**).
   (h) Resuspend ES cells in 1 ml ES medium and plate onto a 50 cm^2 dish that has been gelatinized and pre-plated with MEFs, in a total volume of 10 ml ES medium. Rock the plate, and place in cell culture incubator (*see* **Note 27**).

7. After 24 h, some dead cells will be visible, but the majority of cells will have attached. Remove the medium and add ES cell medium with selection, depending on which resistant cassette was chosen for the generation of the BAC targeting construct [neomycin (Geneticin, G418), 270 μg/ml; puromycin, 1 μg/ml] (*see* **Note 28**).
8. From day 2 until day 14: change medium daily. After the first few days, cell death will be increasing as selection starts. This will be more rapid for puromycin selection, whereas neomycin selection starts more gently. At the end of the first week, small resistant ES cell colonies should be emerging, which will grow larger over the next days (*see* **Notes 29** and **30**).

3.9 Picking Clones in 96-Well Plates for Freezing and DNA Isolation (Timing: 7 Days)

1. Expect to obtain between 50 and 200 clones, which can be picked approximately 2 weeks after electroporation. Emerging clones can be picked and expanded when visible to the naked eye. To pick clones, proceed through the following steps:
   (a) Gelatinize two flat-bottom 96-well plates by adding 50 μl 0.2 % gelatin with a multichannel pipette, and incubate for a minimum of 5 min at room temperature.
   (b) Defreeze a vial of MEFs that is enough for a 50 cm^2 plate, as described in Subheading 3.8, **step 3**. Resuspend the cells in 10 ml of ES medium without selection.
   (c) Remove gelatin from the 96-well plates. Label one plate with "DNA" and the other with "Freezing." Add 10 ml of ES medium to a disposable cell culture reservoir and pipette using the multichannel pipette 100 μl of medium in every well of the "DNA" plate. Add 100 μl of MEF dilution (**step b**) to every well of the "Freezing" plate. Rock the plate to homogenously spread the cells, and put in cell culture incubator until use.
   (d) Take a U-shaped bottom 96-well plate and add 10 μl of cell culture PBS to every well using a reservoir and multichannel pipette.
   (e) Place a microscope in the cell culture hood and clean with ethanol.
   (f) Place the plate with targeted ES cells under the microscope and start picking colonies. Do not remove medium from this plate, as ES cells will stay better in ES medium. To pick a colony, use a P20 pipette and filter-tips to take up a single colony in 2 μl, and transfer this to a well of the U-bottom shaped 96-well plate covered with PBS. It is convenient to pick 96 colonies (if that many colonies are available), as this amount, together with the high targeting efficiency of BAC targeting, will almost certainly guarantee you at least one correctly targeted clone.

(g) When all 96 colonies have been picked, add 25 μl Trypsin–EDTA (pre-warmed at 37 °C) using a reservoir and multichannel pipette, and incubate at 37 °C for 7 min. Shake the plate after 3.5 min, to break up the colonies.

(h) Immediately add 165 μl ES medium to every well using a reservoir and the multichannel pipette to inactivate the Trypsin–EDTA. Set the pipette at 100 μl and pipette up and down a few times to break up the colonies into single cells. Divide the 200 μl in every well over the "DNA" and "Freezing" plate prepared in **step c**. Rock the plates and place in incubator (*see* **Note 31**).

2. Over the next few days, check clones daily and change medium when they turn yellow. Usually, the "DNA" plate cells will grow slower because it is grown without the feeder support. Also differentiation will occur in the "DNA" plates, but that will not affect the quality of DNA (*see* **Note 32**).
3. When the "Freezing" plate becomes confluent (clones will have the same normal ES cell morphology), the plate can be frozen as follows:

(a) Aspirate medium and wash two times with cell culture PBS.

(b) Add 35 μl of Trypsin–EDTA (pre-warmed to 37 °C), and incubate for 7 min at 37 °C. Shake the plate after 3.5 min.

(c) Add 80 μl ES cell medium to each well (using a reservoir and multichannel pipet), and pipette cells up and down a few times to obtain a single cell solution.

(d) Add 115 μl of 2× Freezing solution and carefully pipette up and down a few times using the multichannel pipette.

(e) Close the lid of the plate and wrap the plate with a diaper. Immediately store at −80 °C (*see* **Note 33**).

3.10 DNA Purification in 96-Well Plates and PCR RFLP Analysis to Identify Correct Targeted Clones (Timing: 2 Days)

1. The "DNA" plate should only be harvested when wells are really confluent. In general, this will be 3–4 days later than the "Freezing" plate. Depending on the strategy chosen, at this point it might be required to isolate RNA to ascertain the loss of transcription of a downstream RFLP (SA-tpA strategy, option II, Fig. 2b). To isolate RNA use Trizol reagent, isolate RNA, treat with DNAse, and prepare cDNA using Superscript III (Invitrogen) according to manufacturer's instructions, and proceed to Subheading 3.10, **step 2**. To obtain the DNA:

(a) Remove the medium by turning the plate upside down above the sink. Place the plate upside down on a paper tissue to remove excess medium.

(b) Add 100 μl of Low SDS lysis solution [75] supplemented with proteinase K. Seal the plate using tape or a disposable adhesive and place at 37 °C for a minimum of 4 h.

(c) Precipitate the DNA by adding 100 µl isopropanol, seal the plate and shake or incubate for 10 min on a plate shaker.

(d) Centrifuge the plate at 4,000 × *g* for 30 min at 4 °C.

(e) Remove the supernatant by turning the plate upside down on a paper-tissue. Small DNA precipitations will be visible on the bottom of the wells.

(f) Wash with 100 µl of 70 % ethanol and centrifuge the plate at 4,000 × *g* for 15 min at 4 °C.

(g) Remove the supernatant by turning the plate upside down on a paper-tissue. Let the plate dry until most of the ethanol has evaporated. Dissolve the DNA in 20 µl 10 mM Tris–HCl, pH 7.4. To facilitate dissolving, seal the plate and place at 37 °C for 2 h. Spin the plate down and store at 4 °C.

2. To analyze whether the expected targeting event has occurred, perform a PCR using primers spanning the RFLP. These can be the same primers used to amplify the homologous arms for BAC recombination in bacteria (*see* Subheading 3.3, **step 1**). Digest the PCR product with the RFLP enzyme.

    (a) Perform a PCR using the same conditions as in Subheading 3.3, **step 1**, and run the program from Subheading 3.3, **step 2**. Take 1 µl of clone DNA or cDNA as template (*see* **Note 34**).

    (b) Precipitate the PCR products by adding the following components to the PCR reaction: 50 µl PCR product, 4 µl tRNA (10 mg/ml), 6 µl 3 M NaAc, pH 5.6. Mix and add 150 µl ice cold 100 % ethanol. Seal plate and shake vigorously.

    (c) Incubate the plate at −20 °C for 20 min, and centrifuge at 4,000 × *g* for 30 min at 4 °C.

    (d) Remove the supernatant by turning the plate upside down on a paper-towel. Precipitated DNA and tRNA will be visible at the bottom of the PCR plate.

    (e) Add 100 µl 70 % ethanol and wash by centrifuging at 4,000 × *g* for 15 min at 4 °C.

    (f) Remove the supernatant by turning the plate upside down on a paper-towel. Dry the plate until all ethanol has evaporated.

    (g) Prepare a master mix for the restriction digestion: for each sample, prepare 5 µl of 10× buffer, 0.5 µl 100× BSA (if required), 0.25 µl RNAse (10 mg/ml), and 1 µl of enzyme, in a total volume of 50 µl. Add 50 µl of the master mix to every well of the PCR plate. Seal plate and incubate at 37 °C for 2 h.

    (h) Run the digest on an agarose gel and verify whether the targeted RFLP is lost in the clones (*see* **Notes 35** and **36**).

3. Select the candidate targeted clones based on the allele-specific PCR from Subheading 3.10, **step 2**, and repeat the PCR analysis one more time, or with a different set of primers, to exclude the possibility that the non-digested pattern was obtained due to failure of restriction digestion. If pattern is confirmed, this indicates that clones were correctly targeted.
4. To defreeze the targeted clones, prepare a 24-well plate, coated with gelatin and plated with MEFs in ES medium (0.5 ml per well). Place the "Freezing" plate (*see* Subheading 3.9, **step 3**) from the −80 °C at 37 °C in a cell culture incubator, and wait until wells start to thaw. Pipette the 230 μl content of the well corresponding to the targeted clones into the 24-well plate. Take another 200 μl of ES medium, and pipette up and down in the well of the "Freezing" plate, to collect all remaining cells, and add this to the 24-well plate. Incubate in a cell culture incubator at 37 °C.
5. Change the medium the following day. It will take 4–5 days for the clones to recover. When confluent, split to a 6-well plate, and subsequently expand further and freeze stocks. Isolate new DNA, and confirm the PCR results (*see* **Note 37**).

3.11 Anticipated Results

Using our strategy, we have recently generated targeted alleles for the X-linked genes *Rnf12* and *Xist* [59, 39, 76]. RNF12 is an X-linked, dose dependent activator of X chromosome inactivation, which is an important epigenetic process resulting in the transcriptional shut down of one X chromosome in all mammalian female cells [77, 78]. By generating heterozygous *Rnf12+/−* ES cells, we showed in vitro that RNF12 acts as a crucial, dose dependent activator of *Xist*. By targeting the second allele, generating a homozygous *Rnf12−/−* ES cell line, we obtained evidence that RNF12 is crucial for the initiation of X chromosome inactivation. Furthermore, by generating a targeted deletion of the *Xist* intron 1 region using the same RFLP-based strategy, we showed that pluripotency factor binding to the *Xist* intron 1 region is not essential for *Xist* repression in undifferentiated ES cells or in vivo in mice. In the latter targeting, we designed homologous arms for BAC recombineering in such a way, that a 2.1 kb region of the BAC was replaced by the insert selection cassette, thereby showing that our strategy can also be applied to generate deletions. Although this approach requires an additional cloning step during the generation of the targeting plasmid, it offers an interesting possibility when insertional mutagenesis is not favored.

In our BAC recombineering step, we found that between 70 and 100 % of bacteria colonies are positive for the correct integration of the selection cassette. This high efficiency, combined with the ease of the protocol, allows the generation of multiple constructs for different genes at the same time, within a short time frame.

Using the BAC targeting approach, we obtained higher targeting efficiencies compared to our previous experience with conventional targeting constructs [79]. By targeting four different loci within the *Rnf12* gene, we found that between 3 and 12 % of our obtained ES cell clones are correctly targeted [58]. We are currently applying a similar strategy to target human iPS cells (hLR5 cells [80]), which are transformed in a mouse ES "like" state through induced exogenous expression of the reprogramming factors in the presence of leukemia inhibitory factor. The targeting efficiency of these iPS cells that can be trypsinized is similar to efficiencies obtained with mouse ES cells (our unpublished observations). The generated targeted mouse ES cell lines can be used for in vitro experiments, or can be used for blastocyst injections of diploid or tetraploid blastocysts to generate mouse models [66–70]. If required, the second allele can subsequently be targeted in vitro, or can be intercrossed in the obtained mice. When in vitro targeting is required, a BAC construct is generated using a BAC library isogenic for the second allele. The same targeting arms can be used, but need to be ligated to a different selection cassette (e.g., targeting one allele with neomycin, and the second one with puromycin). The RFLP screen in targeted ES cell clones then consist of detection of complete absence of the PCR product spanning the original RFLP used, since both the first and second targeted allele will be modified by insertion of the targeting cassette.

The hybrid C57Bl/6-Cast/Ei ES cells provide a plethora of available RFLPs for targeting, but might offer disadvantages for certain studies where the genetic background of knockout mouse models is important. Therefore several backcrosses may be required for certain studies. On the contrary, it might even be beneficial to use hybrid ES cells or mice, since polymorphisms can be used to trace parental gene expression. Combined with the many possibilities offered by BAC recombineering, the high efficiency of BAC targeting approaches and our straightforward readout, BAC targeting to generate knockout alleles has become our method of choice to manipulate the mouse genome.

4 Notes

1. The RFLP restriction site should be unique and absent from the cloning vector used to introduce the PCR product in. This facilitates the insertion of the selection cassette into the targeting plasmid in a single cloning step.
2. Do not expose the DNA to UV-light too long as this can introduce nicks and damage the DNA. Always use an UV-transilluminator at a low UV intensity when visualizing DNA that is subsequently used for cloning.

3. If a correct PCR product is not obtained, consider a possible pipetting error, and repeat the PCR reaction. Also, the primers or the PCR conditions could be unsuitable for amplifying the desired product. Try a temperature gradient to find a better annealing temperature for your primers. Ordering new primers might be required.
4. If no colonies are obtained, make sure that you use the proper concentration of kanamycin (50 μg/ml). Alternatively, the clean-up of DNA prior to the Topo reaction might have failed. Check the obtained DNA by NanoDrop, and test an aliquot on gel. Consider a pipetting error for the Topo reaction, and repeat the reaction. Transform a circular control plasmid as well, to make sure that the heat shock procedure works, and to test the bacteria used. Hundreds of colonies should emerge when bacteria are competent.
5. Do not shake vigorously at this step, since this will result in contamination of the plasmid DNA with bacterial genomic DNA.
6. If no DNA was obtained, make sure that the buffers used are made correctly. Did you see protein precipitation? If the expected digestion pattern is not obtained, consider a failure of the digestion. Did you use the right temperature and digestion conditions? Consider repeating reaction with a fresh vial with restriction enzyme. Sometimes a hidden restriction site is present in a DNA construct, which might not be present in the reference sequence. Is the total size of all bands present still the same as you would expect? Alternatively, you might have cloned the wrong PCR fragment. Try a new set of primers, to get a more specific PCR product.
7. Because of exonuclease activity of Klenow enzyme and T4 DNA polymerase, do not incubate longer than 15 min.
8. If no colonies are present, consider the possibility of having used a wrong concentration of antibiotics. Check the DNA concentration of the fragments used, as these might be too low, and increase the concentration when required. You may have used CIP for both the vector, and the insert, which will block ligation of both fragments. Alternatively, if only empty vector sequences were obtained, you may have forgotten to use CIP. Did the blunting reaction fail? Did you use Klenow enzyme, which has been tested, and are your overhangs suitable for Klenow enzyme? Try T4 DNA polymerase as an alternative. If the obtained DNA pattern is wrong, ask yourself whether the correct fragments were used. Check the total size of the obtained plasmid. Is the orientation of the inserted fragment not as expected? Maybe you can still use the obtained orientation. Sometimes, inverted repeats are generated which will be selected against by the bacteria. Therefore, a certain orientation maybe always be obtained.

9. If no colonies were obtained, consider the possibility that a wrong antibiotic plate was used. Replate on new selection plates, otherwise order a new clone.
10. Do not shake vigorously at this point, since this will result in contamination with bacterial genomic DNA.
11. Do not shake vigorously at this point, since this will result in contamination of bacterial DNA in the BAC preparation.
12. Do not freeze BAC DNA, as this will result in fragmentation and degradation of the BAC DNA. Also prevent vigorous pipetting, as this might result in breaks through shearing of DNA.
13. If only high molecular weight bands or no DNA at all were obtained, this could indicate that the BAC has recombined, or only genomic bacterial DNA has been isolated. Repeat the isolation with a fresh culture, and follow the protocol carefully. Growing BACs at 30 °C might prevent spontaneous recombination events. Also the digest might have failed, a too low amount of DNA was digested, or a wrong BAC clone might have been ordered.
14. Recombination proficient bacteria must be grown at 30 °C, to prevent activation of the recombination machinery at higher temperatures.
15. Although in principal one electroporation can be enough to transform the BAC into the recombination proficient bacteria, we found that for several BACs it is rather difficult to be transformed into bacteria, as only few or no colonies are obtained. The use of different concentrations of BAC DNA will increase the chance that one of the electroporations results in positive colonies, thereby saving valuable time.
16. Sometimes electroporation of bacteria results in arcing of the sample, reducing the viability of the cells. Often this indicates that the salt concentration in the DNA solution was too high. Try again with less DNA, or with DNA obtained from a new preparation. Another possibility is that the bacteria may not be electrocompetent enough. You can do a test electroporation using a clean preparation of a small (~3 kb) high-copy plasmid such as pUC19. It is possible to routinely obtain efficiencies up to 1×10^8 CFU/μg DNA, high enough to give several hundreds colonies, even when larger (150–200 kb) BACs are transferred. Otherwise, try preparing competent cells with a new batch of freshly streaked bacterial cells. Also recheck the settings of the electroporator to ensure they are correct—some optimizing might be required.
17. If the BAC was not correctly transformed, repeat the transformation. Some BACs are difficult to transform into bacteria, which may be related to the size and/or sequence. The only solution is to repeat the transformation. If recombined BACs

are obtained, make sure you grow the culture at 30 °C, to prevent activation of the recombination machinery of the bacteria.

18. If no colonies were obtained, make sure you have used the right concentration of antibiotics. Another possibility is that you may have forgotten to induce the recombination machinery by putting the bacteria at 42 °C for 20 min, or there may have been a problem with the electroporation (*see* **Note 16**). If the plate is overgrown with too many colonies, ask yourself whether you used the right selection plates. You may have forgotten to include one of the antibiotics. Maybe the original plasmid containing the targeting fragment is replicating independently in the bacteria (for example undigested targeting plasmid, still carrying an origin of replication (ori), might have contaminated the targeting fragment). Make sure that you have removed the ori of your targeting plasmid prior to use (*see* Subheading 3.3, **step 14**). Maybe try with a different digestion, to isolate an ori-less fragment.
19. Any abnormal, especially high intensity bands, at this stage indicate that something went wrong with the recombination, or that the plasmid containing the targeting fragment is independently replicating in the bacteria, thereby rendering the bacteria resistant to selection.
20. When the expected PCR bands are not obtained, consider a pipetting error. Make a new master mix, and try again. Also the primers might not be optimal. Try a gradient PCR to find a better annealing temperature. Maybe the extension time was too short? Check your PCR program. Maybe the recombination efficiency is very low for your particular BAC. Try to analyze more colonies or repeat the recombination procedure. Also make sure that you are still working with the correct BAC, and no mix-up had occurred.
21. BAC DNA can be very viscous, especially when freshly prepared. Use scissors to cut of the pipette tip to reduce shearing damage of the DNA during pipetting. Linearized BAC DNA should be used soon after the digestion reaction in order to avoid degradation.
22. MEFs can be prepared from day E13.5 mouse embryos [72], or can be purchased commercially.
23. When ES cells are not growing properly, make sure your medium composition is right. Make new medium if you have any doubts. The FCS or the feeder cells may not support the ES cells properly. Try a different batch. Maybe a limited amount of cells was frozen; hence a limited amount of cells will result in few colonies, and the few cells present might suffer from inefficient autocrine and paracrine signaling between the

few ES cells, resulting in poor growth. Sometimes waiting a few days, or trypsinizing slow growing cells with a few colonies might solve the problem. Try to defreeze another vial. Always consider a potential mycoplasma infection. Test, and if positive, discard, or treat your cells (for example using BM Cyclin antibiotics, Roche). Whenever possible, go back to an earlier, non-infected passage.

24. If you observe problems in obtaining a single cell solution, make sure you washed your cells twice with PBS to remove all medium. Pre-warmed Trypsin–EDTA should be used for trypsinization.
25. Precipitation of DNA after PI-SceI digestion, as found in many protocols is not needed, and does not increase targeting efficiency (our unpublished observations), and will lead to introduction of more DNA breaks in the BAC DNA.
26. Different settings might be optimal when using a different pulser. Optimization might be required.
27. MEFs used for the targeting plate and subsequent selection must be resistant against the antibiotic used.
28. If no viable ES cells are left, the electroporation likely killed all of your cells. Optimize the settings of your pulser. Did you by accident plate your cells in media with selection antibiotics? Only start selection after 24 h to allow the cells enough time to produce the resistance proteins. Are there any signs of an infection (e.g., yellowish medium, floating bacteria or yeast)? Always use sterile culture techniques when culturing cells.
29. If MEFs do not resist selection well (e.g., apoptotic cells, holes in the feeder layer), plate out additional MEFs on the selection plate. Sufficient feeder support is crucial to keep ES cell colonies undifferentiated.
30. If too many or too few cells are dying upon selection, you might be using a wrong concentration of antibiotics. A kill curve should be done with wild type ES cells to find the optimal concentration range that kills your cells. Check your media composition to ensure that all components are present in the correct concentration (as indicated in Subheading 2.3, **step 10**).
31. It is crucial to label both replica plates identically at this point, as later on one plate will be frozen and the other one will be used to analyze the DNA. Mixing up at this point will ruin your targeting experiment as you might defreeze the wrong clone later on.
32. If clones are not growing, *see* **Note 23**.
33. It is important not to freeze cells instantly since this will result in higher frequency of cell death. By wrapping in the isolating diaper, the temperature of the plate will decrease more slowly

and increase cell viability. ES cells can be stored for a few months at −80 °C, but for long-term storage they should be placed at −180 °C.

34. Since the amount of DNA obtained might be low, using a high fidelity Taq polymerase can improve the results. Often these enzymes can provide better PCR performance when a limited amount of template is present.
35. In non-targeted clones, PCR amplification of the RFLP fragment followed by restriction digestion will result in three bands: one higher molecular weight band corresponding to the allele which does not contain the RFLP restriction site, and two lower molecular weight bands, which represent the allele with the RFLP. If a correct targeting event has occurred, the RFLP will be destroyed by the insertion of the selection cassette. Therefore, PCR amplification followed by restriction digestion will only reveal the higher molecular weight band, which is undigested because of the absence of a restriction site.
36. If the expected PCR and digestion product were not obtained, consider a failure of DNA isolation. The obtained concentration might be too low, or too high, or too much ethanol might have been left behind, interfering with the enzyme digestion. Check the concentration and DNA quality by NanoDrop measurement. If you have obtained insufficient DNA, defreeze the "Freezing plate" and isolate new DNA at the next passage. Also the PCR reaction might have failed. Repeat with a new master mix, and make sure that the primers are working, and the conditions are suitable for your particular reaction. Was the precipitation of the PCR product successful? Did you use the right concentration of tRNA? Finally, consider the possibility that you did not obtain a correctly targeted clone. Start a new round of targeting, and consider generating a new targeting construct as the chosen genomic region might be resistant to homologous recombination.
37. If the thawed clones do not show the correct PCR bands after the repetition of the analysis to confirm proper targeting of the obtained cell line, consider a failure of the PCR. Try again. Alternatively, a mix-up might have occurred during the freezing, the thawing procedure, or during the initial DNA isolation. In that unlucky case, take a week off, and try a new targeting experiment.

Acknowledgments

We would like to thank all past and present laboratory members for helpful suggestions and fruitful discussions. Work in the Gribnau lab is supported by funding from the Dutch Research Council (NWO-TOP and -VICI and an ERC-starting grant to J.G.).

References

1. Evans MJ, Kaufman MH (1981) Establishment in culture of pluripotential cells from mouse embryos. Nature 292(5819):154–156
2. Martin GR (1981) Isolation of a pluripotent cell line from early mouse embryos cultured in medium conditioned by teratocarcinoma stem cells. Proc Natl Acad Sci U S A 78(12): 7634–7638
3. Bradley A, Evans M, Kaufman MH, Robertson E (1984) Formation of germ-line chimaeras from embryo-derived teratocarcinoma cell lines. Nature 309(5965):255–256
4. Capecchi MR (1989) Altering the genome by homologous recombination. Science 244(4910): 1288–1292
5. Thomas KR, Capecchi MR (1987) Site-directed mutagenesis by gene targeting in mouse embryo-derived stem cells. Cell 51(3):503–512
6. Mansour SL, Thomas KR, Capecchi MR (1988) Disruption of the proto-oncogene int-2 in mouse embryo-derived stem cells: a general strategy for targeting mutations to non-selectable genes. Nature 336(6197):348–352
7. Smithies O, Gregg RG, Boggs SS, Koralewski MA, Kucherlapati RS (1985) Insertion of DNA sequences into the human chromosomal beta-globin locus by homologous recombination. Nature 317(6034):230–234
8. Doetschman T, Gregg RG, Maeda N, Hooper ML, Melton DW, Thompson S, Smithies O (1987) Targeted correction of a mutant HPRT gene in mouse embryonic stem cells. Nature 330(6148):576–578
9. Koller BH, Marrack P, Kappler JW, Smithies O (1990) Normal development of mice deficient in beta 2 M, MHC class I proteins, and CD8+ T cells. Science 248(4960):1227–1230
10. Schwartzberg PL, Goff SP, Robertson EJ (1989) Germ-line transmission of a c-abl mutation produced by targeted gene disruption in ES cells. Science 246(4931):799–803
11. Zijlstra M, Li E, Sajjadi F, Subramani S, Jaenisch R (1989) Germ-line transmission of a disrupted beta 2-microglobulin gene produced by homologous recombination in embryonic stem cells. Nature 342(6248):435–438
12. International Mouse Knockout C, Collins FS, Rossant J, Wurst W (2007) A mouse for all reasons. Cell 128(1):9–13
13. Austin CP, Battey JF, Bradley A, Bucan M, Capecchi M, Collins FS, Dove WF, Duyk G, Dymecki S, Eppig JT, Grieder FB, Heintz N, Hicks G, Insel TR, Joyner A, Koller BH, Lloyd KC, Magnuson T, Moore MW, Nagy A, Pollock JD, Roses AD, Sands AT, Seed B, Skarnes WC, Snoddy J, Soriano P, Stewart DJ, Stewart F, Stillman B, Varmus H, Varticovski L, Verma IM, Vogt TF, von Melchner H, Witkowski J, Woychik RP, Wurst W, Yancopoulos GD, Young SG, Zambrowicz B (2004) The knockout mouse project. Nat Genet 36(9):921–924
14. Auwerx J, Avner P, Baldock R, Ballabio A, Balling R, Barbacid M, Berns A, Bradley A, Brown S, Carmeliet P, Chambon P, Cox R, Davidson D, Davies K, Duboule D, Forejt J, Granucci F, Hastie N, de Angelis MH, Jackson I, Kioussis D, Kollias G, Lathrop M, Lendahl U, Malumbres M, von Melchner H, Muller W, Partanen J, Ricciardi-Castagnoli P, Rigby P, Rosen B, Rosenthal N, Skarnes B, Stewart AF, Thornton J, Tocchini-Valentini G, Wagner E, Wahli W, Wurst W (2004) The European dimension for the mouse genome mutagenesis program. Nat Genet 36(9):925–927
15. Pardo M, Lang B, Yu L, Prosser H, Bradley A, Babu MM, Choudhary J (2010) An expanded Oct4 interaction network: implications for stem cell biology, development, and disease. Cell Stem Cell 6(4):382–395
16. Ciotta G, Hofemeister H, Maresca M, Fu J, Sarov M, Anastassiadis K, Stewart AF (2011) Recombineering BAC transgenes for protein tagging. Methods 53(2):113–119. doi:10.1016/j.ymeth.2010.09.003, S1046-2023(10)00234-3 [pii]
17. Deng C, Capecchi MR (1992) Reexamination of gene targeting frequency as a function of the extent of homology between the targeting vector and the target locus. Mol Cell Biol 12(8):3365–3371
18. Hasty P, Rivera-Perez J, Chang C, Bradley A (1991) Target frequency and integration pattern for insertion and replacement vectors in embryonic stem cells. Mol Cell Biol 11(9): 4509–4517
19. te Riele H, Maandag ER, Berns A (1992) Highly efficient gene targeting in embryonic stem cells through homologous recombination with isogenic DNA constructs. Proc Natl Acad Sci U S A 89(11):5128–5132
20. Zhang P, Li MZ, Elledge SJ (2002) Towards genetic genome projects: genomic library screening and gene-targeting vector construction in a single step. Nat Genet 30(1):31–39
21. Valenzuela DM, Murphy AJ, Frendewey D, Gale NW, Economides AN, Auerbach W, Poueymirou WT, Adams NC, Rojas J, Yasenchak J, Chernomorsky R, Boucher M, Elsasser AL, Esau L, Zheng J, Griffiths JA, Wang X, Su H, Xue Y, Dominguez MG, Noguera I, Torres R, Macdonald LE, Stewart

AF, DeChiara TM, Yancopoulos GD (2003) High-throughput engineering of the mouse genome coupled with high-resolution expression analysis. Nat Biotechnol 21(6):652–659

22. Testa G, Zhang Y, Vintersten K, Benes V, Pijnappel WW, Chambers I, Smith AJ, Smith AG, Stewart AF (2003) Engineering the mouse genome with bacterial artificial chromosomes to create multipurpose alleles. Nat Biotechnol 21(4):443–447
23. Yang Y, Seed B (2003) Site-specific gene targeting in mouse embryonic stem cells with intact bacterial artificial chromosomes. Nat Biotechnol 21(4):447–451
24. Cotta-de-Almeida V, Schonhoff S, Shibata T, Leiter A, Snapper SB (2003) A new method for rapidly generating gene-targeting vectors by engineering BACs through homologous recombination in bacteria. Genome Res 13(9): 2190–2194
25. Song H, Chung SK, Xu Y (2010) Modeling disease in human ESCs using an efficient BAC-based homologous recombination system. Cell Stem Cell 6(1):80–89
26. Testa G, Vintersten K, Zhang Y, Benes V, Muyrers JP, Stewart AF (2004) BAC engineering for the generation of ES cell-targeting constructs and mouse transgenes. Methods Mol Biol 256:123–139
27. Shizuya H, Birren B, Kim UJ, Mancino V, Slepak T, Tachiiri Y, Simon M (1992) Cloning and stable maintenance of 300-kilobase-pair fragments of human DNA in Escherichia coli using an F-factor-based vector. Proc Natl Acad Sci U S A 89(18):8794–8797
28. Ioannou PA, Amemiya CT, Garnes J, Kroisel PM, Shizuya H, Chen C, Batzer MA, de Jong PJ (1994) A new bacteriophage P1-derived vector for the propagation of large human DNA fragments. Nat Genet 6(1):84–89
29. Marra MA, Kucaba TA, Dietrich NL, Green ED, Brownstein B, Wilson RK, McDonald KM, Hillier LW, McPherson JD, Waterston RH (1997) High throughput fingerprint analysis of large-insert clones. Genome Res 7(11): 1072–1084
30. Kelley JM, Field CE, Craven MB, Bocskai D, Kim UJ, Rounsley SD, Adams MD (1999) High throughput direct end sequencing of BAC clones. Nucleic Acids Res 27(6): 1539–1546
31. Osoegawa K, de Jong PJ (2004) BAC library construction. Methods Mol Biol 255:1–46
32. Mozo T, Dewar K, Dunn P, Ecker JR, Fischer S, Kloska S, Lehrach H, Marra M, Martienssen R, Meier-Ewert S, Altmann T (1999) A complete BAC-based physical map of the Arabidopsis thaliana genome. Nat Genet 22(3):271–275
33. Hoskins RA, Nelson CR, Berman BP, Laverty TR, George RA, Ciesiolka L, Naeemuddin M, Arenson AD, Durbin J, David RG, Tabor PE, Bailey MR, DeShazo DR, Catanese J, Mammoser A, Osoegawa K, de Jong PJ, Celniker SE, Gibbs RA, Rubin GM, Scherer SE (2000) A BAC-based physical map of the major autosomes of Drosophila melanogaster. Science 287(5461): 2271–2274
34. Osoegawa K, Tateno M, Woon PY, Frengen E, Mammoser AG, Catanese JJ, Hayashizaki Y, de Jong PJ (2000) Bacterial artificial chromosome libraries for mouse sequencing and functional analysis. Genome Res 10(1):116–128
35. McPherson JD, Marra M, Hillier L, Waterston RH, Chinwalla A, Wallis J, Sekhon M, Wylie K, Mardis ER, Wilson RK, Fulton R, Kucaba TA, Wagner-McPherson C, Barbazuk WB, Gregory SG, Humphray SJ, French L, Evans RS, Bethel G, Whittaker A, Holden JL, McCann OT, Dunham A, Soderlund C, Scott CE, Bentley DR, Schuler G, Chen HC, Jang W, Green ED, Idol JR, Maduro VV, Montgomery KT, Lee E, Miller A, Emerling S, Kucherlapati, Gibbs R, Scherer S, Gorrell JH, Sodergren E, Clerc-Blankenburg K, Tabor P, Naylor S, Garcia D, de Jong PJ, Catanese JJ, Nowak N, Osoegawa K, Qin S, Rowen L, Madan A, Dors M, Hood L, Trask B, Friedman C, Massa H, Cheung VG, Kirsch IR, Reid T, Yonescu R, Weissenbach J, Bruls T, Heilig R, Branscomb E, Olsen A, Doggett N, Cheng JF, Hawkins T, Myers RM, Shang J, Ramirez L, Schmutz J, Velasquez O, Dixon K, Stone NE, Cox DR, Haussler D, Kent WJ, Furey T, Rogic S, Kennedy S, Jones S, Rosenthal A, Wen G, Schilhabel M, Gloeckner G, Nyakatura G, Siebert R, Schlegelberger B, Korenberg J, Chen XN, Fujiyama A, Hattori M, Toyoda A, Yada T, Park HS, Sakaki Y, Shimizu N, Asakawa S, Kawasaki K, Sasaki T, Shintani A, Shimizu A, Shibuya K, Kudoh J, Minoshima S, Ramser J, Seranski P, Hoff C, Poustka A, Reinhardt R, Lehrach H, International Human Genome Mapping C (2001) A physical map of the human genome. Nature 409(6822):934–941
36. Kim UJ, Birren BW, Slepak T, Mancino V, Boysen C, Kang HL, Simon MI, Shizuya H (1996) Construction and characterization of a human bacterial artificial chromosome library. Genomics 34(2):213–218
37. Frengen E, Weichenhan D, Zhao B, Osoegawa K, van Geel M, de Jong PJ (1999) A modular, positive selection bacterial artificial chromosome vector with multiple cloning sites. Genomics 58(3):250–253
38. Frengen E, Zhao B, Howe S, Weichenhan D, Osoegawa K, Gjernes E, Jessee J, Prydz H, Huxley C, de Jong PJ (2000) Modular bacterial

artificial chromosome vectors for transfer of large inserts into mammalian cells. Genomics 68(2):118–126

39. Jonkers I, Barakat TS, Achame EM, Monkhorst K, Kenter A, Rentmeester E, Grosveld F, Grootegoed JA, Gribnau J (2009) RNF12 is an X-Encoded dose-dependent activator of X chromosome inactivation. Cell 139(5):999–1011
40. Hejna JA, Johnstone PL, Kohler SL, Bruun DA, Reifsteck CA, Olson SB, Moses RE (1998) Functional complementation by electroporation of human BACs into mammalian fibroblast cells. Nucleic Acids Res 26(4):1124–1125
41. Jessen JR, Meng A, McFarlane RJ, Paw BH, Zon LI, Smith GR, Lin S (1998) Modification of bacterial artificial chromosomes through chi-stimulated homologous recombination and its application in zebrafish transgenesis. Proc Natl Acad Sci U S A 95(9):5121–5126
42. Muyrers JP, Zhang Y, Benes V, Testa G, Rientjes JM, Stewart AF (2004) ET recombination: DNA engineering using homologous recombination in E. coli. Methods Mol Biol 256:107–121
43. Muyrers JP, Zhang Y, Testa G, Stewart AF (1999) Rapid modification of bacterial artificial chromosomes by ET-recombination. Nucleic Acids Res 27(6):1555–1557
44. Lee EC, Yu D, Martinez de Velasco J, Tessarollo L, Swing DA, Court DL, Jenkins NA, Copeland NG (2001) A highly efficient Escherichia coli-based chromosome engineering system adapted for recombinogenic targeting and subcloning of BAC DNA. Genomics 73(1):56–65
45. Narayanan K, Williamson R, Zhang Y, Stewart AF, Ioannou PA (1999) Efficient and precise engineering of a 200 kb beta-globin human/bacterial artificial chromosome in E. coli DH10B using an inducible homologous recombination system. Gene Ther 6(3):442–447
46. Chan W, Costantino N, Li R, Lee SC, Su Q, Melvin D, Court DL, Liu P (2007) A recombineering based approach for high-throughput conditional knockout targeting vector construction. Nucleic Acids Res 35(8):e64
47. Adams DJ, Biggs PJ, Cox T, Davies R, van der Weyden L, Jonkers J, Smith J, Plumb B, Taylor R, Nishijima I, Yu Y, Rogers J, Bradley A (2004) Mutagenic insertion and chromosome engineering resource (MICER). Nat Genet 36(8):867–871
48. Yang XW, Model P, Heintz N (1997) Homologous recombination based modification in Escherichia coli and germline transmission in transgenic mice of a bacterial artificial chromosome. Nat Biotechnol 15(9):859–865
49. Lalioti M, Heath J (2001) A new method for generating point mutations in bacterial artificial chromosomes by homologous recombination in Escherichia coli. Nucleic Acids Res 29(3): E14
50. Zhang Y, Buchholz F, Muyrers JP, Stewart AF (1998) A new logic for DNA engineering using recombination in Escherichia coli. Nat Genet 20(2):123–128
51. Murphy KC (1998) Use of bacteriophage lambda recombination functions to promote gene replacement in Escherichia coli. J Bacteriol 180(8):2063–2071
52. Sharan SK, Thomason LC, Kuznetsov SG, Court DL (2009) Recombineering: a homologous recombination-based method of genetic engineering. Nat Protoc 4(2):206–223
53. Copeland NG, Jenkins NA, Court DL (2001) Recombineering: a powerful new tool for mouse functional genomics. Nat Rev Genet 2(10):769–779
54. Wu S, Ying G, Wu Q, Capecchi MR (2008) A protocol for constructing gene targeting vectors: generating knockout mice for the cadherin family and beyond. Nat Protoc 3(6):1056–1076
55. Gong S, Kus L, Heintz N (2010) Rapid bacterial artificial chromosome modification for large-scale mouse transgenesis. Nat Protoc 5(10):1678–1696
56. Hofemeister H, Ciotta G, Fu J, Seibert PM, Schulz A, Maresca M, Sarov M, Anastassiadis K, Stewart AF (2011) Recombineering, transfection, Western, IP and ChIP methods for protein tagging via gene targeting or BAC transgenesis. Methods 53(4):437–452. doi: 10.1016/j.ymeth.2010.12.026, S1046-2023 (10)00309-9 [pii]
57. Narayanan K, Chen Q (2011) Bacterial artificial chromosome mutagenesis using recombineering. J Biomed Biotechnol 2011:971296. doi:10.1155/2011/971296
58. Barakat TS, Rentmeester E, Sleutels F, Grootegoed JA, Gribnau J (2011) Precise BAC targeting of genetically polymorphic mouse ES cells. Nucleic Acids Res 39:e21
59. Barakat TS, Gunhanlar N, Pardo CG, Achame EM, Ghazvini M, Boers R, Kenter A, Rentmeester E, Grootegoed JA, Gribnau J (2011) RNF12 activates Xist and is essential for X chromosome inactivation. PLoS Genet 7(1):e1002001
60. Beck JA, Lloyd S, Hafezparast M, Lennon-Pierce M, Eppig JT, Festing MF, Fisher EM (2000) Genealogies of mouse inbred strains. Nat Genet 24(1):23–25
61. Frazer KA, Eskin E, Kang HM, Bogue MA, Hinds DA, Beilharz EJ, Gupta RV, Montgomery J, Morenzoni MM, Nilsen GB, Pethiyagoda CL, Stuve LL, Johnson FM, Daly MJ, Wade CM, Cox DR (2007) A sequence-based variation

map of 8.27 million SNPs in inbred mouse strains. Nature 448(7157):1050–1053

62. Abremski K, Hoess R (1984) Bacteriophage P1 site-specific recombination. Purification and properties of the Cre recombinase protein. J Biol Chem 259(3):1509–1514
63. Abremski K, Hoess R, Sternberg N (1983) Studies on the properties of P1 site-specific recombination: evidence for topologically unlinked products following recombination. Cell 32(4):1301–1311
64. Zhang Z, Lutz B (2002) Cre recombinase-mediated inversion using lox66 and lox71: method to introduce conditional point mutations into the CREB-binding protein. Nucleic Acids Res 30(17):e90
65. Oberdoerffer P, Otipoby KL, Maruyama M, Rajewsky K (2003) Unidirectional Cre-mediated genetic inversion in mice using the mutant loxP pair lox66/lox71. Nucleic Acids Res 31(22):e140
66. Hogan B, Beddington R, Costantini F, Lacy E (eds) (1994) Manipulating the mouse embryo. Cold Spring Harbor Laboratory Press, Cold Spring Harbor, NY
67. Eakin GS, Hadjantonakis AK (2006) Production of chimeras by aggregation of embryonic stem cells with diploid or tetraploid mouse embryos. Nat Protoc 1(3):1145–1153
68. Eggan K, Akutsu H, Loring J, Jackson-Grusby L, Klemm M, Rideout WM 3rd, Yanagimachi R, Jaenisch R (2001) Hybrid vigor, fetal overgrowth, and viability of mice derived by nuclear cloning and tetraploid embryo complementation. Proc Natl Acad Sci U S A 98(11): 6209–6214
69. Eggan K, Rode A, Jentsch I, Samuel C, Hennek T, Tintrup H, Zevnik B, Erwin J, Loring J, Jackson-Grusby L, Speicher MR, Kuehn R, Jaenisch R (2002) Male and female mice derived from the same embryonic stem cell clone by tetraploid embryo complementation. Nat Biotechnol 20(5):455–459
70. Zhao XY, Lv Z, Li W, Zeng F, Zhou Q (2010) Production of mice using iPS cells and tetraploid complementation. Nat Protoc 5(5): 963–971
71. Sambrook J (2000) Molecular cloning: a laboratory manual, 3rd edn. Cold Spring Harbor Laboratory Press, Cold Spring Harbor New York, NY
72. Lin S, Talbot P (2011) Methods for culturing mouse and human embryonic stem cells. Methods Mol Biol 690:31–56. doi:10.1007/978-1-60761-962-8_2
73. Meissner A, Eminli S, Jaenisch R (2009) Derivation and manipulation of murine embryonic stem cells. Methods Mol Biol 482:3–19. doi:10.1007/978-1-59745-060-7_1
74. Nichols J, Ying QL (2006) Derivation and propagation of embryonic stem cells in serum- and feeder-free culture. Methods Mol Biol 329:91–98. doi:10.1385/1-59745-037-5:91
75. Laird PW, Zijderveld A, Linders K, Rudnicki MA, Jaenisch R, Berns A (1991) Simplified mammalian DNA isolation procedure. Nucleic Acids Res 19(15):4293
76. Minkovsky A, Barakat TS, Sellami N, Chin MH, Gunhanlar N, Gribnau J, Plath K (2013) The pluripotency factor-bound intron 1 of Xist is dispensable for X chromosome inactivation and reactivation in vitro and in vivo. Cell Rep 3(3): 905–918. doi:10.1016/j.celrep.2013.02.018, S2211-1247(13)00094-6 [pii]
77. Barakat TS, Gribnau J (2012) X chromosome inactivation in the cycle of life. Development 139(12):2085–2089. doi:10.1242/dev.069328, 139/12/2085 [pii]
78. Barakat TS, Gribnau J (2010) X chromosome inactivation and embryonic stem cells. Adv Exp Med Biol 695:132–154. doi:10.1007/978-1-4419-7037-4_10
79. Monkhorst K, Jonkers I, Rentmeester E, Grosveld F, Gribnau J (2008) X inactivation counting and choice is a stochastic process: evidence for involvement of an X-linked activator. Cell 132(3):410–421
80. Buecker C, Chen HH, Polo JM, Daheron L, Bu L, Barakat TS, Okwieka P, Porter A, Gribnau J, Hochedlinger K, Geijsen N (2010) A murine ESC-like state facilitates transgenesis and homologous recombination in human pluripotent stem cells. Cell Stem Cell 6(6): 535–546.doi: 10.1016/j.stem.2010.05.003, S1934-5909(10)00213-4 [pii]

Chapter 8

Herpesvirus Mutagenesis Facilitated by Infectious Bacterial Artificial Chromosomes (iBACs)

Karl E. Robinson and Timothy J. Mahony

Abstract

A critical factor in the study of herpesviruses, their genes and gene functions is the capacity to derive mutants that harbor deletions, truncations, or insertions within the genetic elements of interest. Once constructed the impact of the introduced mutation on the phenotypic properties of the rescued virus can be determined in either in vitro or in vivo systems. However, the construction of such mutants by traditional virological mutagenesis techniques can be a difficult and laborious undertaking. The maintenance of a viral genome as an infectious bacterial artificial chromosome (iBAC), however, endows the capacity to manipulate the viral genome for mutagenesis studies with relative ease. Here, the construction and characterization of two gene deletion mutants of an alphaherpesvirus maintained as iBAC in combination with an inducible homologous recombination system in *Escherichia coli* is detailed. The methodology is generally applicable to any iBAC and is demonstrated to be a highly efficient and informative approach for mutant virus construction.

Key words Herpesvirus, Bacterial artificial chromosomes, Mutagenesis, Homologous recombination, Gene deletion

1 Introduction

The stable maintenance of infectious herpesvirus genomes as infectious bacterial artificial chromosomes (iBAC) has facilitated unprecedented access to the study of herpesvirus molecular biology. Since the cloning of mouse cytomegalovirus as an iBAC, numerous herpesvirus species that are important to the health and welfare of both human and veterinary animals have been cloned [1–7]. The cloning of a herpesvirus as an iBAC endows numerous advantages over traditional virological techniques used for viral genome manipulation, the most significant being the capacity to readily apply powerful bacterial mutagenesis techniques to the viral genome. Additionally, the maintenance of the herpesvirus genome as an iBAC allows constructed mutants to be fully characterized prior to attempted rescue of infectious virus. This allows for the

Kumaran Narayanan (ed.), *Bacterial Artificial Chromosomes*, Methods in Molecular Biology, vol. 1227,
DOI 10.1007/978-1-4939-1652-8_8, © Springer Science+Business Media New York 2015

identification and evaluation of mutant genotypes that encode severely attenuated virus or that do not facilitate the rescue of infectious virus in traditional mutagenesis studies.

Bovine herpesvirus 1 (BoHV-1), a member of the subfamily *Alphaherpesvirinae*, is the prototype herpesvirus of ruminants and is a globally distributed virus of cattle associated with respiratory and reproductive diseases in production systems [8–10]. The BoHV-1 strain V155 (BoHV-1 V155) has been cloned as the iBAC, pBACBHV37, which facilitates the in vitro rescue of virus indistinguishable from wild-type virus [4]. The pBACBHV37 clone was integral to mutagenesis studies determining the in vitro requirement of the complete gene complement of this virus [11].

Following the construction of the iBAC clone a recombination system then needs to be selected. The pGETrec based recombination system has been highly successful in generating gene specific mutants using pBACBHV37 [4, 11, 12]. It is an inducible homologous recombination system that utilizes the transient expression of *recE* and *recT* genes from lambda bacteriophage to catalyze recombination between linear transgene molecules and BAC constructs in *Escherichia coli* DH10B cells [12]. Initially, the system was developed for the introduction of mutations into a BAC cloned β-globin gene to study hemoglobinopathies in humans [12–14]. The pGETrec system was demonstrated to be a robust and highly efficient recombination technology that can be employed for mutagenesis of BAC maintained constructs, including iBAC of viral genomes.

In this chapter, the methods and materials required for the construction of UL29 and US9 gene deletion mutants of BoHV-1 using the pGETrec system are detailed. The UL29 open reading frame (ORF) encodes for the single-stranded DNA binding protein that is part of the DNA polymerase complex of BoHV-1 and has been designated as being essential to BoHV-1 viability in vitro [11]. The US9 ORF of BoHV-1 encodes for a membrane polypeptide, which is a neuro-virulence factor facilitating reactivation of BoHV-1 from latency [15] and is designated as nonessential for in vitro of replication of BoHV-1 [11]. The methodologies for subsequent characterization of the resultant viral mutants encompassing confirmatory sequencing of the recombination region, BAC restriction endonuclease profiling and the assessment of respective growth properties of rescued virus in vitro are also described. These strategies could be readily applied for the specific mutagenesis of any herpesvirus iBAC clone.

2 Materials

1. Polymerase Chain Reaction (PCR) thermocycler.
2. Oligonucleotide PCR primers (as described in the text and in Table 1).

Table 1
The oligonucleotides used for the generation and characterization of mutants of pBACBHV37 which encodes an infectious genome of bovine herpesvirus 1

Name	Position[a]	Primer sequence[b] 5′-3′
UL29KanF	49235-49285	GGACGCGGCGGCCAAGACGGTAGC CCTCGCGCCGGGGCCCGCGG GCTTCG-ccacgttgtgtctcaaaatctctgatg
UL29KanR	52823-52873	CTATAGCCCCGCCCATGTCGCCGCCG CGCTACACCATGTCAATGGTCAGG-cggttgatgagagctttgttgtaggtg
US9KanF	123552-123602	GAGAGTCCACGCAGCGTCGTCAACGA AAACTATCGAGGCGCTGATGAGGC ccacgttgtgtctcaaaatctctgatg
US9KanR	124350-124400	GGGGAGCCGGAGCTTTGGCCCGCTC GCTCGGCCGGCCCGAATCCTCGGCC cggttgatgagagctttgttgtaggtg
KanME-F		CTCCTTCATTACAGAAACGGC
BHV-F	17128-17146	ATGTTAGCGCTCTGGAACC
BHV-R	17250-17232	CTTTACGTGCGCGAAAAGA
BHV-P	17209-17227	FAM-ACGGACGTGCGCGAAAAGA-BHQ2

[a]Nucleotide position within the BoHV-1 (V155) strain GenBank Accession number AJ004801
[b]Upper case lettering represents sequence homologous to the UL29 or US9 ORF, respectively with lower case lettering representing sequence homologous to the Kan^R selection cassette of EZ::Tn5<KAN>

3. Platinum *Pfx* DNA polymerase and reaction buffers (Life Technologies).
4. 10 mM each of deoxynucleotide triphosphates (dATP, dTTP, dCTP, and dGTP).
5. EZ::TN™ <KAN-2> transposon (Epicentre Technologies).
6. BigDye Terminator v3.1 sequencing mix (Applied Biosciences).
7. Electrophoresis apparatus (Bio-Rad).
8. 10× TBE electrophoresis running buffer: 0.9 M Tris–HCl, pH 8.0, 0.9 M Boric acid, 0.02 M EDTA, pH 8.0.
9. Ethidium Bromide (stock: 10 mg/mL).
10. Agarose, molecular biology grade (Bio-Rad).
11. DNA sample loading buffer (10×): 0.1 M Tris–HCl, pH 8.0, 50 % (v/v) glycerol, 0.1 % (w/v) Bromophenol blue.
12. 1 kb Plus Molecular weight marker (Life Technologies).
13. Gel-Doc UV Transilluminator system (Bio-Rad).
14. 1.5, 15 and 50 mL polycarbonate centrifuge tubes.
15. QIAquick Gel extraction kit (Qiagen).
16. Electroporator, (MicroPulser, Bio-Rad).

17. Electroporation cuvettes (1 mm; Cell Projects).
18. Incubator/shaker platform.
19. Microbiological inoculation loops and plate spreaders.
20. Super optimal broth with catabolite repression (SOC) medium.
21. Luria-Bertani (LB) medium (Liquid and solid) with antibiotics as required; 100 μg/mL Ampicillin (Amp), 12.5 μg/mL Chloramphenicol (CAP), 50 μg/mL of Kanamycin (Kan).
22. L-Arabinose, 20 % (w/v) (Sigma-Aldrich).
23. Sterile 10 % (v/v) glycerol solution.
24. Nucleobond BAC100 BAC extraction kit (Machery-Nagel).
25. Restriction endonuclease *Hind* III and *Dra* I used as detailed by the manufacturer.
26. CRIB-1 cells (ATCC CRL-11883), a clonal derivative of Madin-Darby Bovine Kidney (MDBK) cells that is nonpermissive to bovine viral diarrhea virus infection.
27. Lipofectamine 2000, 1 mg/mL transfection reagent (Life Technologies).
28. Mammalian cell growth media (Maintenance Media): 1× Dulbecco's Modified Eagle Medium (DMEM), 5–10 % fetal calf serum (FCS), 10 mM each of nonessential amino acids (NEAA), HEPES, and sodium pyruvate and supplemented with 200 μM *N′N′*-hexamethylene-bis-acetamide (*N′N′*-HBA).
29. Opti-MEM serum free cell culture medium (Life Technologies).
30. Cell culture flasks and plates (T25cm^3, T75cm^3, T150cm^3, 6- and 24-well plates) (Nunc/Thermo Scientific).
31. Refrigerated microcentrifuge.
32. Phosphate buffered saline (PBS).
33. High Pure Viral DNA Extraction Kit (Roche).
34. TRIS buffer: 10 mM Tris–HCl, pH 8.0.
35. Ethanol: 100 % (v/v) and 70 % (v/v).
36. Liquid Nitrogen used with appropriate handling procedures and equipment.
37. RG-3000 Light Cycler (Qiagen).
38. TaqMan Universal PCR master mix (Life Technologies).

3 Methods

The method outlined here and schematically represented in Fig. 1 is for the complete deletion of specific genes encoded by the BoHV-1 genome and the subsequent characterization of the mutants generated. To achieve the required outcomes several critical steps need to be conducted:

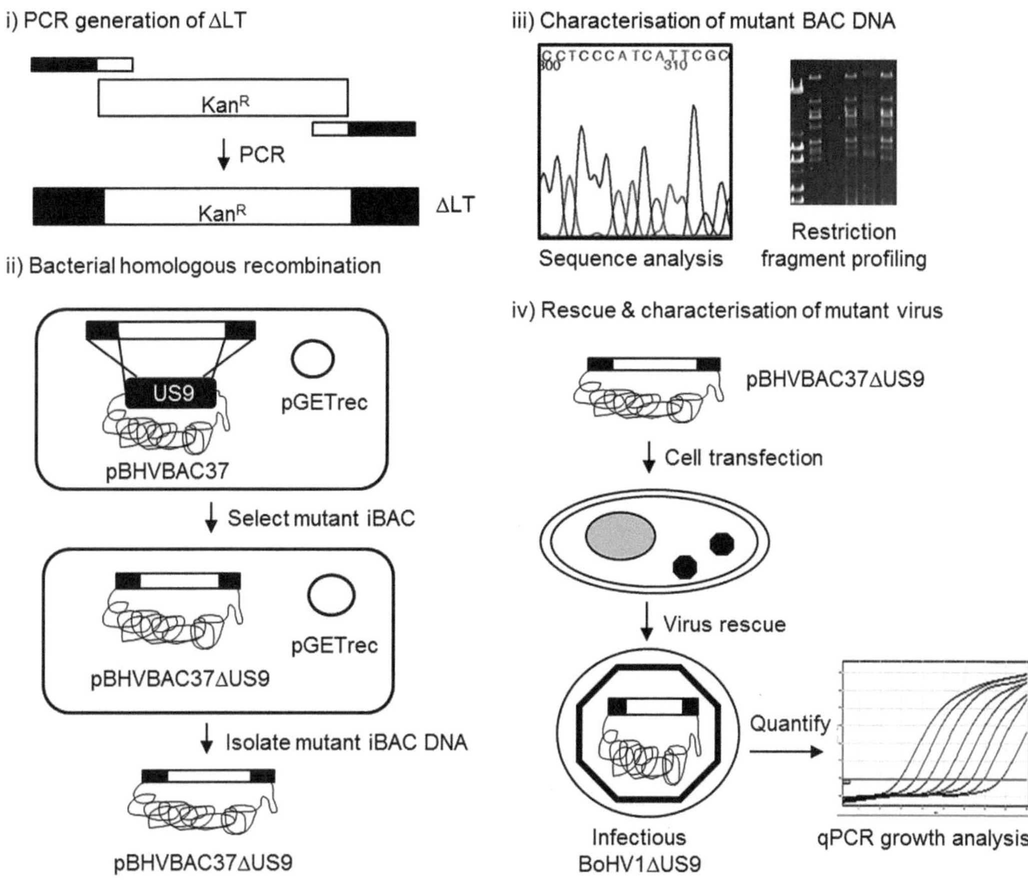

Fig. 1 Schematic representation of the steps required to specifically delete a genetic element from the infectious bacterial artificial chromosome (iBAC) pBACBHV37, which encodes the infectious genome of bovine herpesvirus 1 (BoHV-1). The schematic illustrates the generation and characterization of a US9 deletion mutant as described in the text. (i) Generation of the linear transgene (ΔLT) molecule using PCR and chimeric primers. At the end of the PCR process the ΔLT encodes a bacterial kanamycin resistance cassette (Kan^R) and sequence with identity (black segments) to the 5′ and 3′ of the US9 coding sequence of pBACBHV37. (ii) The ΔLT was electroporated into bacterial cells transiently enabled for homologous recombination. Mutant BAC clones were selected on media containing kanamycin and chloramphenicol. (iii) DNA from putatively recombinant iBAC clones were further characterized by DNA sequence analyses to confirm the presence of the ΔLT. The integrity of the modified iBAC was assessed using restriction endonuclease digestion in comparison to the parental pBACBHV37. (iv) The effect of the introduced mutation on the rescue of infectious virus was determined by transfection of the modified BAC into virus susceptible cells. The replication capacity of the rescued virus was then determined using quantitative PCR (qPCR) relative to the unmodified parental pBACBHV37

1. Amplification of the deletion linear transgene (ΔLT).
2. Generation of the gene-deleted recombinant BAC clone.
3. Confirmation of viral ORF deletion.
4. Characterization of the gene deleted mutant.

3.1 Amplification of Deletion Linear Transgene (ΔLT)

Chimeric primers used for the amplification of the ΔLT incorporate two distinct elements. Firstly, each forward and reverse primer is designed to have 50 bp of homology to the 5′ and 3′ regions, respectively, of the target ORF (other genetic elements can be targeted using the same strategy) in the iBAC. Secondly, both primers contain short (20–27 bp) sequence identity to a prokaryotic selection cassette, e.g., kanamycin, to be inserted into the targeted region of the iBAC.

When used in the PCR process the resulting amplicon (termed ΔLT) contains the selectable marker flanked by the 50 bp of homology to the target sequence (Fig. 1i). These 50 bp elements facilitate the catalyzed homologous recombination between the ΔLT and the target region of the iBAC clone of interest, creating the desired deletion mutant (Fig. 1ii).

1. Here, the UL29 and the US9 ORFs of pBACBHV37 were targeted for deletion. Forward and reverse primers were designed by incorporating 50 bp of identical sequence to the 5′ and 3′ regions of the UL29 and the US9 ORF, respectively of BoHV-1 (*see* **Note 1**). Corresponding primer sequences identical to the 5′ and 3′ regions of the Kanamycin selection cassette (Kan^R) encoded within the EZ::Tn5 <KAN-2> transposon are also incorporated. Primer sequences for the amplification of the LT for the deletion of UL29 (ΔUL29) and US9 (ΔUS9) are shown in Table 1.
2. The PCR components and conditions for the amplification of ΔLT are; 1 U *Taq* DNA polymerase, 10× *Taq* DNA polymerase reaction buffer, 0.4 μM each of dTTP, dATP, dGTP, and dCTP, 1.25 mM $MgCl_2$, 1 fmol/μL EZ::Tn5 <KAN-2> template, and 1 μM each of respective corresponding primers in Table 1.
3. Cycling conditions for the PCR amplification of ΔLT are; initial denaturation, 94 °C, 2 min for 1 cycle, followed by 94 °C, 15 s; 60 °C, 30 s; 68 °C, 90 s for 35 cycles and a final extension of 68 °C and hold at 4 °C (*see* **Note 2**).
4. Resolve PCR amplification products on 0.8–1 % agarose gel at 80 V for 1 h.
5. Image bands on UV transilluminator system.
6. Excise appropriate bands representing the ΔUL29 and ΔUS9 LT at approximately 1.3 kbp from the agarose gel and purify using the QIAquick gel extraction kit (Qiagen) as per manufacturer's conditions. Elute in 30 μL of 10 mM Tris–HCl, pH 8.0. The concentration of the recovered LT should be at least 100 ng/μL before proceeding to next phase of the experiment.
7. Use immediately or store the purified ΔLT at −20 °C until required.

3.2 Generation of the Gene-Deleted Recombinant BoHV-1 Clone

The generation of gene deleted BoHV-1 iBAC clones involves three critical steps:

1. The preparation of recombination competent *E. coli* host strain containing BoHV-1 iBAC and pGETrec.
2. Transformation of recombination competent cells with ΔLT.
3. Recovery of gene-deleted mutants.

3.2.1 Preparation of Recombination Competent E. coli DH10B (pBACBHV37; pGETrec)

Initially the pBACBHV37 iBAC and pGETrec plasmid (a gift from the late Panayiotis Ioannou, The Murdoch Children's Research Institute, Melbourne, Australia) were co-electroporated into DH10B cells. The DH10B cells harboring pBACBHV37 iBAC and the pGETrec plasmid (DH10B: pBACBHV37; pGETrec) were transiently made competent for homologous recombination through inducing expression of recE and recT proteins encoded by pGETrec.

1. Plate out DH10B (pBACBHV37; pGETrec) on LB-CAP/Amp solid media and incubate for approximately 16 h at 37 °C.
2. Transfer an isolated, well-formed bacterial colony to 20 mL of LB-CAP/Amp liquid media using aseptic microbiological technique.
3. Incubate broth culture at 37 °C with vigorous shaking for approximately 16 h.
4. Use 12.5 mL of overnight broth culture to inoculate 200 mL of pre-warmed LB-CAP/Amp liquid media and incubate at 37 °C with vigorous shaking.
5. Monitor the optical density (OD_{600nm}) of the broth culture spectrophotometrically pre- and post-inoculation at 30 min intervals.
6. When the culture OD_{600} is approximately 0.4, induce the production of recombination proteins by the addition of 10 % w/v L-arabinose to a final concentration of 0.2 %.
7. Incubate broth culture at room temperature with vigorous shaking for 40 min.
8. Decant the induced broth culture into four prechilled 50 mL tubes and pellet bacterial cells by centrifugation at 5,000 ×*g* for 10 min at 0 °C (*see* **Note 3**).
9. Discard supernatant and resuspend bacterial pellets in 50 mL of ice-cold 10 % glycerol.
10. Pellet bacterial cells again by centrifugation at 5,000 ×*g* for 10 min at 0 °C.
11. Repeat **steps 8** and **9** a further three times. Finally resuspend all bacterial pellets in total volume of 50 mL 10 % glycerol and pellet cells by centrifugation at 5,000 ×*g* for 10 min at 0 °C.

12. Aspirate and discard the supernatant and resuspend the bacterial pellet in 1 mL ice-cold 10 % glycerol and dispense 50 μL aliquots into prechilled 1.5 mL tubes suitable for use with liquid nitrogen.
13. Use cells immediately or snap-freeze aliquots either in liquid nitrogen or an ethanol/dry-ice bath and store recombination competent DH10B (pBACBHV37; pGETrec) cells at −80 °C until required.

3.2.2 Transformation of Recombination Competent Cells with ΔLT

Here, the ΔLT is introduced into the recombination competent DH10B (pBACBHV37; pGETrec) cells by electroporation. Cells are subsequently allowed to recover without antibiotic selection. Putative gene deleted mutants are then selected by transferring recovered electroporated cells onto solid LB media containing appropriate selective agents.

1. Dispense 450 μL SOC broth into 1.5 mL tubes for each clone to be constructed.
2. Thaw recombination competent DH10B (pBACBHV37; pGETrec) cells on ice immediately prior to use (*see* **Note 3**).
3. Dispense 2–5 μL of the purified ΔLT into thawed, recombination competent DH10B (pBACBHV37; pGETrec) and mix gently and thoroughly by tapping the side of the tube.
4. Transfer the total mixture into the chilled electroporation cuvette taking caution not to introduce bubbles. Gently tap the cuvette on the bench to concentrate the cells at the bottom of the cuvette cell.
5. Place the electroporation cuvette into the cradle, place the holder into the electroporator and pulse (1.8 kV, 100 Ω, 25 μF). Record the actual voltage (kV) and time constant (ms) for future reference.
6. Remove the cuvette from the holder and place immediately on ice for 1 min.
7. Flood the cuvette cell with pre-dispensed 450 μL of SOC, gently aspirate the bacterial cells and transfer to a 1.5 mL centrifuge tube.
8. Incubate recovered cells at 37 °C with gentle shaking for 5 h.
9. Plate recovered cells (200 μL) onto selective agar plates (CAP/Kan) and incubate at 37 °C for at least 16 h. This incubation can be extended to between 24 and 36 h.

3.2.3 Recovery of Gene-Deleted Mutants

Here, putative gene deleted mutant iBAC DNA is extracted from a broth culture by alkaline lysis (Fig. 1ii). Glycerol stock preparations of the putative mutants are also established for long-term storage of the mutant iBAC clones.

1. From the selective plates, pick an isolated, well-formed colony and inoculate 200 mL of LB-CAP/Kan media.
2. Incubate for approximately 16 h at 37 °C with vigorous shaking.
3. Prepare a glycerol stock for long-term storage of the iBAC clones by removing 5 mL of overnight broth culture and pellet bacterial cells by centrifugation at 5,000 × *g* for 5 min at 4 °C.
4. Discard supernatant, wash cells once with 1 mL of ice cold 10 % glycerol and pellet cells by centrifugation at 5,000 × *g* for 5 min at 4 °C.
5. Resuspend bacterial cell pellet in 200 μL of 10 % glycerol and store at −80 °C until required.
6. Decant the remaining broth from the overnight culture into 4 × 50 mL centrifuge tubes and pellet cells by centrifugation at 5,000 × *g* for 10 min at 4 °C.
7. Extract iBAC DNA from pooled cell pellets using the Nucleobond BAC 100 Extraction Kit as per manufacturer's instructions (*see* **Note 4**).
8. Resuspend purified iBAC DNA overnight at 4 °C in 200 μL of 10 mM Tris–HCl, pH 8.0.
9. Store iBAC DNA at 4 °C until required.

3.3 Confirmation of Gene Deleted BoHV-1 iBAC Mutants

The presence of antibiotic resistant colonies to both CAP and Kan selective agents on the agar plates is a preliminary indicator of successful gene deletion (*see* **Note 5**). The modified iBAC is subsequently confirmed by directional sequencing across the two regions where homologous recombination has occurred.

1. Sequencing of gene-deleted mutants utilizes the general primer KanME-F (Table 1), generating a sequencing product that incorporates 330 bp of the 3′ region of the Kan^R cassette, 50 bp of recombination region and run-off sequence into the adjacent ORF or intergenic region of the viral genome.
2. Sequencing reactions (40 μL) are conducted with the following components; 16 μL of BigDye Terminator v3.1 sequencing mix, 5 % (v/v) DMSO, 0.4 μM KanME-F primer, and 2 μg of purified iBAC DNA template.
3. Sequencing reactions are conducted with the following cycling conditions; Initial denaturation, 95 °C, 5 min for 1 cycle, followed by 95 °C, 30 s; 50 °C, 20 s; 60 °C, 4 min for 60 cycles and a completion hold at 4 °C.
4. Precipitate the sequencing reaction products by adding 100 μL of 100 % ethanol and 4 μL of 3 M sodium acetate (pH 4.6).
5. Incubate at room temperature for 15 min in the dark.

Table 2
Analyses of the pBACBHV37 iBAC gene deletion mutants for US9 and UL29

Clone	Site	Region deletion	Infectious	C_t	Requirement
pBACBHV37[a]	NA	Nil	+	18.0	
pBACBHV37ΔUS9	123548-123982	123603-124349	+	23.8	N
pBACBHV37ΔUL29	49233-52844	49285-52822	–	>45	E

The nucleotide location of each gene (site) within the BoHV-1 genome sequence AJ004801, deleted region, recovery and replication fitness of parental and gene deleted BAC clones of BoHV-1 are shown. Recovery of infectious BoHV-1 following transfection of BAC DNA into bovine cells was assessed by the presence (+) or absence (−) of CPE consistent with BoHV-1 infection. The amount of virus released by infected cells was determined by quantitative PCR with the cycle threshold (C_t) value shown. The requirement for the mutated gene for in vitro virus replication was designated as either essential (E) or nonessential (N)

[a]The construction and characterization of this clone has previously been described [4]

6. Pellet the precipitated sequencing products in a microcentrifuge at maximum speed for 15 min.
7. Carefully aspirate and discard the supernatant and wash pellet with 500 μL of 70 % ethanol.
8. Pellet the sequencing products again in a microcentrifuge at maximum speed for 15 min.
9. The DNA sequencing products are then resolved using an appropriate technology platform.
10. Use a compatible computer program for chromatogram visualization and sequence analysis. With respect to our research, sequencing chromatograms were visualized using the 4Peaks program of A. Griespoor and T. Groothuis (www.mekentosj.com). After removal of low quality data points the edited sequence is subject to blast analyses using the BLASTN program [16].
11. Sequencing of the recombination region confirmed the deletion of the target region for both the UL29 and US9 ORF of BoHV-1 (Table 2).

3.4 Characterization of Gene Deleted Viral Mutants

Once generated and the deletion of target ORF is confirmed, subsequent characterization of gene deleted mutants is conducted (*see* **Note 6**, Fig. 1iii). Here, characterization of gene-deleted mutants encompasses:

1. Characterization of iBAC DNA by restriction endonuclease digestion to confirm the molecular constitution of the genome.
2. Determination of the essential or nonessential requirement of the deleted ORF for in vitro viral growth.
3. The establishment of the replication capacity of gene deleted mutants.

3.4.1 Restriction Endonuclease Profiling

Here, the mutant iBAC DNA is digested with restriction endonucleases and the resulting fragment restriction profile compared to the unmodified parental iBAC DNA clone. Profiling extracted iBAC DNA discounts the presence of unmodified parental iBAC DNA in mutant preparations and displays complexity of the genome, indicative of a complete, albeit a slightly modified, genomic molecule.

1. Co-digest approximately 2 μg of iBAC DNA from bacterial host with the restriction endonucleases *Hind* III and *Dra* I as per manufacturer's instructions (*see* **Note** 7).
2. Electrophoretically resolve the digestion products in a 0.7 % agarose gel at 60 V for approximately 16 h in prechilled TAE buffer at 4 °C (*see* **Note 8**).
3. Visualize and photograph the resulting restriction fragment pattern using standard UV transillumination.
4. As can be observed (Fig. 2), complexity comparable to the parental clone is evident in both essential and nonessential deleted clones, suggesting band shifts are a consequence of specific modification of the respective ORF.

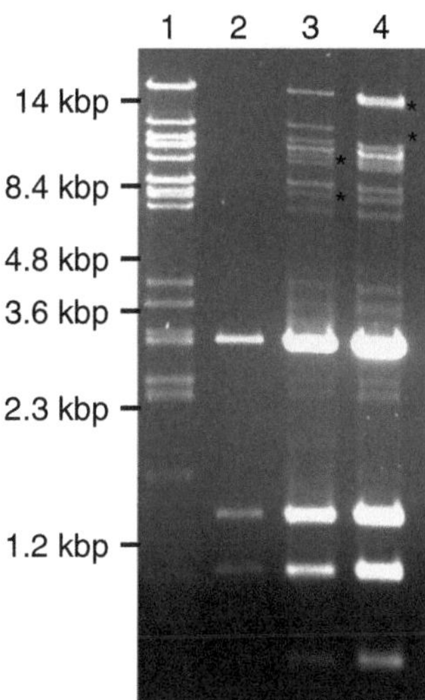

Fig. 2 Restriction endonuclease fragment profiles of pBACBHV37 BAC DNA harboring gene deletion. *Hind* III/*Dra* I digestion of *Lane 1*: Unmodified pBACBHV37 BAC DNA. *Lane 2*: pGETrec. *Lane 3*: Mutant pBACBHV37ΔUS9 BAC DNA. *Lane 4*: Mutant pBACBHV37ΔUL29 BAC DNA. Restriction profiles of gene deletion mutants of pBACBHV37 were digested to assess if the modified genome was comparable in complexity to the parent clone. *Asterisks* highlight specific band shifts that have occurred with respect to the deletion of either the US9 (*Lane 3*) or UL29 (*Lane 4*) open reading frame

3.4.2 *In Vitro Viability of Gene Deleted Mutants*

To determine if the deleted ORF is either essential or nonessential for in vitro virus growth, the mutant BAC is transfected into susceptible cells to rescue infectious virus (Fig. 1iv). After transfection the cells are monitored for the development of cytopathic effect (CPE) consistent with the parental virus. The development of CPE on either the primary transfection or subsequent passages is used to classify the targeted gene as nonessential for in vitro virus growth. Whereas the absence of CPE after primary transfection and three subsequent passages is used to designate the targeted gene as essential for in vitro virus growth.

1. Approximately 24 h prior to requirement, seed CRIB-1 cells in 6-well plates at a density of 5×10^5 cells/well and incubate at 37 °C in an atmosphere of 5 % CO_2.
2. Dilute 8 μL of Lipofectamine 2000 transfection reagent in 92 μL of Opti-MEM serum free medium per transfection.
3. Dilute approximately 500 ng of mutated iBAC DNA and control unmodified iBAC DNA (*see* **Note 9**) in Opti-MEM to a total volume of 100 μL.
4. Add 100 μL of lipofectamine 2000/Opti-MEM to 100 μL iBAC DNA/Opti-MEM and incubate for 45 min at room temperature in the dark.
5. Remove maintenance media from CRIB-1 cells showing approximately 90–95 % confluency and gently wash the monolayer twice with room temperature phosphate buffered saline (PBS) and once with 1 mL of Opti-MEM. Replace with 800 μL of Opti-MEM.
6. Add iBAC DNA/Lipofectamine complexes (200 μL) drop wise to the washed CRIB-1 monolayer and incubate for 16–18 h at 37 °C in a 5 % CO_2 atmosphere.
7. Remove the transfection reaction and wash the monolayer gently with PBS.
8. Dispense 3 mL of pre-warmed (37 °C) 10 % FBS maintenance media supplemented with 200 μM *N′N′*-HBA (*see* **Note 10**) to the monolayer.
9. Incubate plates at 37 °C in a 5 % CO_2 atmosphere for 2–7 days and observe daily for the development of CPE (*see* **Note 11**).
10. Irrespective of the development of CPE, freeze/thaw monolayers and clarify supernatants of cellular debris by centrifugation at 5,000 × *g* for 5 min at 4 °C.
11. Passage 100 μL of respective clarified supernatant on fresh, 90–95 % confluent CRIB-1 cell monolayers.
12. Retain 1 mL of respective clarified supernatant and store at −80 °C until required.
13. Repeat **steps 9–12** a further two times.

14. Observe monolayers for the presence of active infection upon initial transfection and subsequent passages. The development of CPE suggests that the deletion of the respective genetic element is nonessential for in vitro viability of BoHV-1. The absence of CPE on initial transfection or subsequent passage suggests the ORF is essential for BoHV-1 in vitro viability (Table 2).

3.4.3 In Vitro Replication Capacity of Gene Deleted Mutants

Once the requirement of specific ORFs has been established by the presence or absence of CPE in vitro, the effect of ORF deletion with respect to in vitro replication capacity of virus rescued from the mutant clone can be assessed. This is achieved by quantitative real time-PCR (qPCR) of the cell culture supernatants after three passages (Fig. 1iv). The in vitro replication capacity is indirectly assessed by establishing the cycle threshold (C_t) number of the mutant virus harboring a gene deletion and compared against a standard curve generated from the replication capacity of the unmodified pBACBHV37 clone.

1. To define replication capacity, extract total DNA from 200 μL of clarified, passage 3 supernatants using the High Pure Viral DNA extraction kit (Roche) as per the manufacturer's instructions.
2. Conduct qPCR reactions in triplicate in a volume of 20 μL. Reaction components include 10 μL Taqman Universal PCR master mix, 0.6 μM each of primers BHV-F and BHV-R, 0.2 μM probe (BHV-P) labelled with 6-carboxy-fluorescein (FAM) (Table 1), and 1 μL of template viral DNA [17].
3. The qPCR thermal cycling conditions are; initial template incubation of 60 °C for 2 min and an enzyme activation step at 95 °C for 10 min followed by; 40 cycles of 95 °C for 15 s and 60 °C for 1 min. A C_t value of 0.2 was used for data acquisition.
4. After qPCR determine the impact of the introduced mutations on in vitro replication by comparing the C_t values for the various supernatant extracts to similar extracts from virus rescued from the parental iBAC. For qualitative comparisons of detectable virus in the supernatant, a C_t value <45 suggests the modified element is not essential for in vitro growth while no determined C_t value (i.e., >45) suggests the modified element is essential for in vitro growth (Table 2).

Here, the materials and methods required for the mutagenesis of an alphaherpesvirus viral genome maintained as an iBAC have been described (Fig. 1). The methodologies were demonstrated by the specific deletion of the UL29 and US9 ORF of a BoHV-1 genome maintained as an iBAC. Further, the strategies to characterize the generated mutants were detailed including, nucleotide sequencing to confirm the deletion of genes, restriction endonuclease profiling and assessment of the phenotypic effect of deleting the targeted genes on viral growth in vitro. The results of the transfection experiments to determine the replication capacity of pBACBHV37ΔUS9 and pBACBHV37ΔUL29 mutants showed the US9 deletion clone to have a reduced replication capacity when compared to the replication capacity of the parental clone. As indicated from the viability studies of the UL29 deletion clone, no indicative CPE and no DNA was detectable in passage 3 supernatants suggesting the mutated iBAC was unable to facilitate the recovery of infectious BoHV-1. Therefore UL29 ORF was designated essential for viability of BoHV-1 (Table 2). The pGETrec facilitated inducible homologous recombination methodologies presented can be readily applied to the mutation and deletion of specific genetic elements or the introduction of exogenous DNA sequences into any herpesvirus iBAC clone of interest.

4 Notes

1. When deleting genes a decision of anchoring the ΔLT within the ORF or outside of the ORF adjacent to the start codon needs to be made. Anchoring the primer inside the gene results in the first 50 bp and last 50 bp of the ORF remaining although the majority of the ORF is deleted, in some smaller ORFs or other genetic elements, this residual sequence may affected the phenotype determination of the virus. Anchoring recombination regions outside the gene of interest will result in the regulatory elements still being present however, complete deletion of the ORF including start codons is achieved. The overlapping nature of viral genetic elements should also be considered to ensure the introduced mutations only affect the element of interest.
2. To recover an appropriate amount of linear transgene from the gel purification step it is appropriate to run several PCRs. Here we utilized two 50 μL PCR reactions that were resolved on a 1 % agarose gel. The gel slices taken for gel purification were pooled and amplified products purified on the same column. Transgenes were eluted in a total vol of 30 μL pre-warmed (50 °C) elution buffer.
3. The maintenance of the cold chain is critical in reducing the breakdown of recE and recT proteins. It is advised to chill all apparatus and consumables on ice for the entire process. As a result of this, caution should be used in the use of the electroporation equipment to ensure condensation does not affect the performance of the equipment or present a danger to the operator.
4. Several BAC extraction kits are available utilizing either liquid or affinity column extraction methods. Standard alkaline lysis

and phenol–chloroform extraction has also been successfully utilized [18]. It should be noted that commercial BAC DNA extraction methods typically result in lower yields compared alkaline lysis. However the lower purity of alkaline lysis preparations can negatively affect downstream analyses particularly the rescue of infectious virus. In this study, moderate yields with high purities were regularly achieved utilizing the Qiagen midi prep kit for extraction of pBACBHV37. However the Nucleobond BAC 100 kit gave the best combination of yield and purity of the recovery of pBACBHV37 DNA, as the DNA was suitable for sequencing, transfection and restriction profile characterization of mutants. The user may be required to empirically investigate the best BAC isolation methodology for the iBAC under investigation.

5. In this laboratory, the development of bacterial colonies on dual selection plates after the recombination process has yielded a gene deleted BoHV-1 iBAC with 100 % efficiency. If there is any doubt over the phenotype of the colonies, a PCR assay can be designed that amplifies from the recombined ΔLT across the iBAC insertion point into the viral genome. Utilizing a primer set that incorporates one primer anchored inside the selection cassette of the linear transgene and one primer that is anchored adjacent to the recombination region, modified iBACs can be effectively identified. A positive PCR (at the estimated size) is indicative of a recombinant clone. A negative PCR result is indicative of a failure or non-target recombination between the iBAC and the linear transgene.
6. The characterization methods required are dictated by the nature of the work and the ultimate requirements of the project. With respect to the studies in this laboratory, the aims were to assess the requirements of the virally encoded ORFs as possible targets for replacement with exogenous DNA and as such, to characterize resulting mutants that harbored nonessential gene deletions and could replicate at a comparable rate to the unmodified parent virus.
7. The choice of restriction endonuclease is important as the enzyme should generate a fragment profile that allows for an assessment of the integrity of the remainder of the iBAC DNA. This is best achieved with restriction endonucleases that digest the BAC DNA approximately 15 to 30 times and generates a consistent fragment profile. Following electrophoretic resolution of the digestion fragments, a number of bands will be static and directly comparable to the parental iBAC. However, a small number will show electrophoretic band shifts corresponding to changes in restriction sites due to the deletion of an ORF. In some cases depending on the relative location of the inserted transgene and restriction endonuclease

sites no specific band shifts may be evident. Even in these cases restriction endonuclease profiles are useful to ensure that no unintended modifications have occurred to the iBAC during the various manipulations.

In this laboratory the recovered DNA from the BAC preparations is also double digested with *Dra* I, which cuts pGETrec into three fragments and *Hind* III, which is the enzyme of choice used for biotyping and strain differentiation of BoHV-1 (and pBACBHV37) (Fig. 2). The purpose of this is to ensure that the high quantity of pGETrec DNA that is present in the bacterial cells in higher copy number compared to the iBAC does not prevent visualization of the iBAC fragments due to the higher total fluorescence.

It has been reported that without antibiotic selection (ampicillin) the pGETrec plasmid is spontaneously lost during the recovery and subsequent culture processes however, the presence of pGETrec was not a concern in our studies [12]. To remove pGETrec it is suggested five or more rounds of subculture of iBAC in host cells on growth medium in the absent of ampicillin can facilitate the loss of the pGETrec plasmid.

A second approach resulting in the removal of the pGETrec plasmid is the rescue of infectious virus by transfection of the iBAC DNA into mammalian cells. After a minimum of two passages in mammalian cells, the iBAC can be recovered from infected cell DNA extracts by electroporation into bacterial host cells. Depending on the project requirements, either approach is suitable for removal of pGETrec from host cells.

8. Running electrophoretic separation of digested DNA at lower temperatures reduces heat-induced diffusion of the digestion fragments and results in increased banding sharpness. Chilled buffer can be used in a reticulated cooling system using a peristaltic pump apparatus; tubing and an ice bath or electrophoresis can be conducted in a cold room.

9. The Lipofectamine 2000 transfection method is forgiving with respect to both DNA quantity and quality, however, iBAC DNA prepared by using the Nucelobond BAC100 kit is suitable for this protocol. The Lipofectamine 2000 reagent is marginally toxic to the cells and to prevent excessive cell death in the monolayer, lower concentrations of DNA can be used (e.g., 500 ng). Also, maintenance media should be warmed to at least room temperature before applying to the transfected monolayer.

10. The addition of 2 mM *N′N′*-HBA to the cell culture media has been shown to promote viral gene transcription from the BoHV-1 genome [19]. This component may not be required for non-BoHV-1 and nonviral applications in BAC transfection studies.

11. The development of CPE typical of the parent virus may occur between 2 and 7 days post transfection. However, in this

laboratory several iBAC mutants, not described in this study, have been observed that produce a non-lytic CPE. Thus the transfection of cells with mutated iBAC clones may produce atypical CPE in transfection studies. Therefore some consideration should be given to the criteria used to assess the effects of the introduced mutations prior to commencing these types of studies.

References

1. Messerle M, Crnkovic I, Hammerschmidt W, Ziegler H, Koszinowski UH (1997) Cloning and mutagenisis of a herpesvirus genome as an infectious bacterial artificial chromosome. Proc Natl Acad Sci U S A 94:14759–14763
2. Borst E, Hahn G, Koszinowski UH, Messerle M (1999) Cloning of the human cytomegalovirus (HCMV) genome as an infectious bacterial artificial chromosome in *Escherichia coli*: a new approach for construction of HCMV mutants. J Virol 73:8320–8329
3. Schumacher D, Karsten Tischer B, Fuchs Wosterrieder N (2000) Reconstitiution of Mareks disease virus serotype 1 (MDV-1) from DNA cloned as a bacterial artificial chromosome and characterisation of a glycoprotein B-negative MDV-1 mutant. J Virol 74:11088–11098
4. Mahony TJ, McCarthy FM, Gravel JL, West L, Young PL (2002) Construction and manipulation of an infectious clone of the bovine herpesvirus 1 genome maintained as a bacterial artificial chromosome. J Virol 76:6660–6668
5. Chang WLW, Barry PA (2003) Cloning of the full-length Rhesus cytomegalovirus genome as an infectious and self-excisable bacterial artificial chromosome for analysis of viral pathogenesis. J Virol 77:5073–5083
6. Gillet L, Daix V, Donofrio G, Wagner M, Koszinowski UH, China B, Ackermann M, Markine-Goriaynoff N, Vanderplasschen A (2005) Development of bovine herpesvirus 4 as an expression vector using bacterial artificial chromosome cloning. J Gen Virol 86: 907–917
7. Hansen K, Napier I, Koen M, Bradford S, Messerle M, Bell E, Seshadri L, Stokes HW, Birch D, Whalley JM (2006) In vitro transposon mutagenesis of an equine herpesvirus 1 genome cloned as a bacterial artificial chromosome. Arch Virol V151:2389–2405
8. Schwyzer M, Ackermann M (1996) Molecular virology of ruminant herpesviruses. Vet Microbiol 53:17–29
9. Gibbs EPJ, Rweyemamu MM (1977) Bovine herpesvirus. Part 1. Vet Bull 47:317–343
10. Muylkens B, Thiry J, Kirten P, Schynts F, Thiry E (2007) Bovine herpesvirus 1 infection and infectious bovine rhinotracheitis. Vet Res 38:181–209
11. Robinson KE, Meers J, Gravel JL, McCarthy FM, Mahony TJ (2008) The essential and non-essential genes of Bovine herpesvirus 1. J Gen Virol 89:2851–2863
12. Narayanan K, Williamson R, Zhang Y, Stewart AF, Ioannou PA (1999) Efficient and precise engineering of a 200 kb β-globin human/bacterial artificial chromosome in *E. coli* DH10B using an inducible homologous recombination system. Gene Ther 6:442–447
13. Jamsai D, Nefedov M, Narayanan K, Orford M, Fucharoen S, Williamson R, Ioannou PA (2003) Insertion of common mutations into the human β-globin locus using GET reecombination and an *Eco* RI endonuclease counterselection cassette. J Biotech 101:1–9
14. Nefedov M, Williamson R, Ioannou PA (2000) Insertion of disease-causing mutations in BACs by homologous recombination in *Escherichia coli*. Nucleic Acids Res 28:e79
15. Butchi NB, Jones C, Perez S, Doster A, Chowdhury SI (2007) Envelope protein US9 is required for the anterograde transport of bovine herpesvirus type 1 from trigeminal ganglia to nose and eye upon reactivation. J Neurovirol 13:384–388
16. Altschul SF, Madden TL, Schäffer AA, Zhang J, Zhang Z, Miller W, Lipman DJ (1997) Gapped BLAST and PSI-BLAST: a new generation of protein database search programs. Nucleic Acids Res 25:3389–3402
17. Lovato L, Inman M, Henderson G, Doster A, Jones C (2003) Infection of cattle with a bovine herpesvirus 1 strain that contains a mutation in the latency-related gene leads to increased apoptosis in trigeminal ganglia during the transition from acute infection to latency. J Virol 77:4848–4857
18. Sambrook J, Russell DW (2001) Molecular cloning: a laboratory manual. Cold Springs Harbour Laboratory Press, New York
19. Stavropoulos TA, Strathdee CA (1998) An enhanced packaging system for helper-dependent herpes simplex virus vectors. J Virol 72:7137–7143

Chapter 9

Conversion of BAC Clones into Binary BAC (BIBAC) Vectors and Their Delivery into Basidiomycete Fungal Cells Using *Agrobacterium tumefaciens*

Shawkat Ali and Guus Bakkeren

Abstract

The genetic transformation of certain organisms, required for gene function analysis or complementation, is often not very efficient, especially when dealing with large gene constructs or genomic fragments. We have adapted the natural DNA transfer mechanism from the soil pathogenic bacterium *Agrobacterium tumefaciens*, to deliver intact large DNA constructs to basidiomycete fungi of the genus *Ustilago* where they stably integrated into their genome. To this end, Bacterial Artificial Chromosome (BAC) clones containing large fungal genomic DNA fragments were converted via a Lambda phage-based recombineering step to *Agrobacterium* transfer-competent binary vectors (BIBACs) with a *Ustilago*-specific selection marker. The fungal genomic DNA fragment was subsequently successfully delivered as T-DNA through *Agrobacterium*-mediated transformation into *Ustilago* species where an intact copy stably integrated into the genome. By modifying the recombineering vector, this method can theoretically be adapted for many different fungi.

Key words Recombineering, Binary vector, Bacterial artificial chromosome, Basidiomycete, *Ustilago*, *Agrobacterium*-mediated transformation

1 Introduction

Genetic complementation is required to prove the function of genes identified either by homologous search or by map-based cloning. To functionally analyze entire genes with regulatory elements, in their native genomic setting, or complete clusters of genes, it is highly desirable to be able to transfer these large DNA fragments to recipient cells. Large genome fragments, represented by bacterial artificial chromosome (BAC) clones, are routinely used for positional cloning, physical map construction, and whole genome sequencing because of their capability to replicate stably in *E. coli* [1, 2]. Typically, genomic DNA fragments contained in BAC clone libraries range in size from 50 to 150 kb. However, cloning of such BAC-size fragments into plasmid constructs that

Kumaran Narayanan (ed.), *Bacterial Artificial Chromosomes*, Methods in Molecular Biology, vol. 1227,
DOI 10.1007/978-1-4939-1652-8_9,

can be used for genetic transformation, using conventional cloning methods such as restriction enzyme digestion and ligation, are often time-consuming and inefficient because of the big sizes and few convenient restriction sites available for cloning in such fragments. An alternative approach to manipulate large DNA fragments uses a yeast-based recombineering system [3] or recombineering technology based on Lambda (λ) phage biology [4–7]. In addition, genetic transformation using chemical procedures or electroporation, in particular for large DNA constructs, is not efficient for some fungi. An alternative transformation procedure using *Agrobacterium tumefaciens*, shown initially to be able to genetically transform fungi [8], was further developed in our laboratory for transformation of *Ustilago* species [9]. Previously it had been shown that large genomic DNA fragments could be delivered by *A. tumefaciens* into tobacco plants by developing a BAC vector suitable for maintenance in and transfer from this bacterium as a so-called binary vector [10] which was termed a BIBAC vector; this also required the proper placement of transfer (T)-DNA border sequences [11].

We combined different techniques and made use of the λ phage-based in vivo genetic engineering technology to convert BAC clones with *Ustilago*-specific selection markers into binary BAC (BIBAC) constructs by homologous recombination in *E. coli*. This recombineering technology is based on three λ RED-encoded genes (*Exo*, *Beta*, and *Gamma*) that have been transferred to the chromosome of the recombineering-proficient *E. coli* strain SW102 by integrating a λ prophage [12]. The Exo protein is a 5′ to 3′ exonuclease that binds to the end of introduced linear dsDNA and degrades one strand to make a 3′ ssDNA overhang [6]. Beta is a ssDNA binding protein that binds to the 3′ overhang and promotes its annealing and homologous recombination with the complementary DNA strand of the BAC clone [6, 13]. Gamma encodes an inhibitor protein of *E. coli* RecBCD and SbcCD nuclease activities to protect the linear DNA and make it available for recombination with the BAC clone [6, 14, 15].

To convert BAC clones into BIBAC constructs that can be delivered by *A. tumefaciens* to the host cells (fungi), some specific functions such as a bacterial selectable marker, T-DNA specific border sequences (a left border LB and a right border RB sequences) and a broad-host range origin of replication need to be introduced. These essential functions are supplied by a linear DNA fragment produced by the Polymerase Chain Reaction from the recombineering vector (Fig. 1). A crucial part of this linear REC fragment is the presence of two 40–50 bp sequences on each end that are complementary to the selection marker in the BAC clone such as the chloramphenicol acetyl transferase gene (CAT resistance gene) present in most BAC vectors/clones (Fig. 1, Cat-1 and Cat-2). This provides the necessary homologous sequences for the λ RED

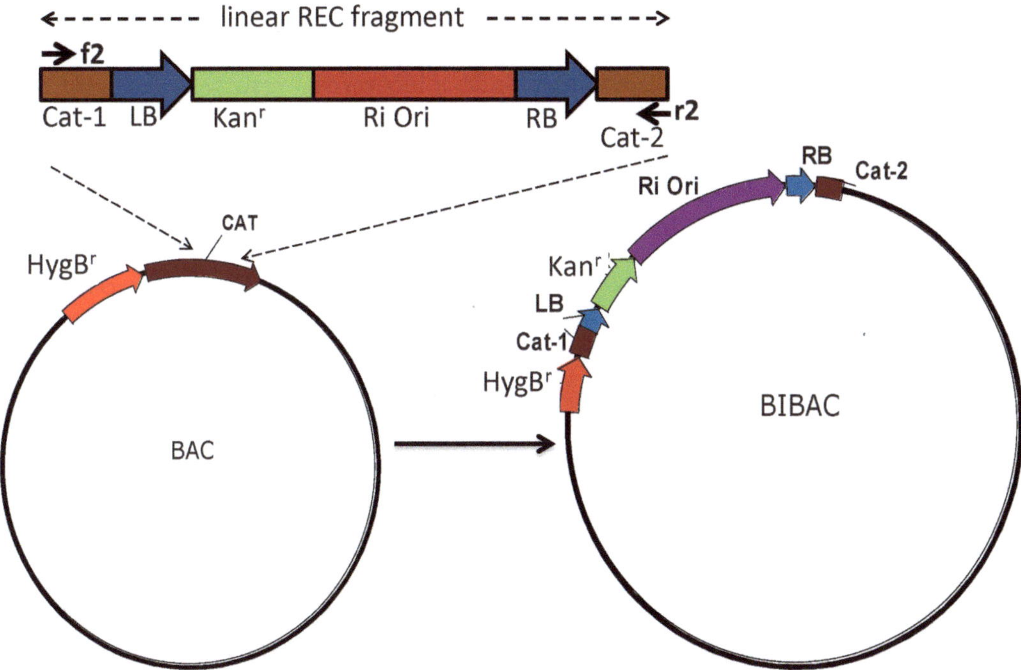

Fig. 1 Schematic representation of the BAC to BIBAC conversion method. Conversion of a Bacterial Artificial Chromosome vector, BAC, harboring genomic inserts (*black circle*), into a binary BAC (BIBAC) vector using a linear DNA fragment (linear REC fragment) amplified by PCR with primers f2 and r2 from recombineering vector pFT41 [16]. The REC fragment recombines into the chloramphenicol resistance gene (CAT) present on the BAC clone, using homologous regions present on both the left (Cat-1) and the right end (Cat-2) of the REC fragment (*dashed arrows*). Recombinants are selected for kanamycin resistance (Kan^r) present on the REC fragment and screened for loss of chloramphenicol resistance indicating proper integration. The resulting BIBAC constructs are then transformed into a suitable *A. tumefaciens* strain for subsequent transformation into *Ustilago*. Any DNA present between the right border (RB) and *left border* (LB) sequence elements in BIBAC clones functions as T-DNA and is transferred by the *A. tumefaciens* machinery to a host cell where it is integrated into the genome. *Ustilago* transformants were selected on hygromycin B; a *Ustilago*-specific hygromycin B phosphotransferase cassette present on the BAC vector (HygB^r, *red arrow*) placed this selection cassette after recombination with the REC fragment on the T-DNA of the BIBAC construct. The figure is not drawn to scale

proteins to integrate the linear REC fragment into the BAC clone, but also provides counter selection by conversion to chloramphenicol sensitivity indicating proper recombination.

Such a Rec vector was developed by Takken et al. [16] for use in ascomycete fungi by introducing a fungal-specific selectable marker cassette to allow *Agrobacterium*-mediated transformation (AMT) of *Fusarium* and *Aspergillus* species. We used two different types of REC vectors to convert BAC clones into BIBAC constructs. For one kind of BAC conversion, we used the pFT41 backbone [16] to convert the previously developed *Ustilago* specific BAC vector, pUsBAC5 [17]. This BAC clone already has a *Ustilago*-specific hygromycin B (HygB^r) selection cassette under control of the HSP70 promoter and terminator signals [18] and therefore no fungal selection marker was required in the REC vector [9]. In the second BAC conversion approach, we replaced the Ascomycete-specific phleomycin resistance cassette from pFT41 with the *Ustilago*-specific HygB resistance cassette (S. Ali, J. Xu, and G. Bakkeren, unpublished). This REC vector can be used for conversion of any BAC clone that does not have a *Ustilago* (fungal) selectable marker.

We used *A. tumefaciens* to deliver these BIBAC clones into *U. hordei* and *U. maydis* and showed that *Agrobacterium* can efficiently deliver large DNA fragments into these basidiomycete fungi [9]. AMT has proved very efficient for delivery of foreign DNA into several fungal species that are refractory to genetic transformation with conventional transformation methods [19–23]. In addition to the higher efficiency of AMT in different fungi, it also usually results in single-copy T-DNA integration of mostly intact DNA fragments into the host genome at random sites [24–26]. This method has been used to transfer large DNA fragments into different fungi and plant cells [9, 16, 27]. In this chapter we will describe in detail the method for BAC to BIBAC conversion and their subsequent transfer into the basidiomycete fungus *Ustilago* by *Agrobacterium*.

2 Materials

Prepare all solution using ultrapure deionized water and analytical grade reagents.

2.1 Bacterial and Fungal Strains, Growth Media, and Culture Conditions

1. *E. coli* recombineering strain SW102 (a DY380 derivative with genotype *λcI*857, no *cro-BioA* but *tet*) can be requested from Dr. Neal Copeland (National Cancer Institute, Frederick, MD) [4, 12]. Supervirulent *Agrobacterium tumefaciens* strain COR309 is a recA-deficient C58 nopaline strain UIA143 harboring disarmed pTiB6 derivative plasmid pMOG101 [10] and a special vir helper plasmid pCH32, which provides extra copies of the *virA* and *virG* two-component signaling genes. *A. tumefaciens* strain COR308 is similar to COR309 except that it has disarmed pTi derivative plasmid pMP90 instead of pMOG101; they can be obtained from Cornell University (http://www.biotech.cornell.edu/BIBAC/BIBACHomePage.html). *Ustilago hordei* haploid strain Uh4857-4 (alias Uh364, *MAT-1*) has been described [17] and *U. maydis* haploid strain 324 (mating type *a2b2*) is Um521 [28].
2. Luria–Bertani (LB) medium: 1 % NaCl w/v, 1 % Bacto Tryptone w/v, 0.5 %, Bacto Yeast Extract w/v. LB broth and agar plates supplemented with appropriate antibiotics for plasmid selection in *E. coli* and *Agrobacterium*.

3. SOC medium: 2 % Bacto Tryptone w/v, 0.5 %, Bacto Yeast Extract w/v, 10 mM NaCl, 2.5 mM KCl, 10 mM $MgCl_2$, 10 mM $MgSO_4$, 20 mM glucose, pH 7.0.
4. LC medium: 0.8 % NaCl w/v, 1 % Bacto Tryptone w/v, 0.5 % Bacto Yeast Extract w/v.
5. Induction medium: 10 mM K_2HPO_4, 10 mM KH_2PO_4, 2.5 mM NaCl, 2 mM $MgSO_4{\cdot}7H_2O$, 0.7 mM $CaCl_2{\cdot}2H_2O$, 9 μM $FeSO_4$, 4 mM $(NH_4)_2SO_4$, 10 mM glucose supplemented with tetracycline (5 μg/mL), kanamycin (50 μg/mL), 40 mM MES, and 0.5 % glycerol.
6. Complete medium (CM): per liter, 5 g Casamino acids, 1.5 g Ammonium nitrate (18 mM end concentration), 10 g Yeast extract, 60 mL *Ustilago* macro salts (see below), set pH 7.0 with NaOH; (for plates, add 15 g agar). After autoclaving, add 20 mL sterile 50 % Glucose to 1 %.
7. *Ustilago* macro salts: per liter, 16 g KH_2PO_4, 4 g Na_2SO_4, 8 g KCl, 2 g $MgSO_4{\cdot}7H_2O$, 1 g $CaCl_2{\cdot}2H_2O$, 8 mL Trace Elements Solution.
8. Trace Elements Solution: per liter, 60 mg H_3BO_3 (boric acid), 140 mg $MnCl_2$, 400 mg $ZnCl_2$, 40 mg $Na_2Mo_4{\cdot}2H_2O$, 100 mg $FeCl_3$, 400 mg $CuSO_4$.
9. Selection medium: CM supplemented with Cefotaxime 200 μg/mL to kill off *A. tumefaciens* and 300 μg/mL hygromycin B to select for *Ustilago* transformants.
10. Acetosyringone (AS) 200 mM stock in DMSO (*see* **Note 1**).
11. Incubator and shaker at 20, 24, 28, and 30 °C.
12. Shaking water bath at 42 °C.
13. High speed refrigerated centrifuge with swing out rotor for 50 and 15 mL Falcon tube and refrigerated microcentrifuge for small tubes.
14. ME-25 filters (Schleicher and Schuell, 0.45 μm pore size, 47 mm diameter).

2.2 DNA Extraction

1. Sterilized acid-washed glass beads of 0.5 mm in size.
2. Lysis buffer: 0.5 M NaCl, 2 M Tris–HCl pH 7.5, 10 mM EDTA pH 8.0, 1 % SDS (*see* **Note 2**).
3. Phenol–Chloroform–Isoamylalcohol 24:24:1.
4. Ethanol 70 %.
5. Isopropanol.
6. TE pH 8.0 (10 mM Tris–HCl pH 7.5, 1 mM EDTA).
7. Prechilled mortars and pestles.

2.3 DNA Blotting, Probe Labeling, and Hybridization

1. 20× SSC buffer: 0.3 M sodium chloride, 0.03 M sodium citrate, pH 7.0, and 0.1 % w/v sodium dodecyl sulfate (SDS).
2. 1× Tris–Borate–EDTA (TBE) buffer: 89 mM Tris base, 89 mM boric acid, 2 mM EDTA, pH 8.0.
3. Saran Wrap, X-ray film (e.g., Kodak Biobax film, Kodak Canada, Toronto, ON, Canada), and developing reagents.
4. Agarose gel apparatus.
5. Transfer tray and alkaline transfer buffer. Alkaline transfer buffer is composed of 0.4 N Sodium hydroxide (NaOH) and 1 M sodium chloride (NaCl).
6. Hybond-N^+ nylon membrane (Amersham Biosciences/GE healthcare) or similar product from other suppliers.
7. αP^{32}-dCTP for radioactive labeling of probes.
8. Rediprime II Random Prime Labeling system (Amersham Biosciences/GE healthcare) or similar product from other suppliers.
9. Hybridization buffer, ULTRA-hyb buffer (Ambion) or similar product from other suppliers.
10. Hybridization oven and hybridization tube.

3 Methods

3.1 Preparation and Transfer of the Target BAC Clones in Recombineering Proficient E. coli SW102 Cells

1. To prepare the BAC plasmid, streak a BAC clone for single colony selection from a glycerol stock at −80 °C on LB plates with appropriate antibiotic (Cm).
2. Transfer a single colony to 5 mL LB medium in a 13 mL snap-cap tube with appropriate selective antibiotic, incubate overnight at 37 °C with vigorous shaking at 200–230 rpm.
3. Take 500 μL of the starter culture and inoculate into 150 mL LB media with selective antibiotic in a 500 mL conical flask and grow at 37 °C for 14–16 h with vigorous shaking.
4. Harvest cells at 3,700 ×*g* for 15 min at 4 °C and follow the protocol for plasmid midi preparation (e.g., QIAGEN's QIAprep Spin Midi Kit and Protocol or a similar product from other suppliers (*see* **Note 3**).
5. Quantify the BAC DNA using a spectrophotometer; pure DNA should have an OD_{260}/OD_{280} ratio of >1.8. Use 1 μL of the BAC DNA solution for PCR analysis to confirm the plasmid, and use about 3–5 μg for restriction analysis.
6. Transfer 1–5 μL (0.2–2 μg) of freshly prepared target BAC constructs (*see* **Note 4**) into 50 μL electro-transformation-competent recombineering proficient *E. coli* SW102 cells

(*see* Subheading 3.3), on ice. As a negative control for plasmid transformation, use 5 μL of ddH_2O for 50 μL cells.

7. Pipette the *E. coli* cells with the BAC DNA into a prechilled 2 mm gap electroporation cuvette.
8. Set the electroporator to 2.5 kV/cm, 25 μ FD, 200 Ω. This setting is used for a Gene Pulser (Bio-Rad) using an electroporation cuvette with 2 mm gap. For 1 mm gap cuvettes, change the voltage to 1.2–1.6 kV/cm. Other electroporation apparatuses can be used following the manufacturer's instructions.
9. Flick the cuvette gently to mix and knock on the table to remove air bubbles. Dry the cuvette with a tissue and place into position in the electroporator. Proper transformation normally occurs when no arching is observed and a time constant of 4–5 is obtained.
10. Immediately remove the cuvette from the chamber and add 1 mL SOC medium without antibiotics to the cuvette. Remove all the contents of the cuvette using a sterile Pasteur pipette and transfer it to a 13 mL snap-cap tube.
11. Incubate the cultures at 28 °C to avoid premature induction of the phage recombineering (RED) genes in a shaking incubator for 60 min.
12. Plate out 50–100 μL of the cells on LB agar plates with appropriate antibiotic (Cm) and incubate overnight at 28 °C. No-DNA controls are treated the same and should be included.
13. Colonies should appear on selection plates where cells received BAC DNA whereas controls should not. Select a single colony and perform colony PCR with specific primers to confirm proper transformation with the target BAC clone. Make a glycerol stock of the confirmed colony and store at −80 °C for future use.

3.2 Generation of the REC Fragment by Polymerase Chain Reaction (PCR)

1. Design primers to generate a PCR product with 40–50 bp termini homologous to a selectable marker in the BAC construct. In REC vector pFT41, the CAT gene was targeted and primers cat-f2 (CCGTTGATATATCCCAATGGC) and cat-r2 (ACAAACGGCATGATGAACCT) were used.
2. Use a proof reading polymerase that is optimized for the generation of long PCR products such as *TaKaRa LA Taq* polymerase (TAKARA Bio INC) or a 18:1 blend of SuperTaqPlus (Ambion) and Pfu (Promega) or similar products from other suppliers.
3. For the amplification of long PCR products for recombineering, use the following program on a thermal cycler: an initial denaturation step of 5 min at 95 °C, followed by 35 cycles of

30 s at 95 °C, 40 s at 58 °C (depending on the annealing temperature of the primers; this should be empirically determined) and 8 min at 68 °C, with a final extension at 68 °C for 15 min (*see* **Note 5**).

4. Separate the generated PCR molecules from the template REC plasmid DNA, primers and dNTPs by agarose gel electrophoresis. PCR purification kits can also be used for cleaning up the PCR product for the next enzymatic reaction (*see* **Note 6**). Approximately 150 ng of the final, clean PCR product is needed per transformation.
5. Careful removal (destruction) of the template DNA (the REC substrate plasmid) is required to eliminate background transformation. Digest the PCR product with restriction enzyme Dpn1, which recognizes a 4-base pair site (GATC), and only targets methylated DNA. Template DNA plasmid is usually methylated as most commonly used bacterial strains are dam^+ while the linear PCR product is usually not methylated. Dpn1 works well in PCR buffer and **step 4** can be eliminated by directly adding 2–4 μL of Dpn1 in 50 μL PCR reaction and incubating for 1 h at 37 °C. Scale up this digestion reaction when using larger volumes of PCR reaction.
6. Separate the PCR-amplified fragments from the digested template DNA by agarose gel electrophoresis and isolate the fragment as in the **step 4** above. Elute the DNA in sterile ddH_2O to avoid addition of any salts which will interfere in the electrotransformation step. If the concentration of eluted DNA is low, precipitate the DNA by adding 1/5 volume of 3 M sodium acetate and 2 volumes of 96–100 % ethanol. Mix and incubate the tube on ice for 30 min and centrifuge for 15 min at 20,000 × *g* at 4 °C. Discard the supernatant and wash the pellet with 70 % ethanol and dissolve in a small volume of ddH_2O.

3.3 Preparation of Electro-competent Recombineering-Proficient *E. coli* SW102 Cells

Before starting the electro-competent cell preparation, cool down sterile ddH_2O, 10 % glycerol, and the electroporation cuvettes on ice for 1–2 h.

Remember to grow this strain at 28 °C because you do not want to induce the λ RED genes which are under the control of the temperature-sensitive λ-repressor, prematurely.

1. To prepare the electro-competent cells, streak *E. coli* strain SW102 or SW102 strains previously transformed with target BAC clones, from glycerol stocks for single colony selection on LB plates (with appropriate selection). To avoid any contamination, use tetracycline at 12.5 μg/mL for SW102 and use additional appropriate antibiotics for strains that harbor target BAC clones.

2. Transfer a single colony to 5 mL LB medium in a 13 mL snap-cap tube with appropriate selective antibiotic, incubate overnight at 28 °C with vigorous shaking at 200–230 rpm.
3. Take 500 μL of the starter culture and inoculate into 50 mL LB media with selective antibiotic in a sterile 250 mL conical flask. Incubate for 3–6 h at 28 °C in a shaking incubator until the density reaches an OD600 of 0.4–0.5. To monitor cell growth, take a sample for measuring the OD after 3 h and repeat each 30 min; use LB media as a reference when measuring the OD.
4. When the desired OD is reached, cool down the flasks containing the bacteria for 5 min on an ice slurry and transfer 40 mL of the culture into precooled 50 mL Falcon tubes.
5. Harvest the cells at 3,700 × *g* for 10 min at 0 °C and resuspend the cell pellet in 2 mL ice cold 10 % glycerol by gently shaking the tube in the ice slurry. When the cells are fully resuspended, add 13 mL 10 % glycerol to a final volume of 15 mL.
6. Invert the tube several times and centrifuge the cells at 3,700 × *g* for 5 min at 4 °C.
7. Remove the supernatant, resuspend the cells as in **step 5**, and spin once more as in **step 6**.
8. Remove all supernatant gently by inverting the tube. One needs to be very careful at this step to not lose the pellet, which should be kept on ice all the time.
9. Resuspend the cells gently on ice in 120–160 μL sterile ddH_2O or 10 % glycerol (use 10 % glycerol if the cells need to be stored at −80 °C for future use). This should give approximately 10^9 cells per mL. Aliquot 40 μL of the cells in a 1.5 mL precooled Eppendorf tube on ice.
10. Freeze remaining cells in 50 μL aliquots in liquid N_2 and store at −80 °C.

3.4 Heat Shock Induction of the λ RED Recombination Proteins and Recombineering Step

1. For induction of the λ RED proteins, grow strains that contain the target BAC clones at 28 °C to avoid premature induction. When the OD600 of the bacterial culture reaches 0.4–0.6, transfer 30 mL into a 250 mL conical flask and incubate in a shaking incubator at 42 °C for 15 min to induce the phage recombination RED genes, the expression of which is under control of a temperature-sensitive λ-repressor. The rest of the culture remains at 28 °C as non-induced control.
2. After 15 min, put both flasks containing induced and non-induced bacterial culture on ice for 20 min to cool them down and pour into 50 mL prechilled Falcon tubes; keep on ice.
3. Prepare electro-competent cells of both heat-induced and non-induced bacterial cultures as described above, starting

with **step 5** in Subheading 3.3, until 40 μL aliquots with a cell density of approximately 10^9 cells per mL are obtained in a 1.5 mL precooled Eppendorf tubes on ice (**step 9** in Subheading 3.3).

4. Add 150 ng of the clean, Dpn1-digested REC PCR product from **step 6** in Subheading 3.2 and pipette the *E. coli* cells with the DNA into a prechilled 2 mm gap electroporation cuvette.
5. Set the electroporator to 2.5 kV/cm, 25 μ FD, 200 Ω. This setting is used for a Gene Pulser (BIO-RAD) using an electroporation cuvette with 2 mm gap. For 1 mm gap cuvettes, change the voltage to 1.2–1.6 kV/cm. Other electroporation apparatuses can be used following the manufacturer's instructions.
6. Flick the cuvette gently to mix and knock on the table to remove air bubbles. Dry the cuvette with a tissue and place into position in the electroporator. Proper transformation normally occurs when no arching is observed and a time constant of 4–5 is obtained.
7. Immediately remove the cuvette from the chamber and add 1 mL SOC medium without antibiotics to the cuvette. Remove all the contents of the cuvette using a sterile Pasteur pipette and transfer it to a 13 mL snap-cap tube.
8. Incubate the cultures at 28 °C to avoid premature induction of the phage recombineering (RED) genes in a shaking incubator for 60 min.
9. Plate out 100–200 μL of the cells on LB plates with 50 μg/mL kanamycin and incubate at 28 °C. In the non-induced cells, no or very few colonies will appear while on the plate with induced cells, 100–200 colonies will grow.
10. Test at least 100 colonies for sensitivity to chloramphenicol and resistance to kanamycin. This can be done by taking a colony with a sterilized tooth pick and streaking it first on a LB + Cm plate and then, using the same toothpick, on a LB + Kan plate. The whole plate can also be replica-plated using sterilized pieces of velvet.
11. Colonies that are resistant to Kan and sensitive to Cm likely contain properly recombined BIBAC clones as correct recombination of the Rec vector into the BAC clone will have disrupted the chloramphenicol acetyltransferase (CAT) gene. This should be further verified with PCR and restriction enzyme digestion. However, there will be some colonies resistant to both Kan and Cm and could be the result of REC DNA integration into chromosomal DNA or non-homologous integration of the REC fragment into the BAC clone at other locations than the proper CAT gene.

3.5 Fungal Transformation

1. Transfer of desired BIBAC constructs into electro-competent *A. tumefaciens* strains is done by standard electroporation method as described for *E. coli* above, except with a longer incubation time (90–120 min) before plating on selection medium. It is important to use an *A. tumefaciens* strain that contains an extra copy of *virA* and *virG* genes since these have been shown to be important for the transfer of large T-DNA fragments to plant cells [27]. In our experiments we have used strains COR308 and COR309 that are recA deficient and contain extra copies of *virA* and *virG* on a helper plasmid pCH32. To maintain the helper plasmid, grow these strains under 5 μg/mL tetracycline selection at all times (*see* **Note 7**).
2. Streak the *A. tumefaciens* strains with the BIBAC construct from glycerol stocks on LB media with appropriate antibiotic and incubate at 28 °C.
3. Inoculate a single colony into 10 mL LC medium with selection for both the helper and BIBAC plasmid and incubate overnight.
4. Spin down 10 mL of BIBAC-containing *A. tumefaciens* culture for 10 min at 3,700 × *g*, at 4 °C, remove the supernatant and resuspend the pellet in 10 mL induction medium (IM) supplemented with 40 mM MES, 0.5 % glycerol, and 0.2 % glucose.
5. Resuspend the pellet to an OD_{600} of 0.4 in 5 mL IM and supplemented with 200 μM AS.
6. Incubate the *Agrobacterium* at 28 °C for 6–8 h to reach OD_{600} of 0.5–0.6. Incubate control cells in the same medium but without AS.
7. Grow fungi that you want to transform in liquid medium. In our laboratory, we grow *Ustilago* species in CM [29]; *U. hordei* at 22 °C and *U. maydis* at 28 °C. Re-inoculate 20 mL fresh CM to an OD_{600} of 0.15 and subsequently grow these cultures until an OD_{600} of 0.5 is reached.
8. Dilute the fungal cell cultures tenfold in IM and mix with an equal volume of AS-induced *A. tumefaciens.* This will result in an approximate ratio of 10:1 *Agrobacterium* to fungal cells (*see* **Note 8**).
9. 20 min after mixing the two cell cultures, inoculate 100–200 μL of the mixture onto ME-25 filters (0.45 μm pore size, 47 mm diameter) and place them on co-cultivation medium (IM, but with 0.1 % glucose and 200 μM AS). IM without AS can be used as a negative control for co-cultivation medium.
10. Air-dry the membranes for 30 min and incubate at 20–24 °C for 2–5 days or till individual colonies appear (*see* **Note 9**).

11. To select for *Ustilago* cells properly transformed with the T-DNA delivered from the BIBAC vectors by *A. tumefaciens*, transfer the membrane to selection plates containing CM medium supplemented with antibiotics such as HygB for which the resistance cassette is present on the T-DNA. Use cefotaxime at 200 μg/mL in the selection medium to kill off the *A. tumefaciens*. After 4 days, transfer individual colonies to CM with selective antibiotic and cefotaxime as above.

3.6 DNA Isolation from Fungal Cells

3.6.1 DNA Isolation for PCR Analysis

1. For PCR analysis, genomic DNA can be isolated using a mini prep protocol for *Ustilago* that is optimized in our laboratory and modified from a protocol used for yeast [30].
2. Inoculate transformed as well as untransformed control fungal cells from a fresh CM plate (*see* **Note 10**) in 5 mL CM with appropriate antibiotic in 13 mL snap-cap tubes until late-log growth, at an approximate OD_{600} of 1.5–2 (*see* **Note 11**).
3. Transfer 2 mL of the culture to a 2 mL Eppendorf tube and centrifuge for 2 min at 20,000 × *g* and aspirate off all the supernatant using regular laboratory vacuum.
4. Add 0.2–0.5 g of sterilized acid-washed glass beads of 0.5 mm in size to each tube. In our laboratory we are using a scoop made from cap of a microfuge tube and add one scoop to each tube (*see* **Note 12**).
5. Add 500 μL of lysis buffer and 250 μL of Phenol–Chloroform–Isopropanol (PCI) and vortex at high speed for 3 min. Do this step in a fume hood.
6. Centrifuge the tube for 3 min at 20,000 × *g* to separate the phases. Carefully transfer 450 μL of the upper aqueous phase to a new tube, avoiding touching and taking along the cell debris at the interface.
7. Add 250 μL PCI to the aqueous phase, vortex for 30 s, and centrifuge for 1 min at 20,000 × *g*. Carefully transfer 400 μL of the upper aqueous phase to a new tube.
8. To precipitate the genomic DNA, add 0.6 volumes of isopropanol, mix and keep on ice or at −20 °C for 30 min or more. Overnight storage at −20 °C works better.
9. Centrifuge for 15 min at 20,000 × *g* at room temperature; a white pellet should be visible which contains genomic DNA and a large amount of RNA. Remove supernatant by pipette or inverting.
10. Add 1 mL of 70 % room-temperature ethanol, centrifuge for 5 min at 20,000 × *g* at room temperature and remove supernatant. Be careful, since the pellet may be loose. Invert the tube on a paper towel for 10–20 min or until the pellet is dry. The pellet may also be dried in a speedVac but avoid over-drying since the pellet will be difficult to dissolve.

11. Dissolve in 100 μL TE buffer pH 8. This protocol yields both DNA and RNA. The quality and yield of the DNA can be checked by running 5 μL of the sample on 1 % agarose mini gel.

3.6.2 DNA Isolation for Gel Blot Analysis

To get a higher yield and a more pure quality of genomic DNA for gel blot analysis, isolate the DNA using the following protocol:

1. Inoculate a single colony of the transformed as well as untransformed fungi into 1 mL CM medium with appropriate antibiotic and incubate overnight. Transfer this 1 mL culture into 100 mL CM medium in 500 mL conical flasks and grow until late-log growth, at an approximate OD_{600} of 1.5–2.
2. Harvest the fungal cell culture in 50 mL Falcon tubes at 3,700 × *g* for 15 min at 4 °C. Remove the supernatant and freeze the pellet rapidly in liquid nitrogen.
3. Grind frozen cell pellets with prechilled mortars and pestles using liquid nitrogen and isolate the genomic DNA using the DNeasy Plant Maxi Kit (QIAGEN) following the manufacturer's instructions. Similar kits from other suppliers can also be used.

3.7 Verification of Fungal Transformation by Molecular Analysis

Standard molecular biological techniques such as PCR and genomic DNA blot analysis can be used to confirm proper transformation of fungal cells with T-DNA as expected from the BIBAC constructs.

1. To verify integration of intact T-DNA, perform PCR analysis for the presence of left border, right border, and selection marker.
2. For DNA blot analysis, use two probes one for the left border and one for the right border. For *Ustilago*, digest 8–10 μg of genomic DNA with appropriate restriction enzymes that cut the T-DNA not more than 2 kb from the border.
3. Run the digested DNA slowly on a 0.8 % (w/v) agarose gel in 1× Tris–borate–EDTA (TBE) buffer. Stain the gel with ethidium bromide, photograph with a ruler alongside, making sure size standards are matched up with the ruler. Treat the gel with a 0.25 M HCl solution by gently shaking for 15 min to reduce the sizes of the genomic DNA fragments and immediately transfer the gel to a 0.4 N NaOH solution and shake gently for 15 min. Transfer the DNA overnight from the gel to a Hybond-N^+ nylon membrane using a 0.4 N NaOH solution. Use a standard setup. After transfer, dismount the assembly and float the membrane carefully on a 6× SSC solution for 5 min, then air-dry on filter paper.
4. To detect specific fragments in the genome by hybridizations, label probes with a αP^{32}-dCTP, using the Rediprime II Random Prime Labeling system (Amersham Biosciences/GE healthcare) or a similar product from other suppliers. Nonradioactive methods can also be used for labeling the probes and several kits from different suppliers are available.

5. Hybridize labeled and denatured probes to the membrane using ULTRA-hyb buffer (Ambion) at 42 °C according to the manufacturer's instructions.
6. Wash the blot twice for 5 min each with 2× SSC plus 0.1 % w/v sodium dodecyl sulfate (SDS) solution, followed by two washes each of 15 min with 0.1× SSC plus 0.1 % SDS solution. Do all the washes at 42 °C.
7. Depending on signal strength, expose the blot to Kodak BioMax film (we used Kodak Canada, Toronto, ON, Canada) or other type film for 5–8 h. If the signals are weak, overnight exposure or the use of an intensifying screen and exposure at −80 °C might give better results. Detection using phosphor-imaging screens and laser detection can also be used.

4 Notes

1. Make a 200 mM acetosyringone solution in DMSO, filter-sterilize and dispense in aliquots of 500 μL in 1.5 mL microtubes and store in a −20 °C freezer until use.
2. Make the lysis buffer fresh each time. To make 10 mL of lysis buffer, mix together 6.8 mL of ddH_2O, 1 mL of a 5 M NaCl stock solution, 1 mL of a 2 M Tris–HCl (pH 7.5) solution, 0.2 mL of a 500 mM EDTA (pH 8.0) solution, and 1 mL of 10 % SDS solution (all autoclaved separately).
3. It is important to use midi or maxi kits designed for BAC plasmid preparation because of their big sizes and low copy numbers. The mini plasmid kits yield too little DNA and not of sufficient quality to be used for transformation in this protocol; larger-scale preparations normally work better.
4. It is important to use freshly prepared BAC DNA since large constructs deteriorate over time; older DNA preparations stored at 4 °C transform with less efficiency.
5. The extension time depends on type of polymerase used, however use 1 min per kb as general rule or follow the manufacturer's protocol provided with the polymerase.
6. Direct PCR purification should be used only when a clean single band of the expected size is amplified. Standard procedures can be used for the isolation of the PCR product from agarose gels. Different DNA-fragment isolation kits are commercially available that can be used according to the manufacturer's instructions. Depending on the method used, DNA loss can be substantial and several PCR reactions combined may be required.
7. *Agrobacterium tumefaciens* strains COR308 and COR309 were used for the transformation of BIBAC into *Ustilago* species.

Both COR308 and COR309 are recombination (rec) deficient strains and contain extra copies of two virulence genes: *virA* and *virG* on a helper plasmid pCH32. The rec deficient strains reduce instability of larger DNA fragments in *A. tumefaciens* and extra copies of these virulence genes, the two-component acetosyringone-sensing and virulence region inducers, increase its virulence and transformation efficiency into the host cells [10, 28].

8. Here the ratio of *Agrobacterium* to fungal cells is 10:1; however, to optimize the protocol for other fungi, different ratios of *Agrobacterium* to fungal cells, such as 1:1, 2:1, 5:1 and 20:1, can be tested.
9. Drying the membrane for different times, changes the transformation efficiency of different fungi [31]. To get the optimal transformation efficiency, different drying times, such as 10, 20, 30, 40, 50, and 60 min, can be tested.
10. Take the inoculum from a fresh agar plate not more than 7–10 days old. Inoculation from old agar plate stored for few weeks at 4 °C will fail to grow or will grow very slowly.
11. For *U. hordei*, it usually takes 2 days to reach the desired OD_{600} of 1.5–2 when growing at 22 °C, while *U. maydis* grows faster at 28 °C and may take only 24 h to reach the desired OD. Incubating *U. maydis* for more than 30 h will yield DNA that will be challenging to perform PCR with, or to cut with restriction enzymes.
12. Make sure glass beads do not stick near the rim of microfuge tube; failing to do so makes the tube not to close properly and leakage of phenol will happen during the vortexing step.

Acknowledgment

We thank Dr. Frank Takken, University of Amsterdam, for the plasmid pFT41 and pioneering work [16] and Dr. Neal Copeland, National Cancer Institute, Frederick, MD, for *E. coli* strain SW102 [4]. This work was supported by a Natural Sciences and Engineering Research Council of Canada grant to G. Bakkeren.

References

1. Shizuya H, Birren B, Kim UJ, Mancino V, Slepak T, Tachiiri Y, Simon M (1992) Cloning and stable maintenance of 300-kilobase-pair fragments of human DNA in *Escherichia coli* using an F-factor-based vector. Proc Natl Acad Sci U S A 89:8794–8797
2. Hosoda F, Nishimura S, Uchida H, Ohki M (1990) An F factor based cloning system for large DNA fragments. Nucleic Acids Res 18:3863–3869
3. Nagano Y, Takao S, Kudo T, Iizasa E, Anai T (2007) Yeast-based recombineering of DNA

fragments into plant transformation vectors by one-step transformation. Plant Cell Rep 26: 2111–2117

4. Lee EC, Yu D, Martinez de Velasco J, Tessarollo L, Swing DA, Court DL, Jenkins NA, Copeland NG (2001) A highly efficient *Escherichia coli*-based chromosome engineering system adapted for recombinogenic targeting and subcloning of BAC DNA. Genomics 73:56–65
5. Yu D, Ellis HM, Lee EC, Jenkins NA, Copeland NG, Court DL (2000) An efficient recombination system for chromosome engineering in *Escherichia coli*. Proc Natl Acad Sci U S A 97:5978–5983
6. Muniyappa K, Radding C (1986) The homologous recombination system of phage lambda. Pairing activities of beta protein. J Biol Chem 261:7472–7478
7. Copeland NG, Jenkins NA, Court DL (2001) Recombineering: a powerful new tool for mouse functional genomics. Nat Rev Genet 2:769–779
8. Bundock P, den Dulk-Ras A, Beijersbergen A, Hooykaas PJ (1995) Trans-kingdom T-DNA transfer from *Agrobacterium tumefaciens* to *Saccharomyces cerevisiae*. EMBO J 14: 3206–3214
9. Ali S, Bakkeren G (2011) Introduction of large DNA inserts into the barley pathogenic fungus. Ustilago hordei, via recombined binary BAC vectors and Agrobacterium-mediated transformation. Curr Genet 57:63–73
10. Hamilton CM, Frary A, Lewis C, Tanksley SD (1996) Stable transfer of intact high molecular weight DNA into plant chromosomes. Proc Natl Acad Sci U S A 93:9975–9979
11. Lee L-Y, Gelvin SB (2008) T-DNA binary vectors and systems. Plant Physiol 146:325–332
12. Warming S, Costantino N, Court DL, Jenkins NA, Copeland NG (2005) Simple and highly efficient BAC recombineering using galK selection. Nucleic Acids Res 33:e36
13. Li Z, Karakousis G, Chiu S, Reddy G, Radding C (1998) The beta protein of phage λ promotes strand exchange. J Mol Biol 276: 733–744
14. Kulkarni SK, Stahl FW (1989) Interaction between the *sbcC* gene of *Escherichia coli* and the *gam* gene of phage lambda. Genetics 123:249–253
15. Murphy KC (2007) The λ Gam protein inhibits RecBCD binding to dsDNA ends. J Mol Biol 371:19–24
16. Takken FL, Van Wijk R, Michielse CB, Houterman PM, Ram AF, Cornelissen BJ (2004) A one-step method to convert vectors into binary vectors suited for *Agrobacterium*-mediated transformation. Curr Genet 45: 242–248
17. Linning R, Lin D, Lee N, Abdennadher M, Gaudet D, Thomas P, Mills D, Kronstad JW, Bakkeren G (2004) Marker-based cloning of the region containing the *UhAvr1* avirulence gene from the basidiomycete barley pathogen *Ustilago hordei*. Genetics 166:99–111
18. Wang J, Holden DW, Leong SA (1988) Gene transfer system for the phytopathogenic fungus *Ustilago maydis*. Proc Natl Acad Sci U S A 85:865–869
19. Chen X, Stone M, Schlagnhaufer C, Romaine CP (2000) A fruiting body tissue method for efficient *Agrobacterium*-mediated transformation of *Agaricus bisporus*. Appl Environ Microbiol 66:4510–4513
20. Mikosch TSP, Lavrijssen B, Sonnenberg ASM, van Griensven LJLD (2001) Transformation of the cultivated mushroom *Agaricus bisporus* (Lange) using T-DNA from *Agrobacterium tumefaciens*. Curr Genet 39:35–39
21. Meyer V, Mueller D, Strowig T, Stahl U (2003) Comparison of different transformation methods for *Aspergillus giganteus*. Curr Genet 43:371–377
22. Degefu Y, Hanif M (2003) *Agrobacterium-tumefaciens*-mediated transformation of *Helminthosporium turcicum*, the maize leaf-blight fungus. Arch Microbiol 180:279–284
23. Michielse CB, Hooykaas PJJ, van den Hondel CAMJJ, Ram AFJ (2005) *Agrobacterium* -mediated transformation as a tool for functional genomics in fungi. Curr Genet 48:1–17
24. Mullins ED, Chen X, Romaine P, Raina R, Geiser DM, Kang S (2001) *Agrobacterium*-mediated transformation of *Fusarium oxysporum*: an efficient tool for insertional mutagenesis and gene transfer. Phytopathology 91: 173–180
25. Combier JP, Melayah D, Raffier C, Gay G, Marmeisse R (2003) *Agrobacterium tumefaciens*-mediated transformation as a tool for insertional mutagenesis in the symbiotic ectomycorrhizal fungus *Hebeloma cylindrosporum*. FEMS Microbiol Lett 220:141–148
26. Takahara H, Tsuji G, Kubo Y, Yamamoto M, Toyoda K, Inagaki Y, Ichinose Y, Shiraishi T (2004) *Agrobacterium tumefaciens*-mediated transformation as a tool for random mutagenesis of *Colletotrichum trifolii*. J Gen Plant Pathol 70:93–96
27. Frary A, Hamilton CM (2001) Efficiency and stability of high molecular weight DNA transformation: an analysis in tomato. Transgenic Res 10:121–132

28. Kronstad JW, Leong SA (1989) Isolation of two alleles of the *b* locus of *Ustilago maydis*. Proc Natl Acad Sci U S A 86:978–982
29. Holliday R (1974) Ustilago maydis. In: King RC (ed) Handbook of genetics. Plenum, New York, pp 575–595
30. Elder RT, Loh EY, Davis RW (1983) RNA from the yeast transposable element Ty1 has both ends in the direct repeats, a structure similar to retrovirus RNA. Proc Natl Acad Sci U S A 80:2432–2436
31. Almeida AJ, Carmona JA, Cunha C, Carvalho A, Rappleye CA, Goldman WE, Hooykaas PJ, Leao C, Ludovico P, Rodrigues F (2007) Towards a molecular genetic system for the pathogenic fungus *Paracoccidioides brasiliensis*. Fungal Genet Biol 44:1387–1398

Chapter 10

Infectious Delivery of Alphaherpesvirus Bacterial Artificial Chromosomes

Kurt Tobler and Cornel Fraefel

Abstract

Bacterial artificial chromosomes (BACs) can accommodate and stably propagate the genomes of large DNA viruses in *E. coli*. As DNA virus genomes are often per se infectious upon transfection into mammalian cells, their cloning in BACs and easy modification by homologous recombination in bacteria has become an important strategy to investigate the functions of individual virus genes. This chapter describes a strategy to clone the genomes of viruses of the *Alphaherpesvirinae* subfamily within the family of the *Herpesviridae*, which is a group of large DNA viruses that can establish both lytic and latent infections in most animal species including humans. The cloning strategy includes the following steps: (1) Construction of a transfer plasmid that contains the BAC backbone with selection and screening markers, and targeting sequences which support homologous recombination between the transfer plasmid and the alphaherpesvirus genome. (2) Introduction of the transfer plasmid sequences into the alphaherpesvirus genome via homologous recombination in mammalian cells. (3) Isolation of recombinant virus genomes containing the BAC backbone sequences from infected mammalian cells and electroporation into *E. coli*. (4) Preparation of infectious BAC DNA from bacterial cultures and transfection into mammalian cells. (5) Isolation and characterization of progeny virus.

Key words Alphaherpesvirus, *Cre-lox*P, GFP, BoHV-1, BoHV-5, FeHV-1, BAC cloning

1 Introduction

The cloning of the genomes of large DNA viruses in bacterial artificial chromosomes (BAC) has become an important strategy to investigate the roles of individual virus genes and the molecular mechanisms of virus–cell interactions. Specifically, it supports the efficient and precise manipulation of the virus genome by homologous recombination in *E. coli* and subsequent analysis of the mutant virus in mammalian cells [1]. BACs have proven particularly suitable for accommodating the 120–230 kbp genomes of the *Herpesviridae*, a family of large DNA viruses that establish lytic and latent infections in almost all animal species, including humans [2–4]. On the basis of their biological properties, the *Herpesviridae* can be subdivided into three subfamilies, *Alpha*-, *Beta*-, and

Kumaran Narayanan (ed.), *Bacterial Artificial Chromosomes*, Methods in Molecular Biology, vol. 1227,
DOI 10.1007/978-1-4939-1652-8_10, © Springer Science+Business Media New York 2015

Gammaherpesvirinae [5, 6]. The *Alphaherpesvirinae* are characterized by a variable host range and a rapid replication cycle; these viruses establish latency primarily but not exclusively in neurons [7]. Three of the eight human herpesviruses identified to date, including the herpes simplex viruses type 1 and type 2 (HSV-1, HSV-2) and varicella zoster virus (VZV), belong to the *Alphaherpesvirinae.* The *Betaherpesvirinae* include three members of the human herpesviruses, human cytomegalovirus (HCMV) and the human herpesviruses 6 and 7 (HHV-6, HHV-7). The *Gammaherpesvirinae* comprise two members of the human herpesviruses, Epstein–Barr virus (EBV) and human herpesvirus 8 (HHV-8). Herpesviruses are also important pathogens in veterinary medicine, in particular bovine (BoHV-1, BoHV-2, BoHV-5), felid (FeHV-1), equid (EHV-1), and suid (SuHV-1) herpesviruses can cause substantial economic losses [8, 9].

To identify the roles of specific virus genes, herpesvirus genomes have conventionally been mutated by homologous recombination in mammalian cells, a procedure that is time-consuming and can generate undetected second-site mutations. Moreover, genes that are essential for virus replication in cell culture can be mutated only if cells are available that complement the essential virus gene. Because of their large size, herpesvirus genomes cannot be cloned into standard multi-copy bacterial plasmids. By contrast, bacterial artificial chromosomes (BACs) can accommodate at least 300 kbp of DNA and are genetically stable because they are present as a single copy in bacteria, thereby reducing the risk of recombination events that could lead to rearrangement of the viral genome. The first herpesvirus genome cloned as a BAC was the 230 kbp mouse cytomegalovirus DNA [10]. Since then, all of the human herpesviruses except for HHV-7, and many of the animal herpesviruses have been cloned using a multitude of different strategies [11–19].

The protocols presented in this chapter were written to summarize the experience and expertise gained by cloning of BoHV-1, BoHV-5 and FeHV-1 as BACs [20, 21]. The procedures may be applicable on alphaherpesviruses in general. The strategy is based on the homologous recombination of co-transfected genomic and plasmid DNA. Such plasmid DNA, termed as transfer plasmid, combines several features (*see* Fig. 1a), which are required or convenient for the BAC cloning. They carry (1) a high copy origin of replication for efficient propagation and preparation, (2) a low copy origin of replication for the propagation of large DNA as BAC in *E. coli*, (3) a chloramphenicol resistance gene for the selection of BAC containing *E. coli*, (4) herpesvirus derived homologous sequences, which are flanked by *lox*P sites for removal of the plasmid backbone if required, and (5) a GFP expression cassette for screening of recombinant viruses in mammalian cell culture.

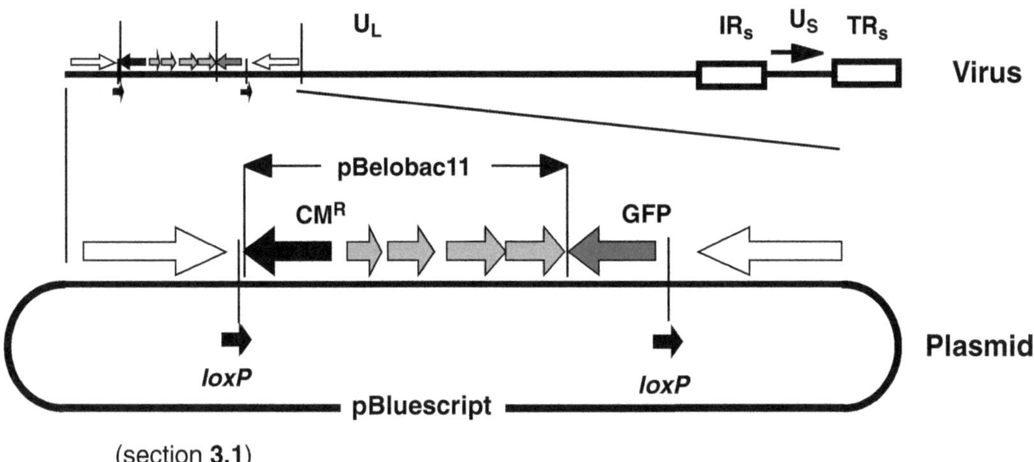

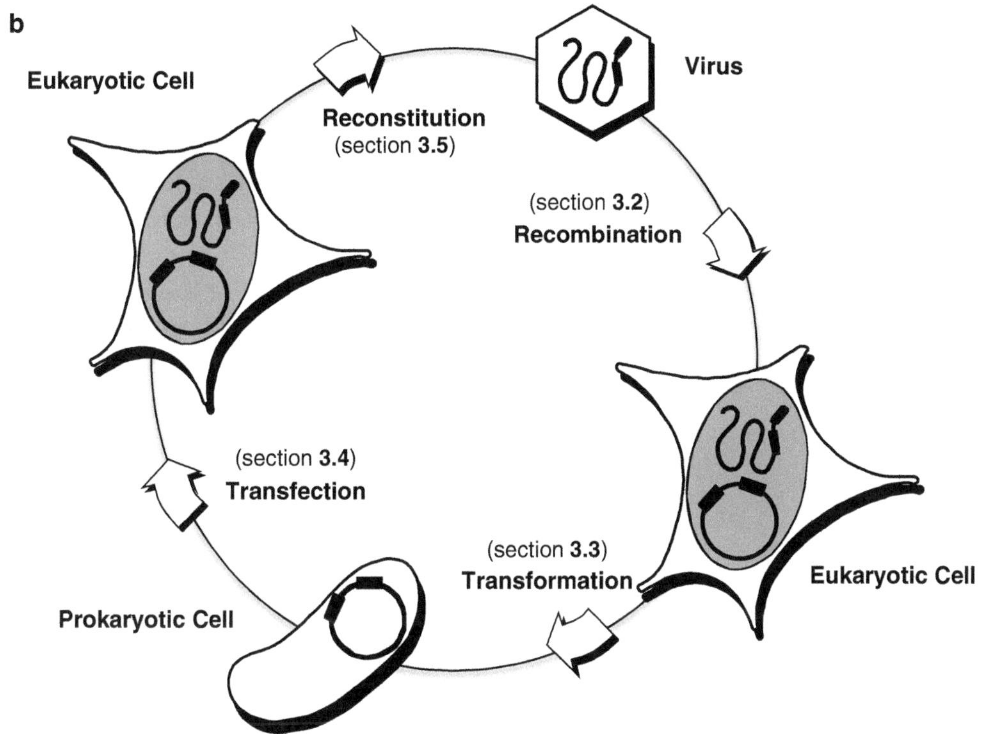

Fig. 1 (**a**) Schematic representation (not to scale) of recombinant herpesvirus DNA (*upper panel*) and transfer plasmid DNA (*lower panel*) referring to the Subheading 3.1 (*see* **Note 22**). The BoHV-1 genome segments unique long (U_L), unique short (U_S), inverted repeat short (IR_S), and terminal repeat (TR_S) are shown. The altered part of the herpesvirus genome—virtually the inserted transfer plasmid (enlarged)—carries the genetic elements required for DNA replication (pBeloBac11) and selection (CM^R) of plasmid DNA in prokaryotic cells and visual screening of recombinant virus progeny (GFP). The herpesvirus-derived homologous sequences are shown as white arrows. The transfer plasmid backbone derived from pBluescript is shown. This part of the transfer plasmid will be lost upon homologous recombination with viral DNA. Likewise, the genetic elements required for propagation as BAC DNA in prokaryotic cells, which are flanked by *lox*P sites, might be removed in eukaryotic cells by *Cre*-mediated recombination, leaving a virus DNA which differs from the parent virus only in one single *lox*P site. (**b**). Schematic presentation of the steps described in this chapter referring to the Subheadings 3.2–3.5

The cloning procedure of the herpesvirus genome may be divided into five consecutive steps (*see* Fig. 1). (1) Preparation of the transfer plasmid and virus DNA (*see* Subheading 3.1). (2) The generation and selection of recombinant viruses carrying the genetic elements required for DNA propagation in bacteria by recombination (*see* Subheading 3.2). (3) Preparation of infectious herpesvirus BAC DNA by transformation (*see* Subheading 3.3). (4) Recovery and propagation of progeny virus by transfection (*see* Subheading 3.4). (5) Characterization of progeny virus upon reconstitution (*see* Subheading 3.5).

2 Materials

2.1 Materials for Working with Prokaryotes

2.1.1 Bacterial Strain and Plasmids

1. *E. coli* electro-competent cells (e.g. ElectroMAX™ DH10B™, Invitrogen).
2. pBeloBACModified (Genbank #AY665170) (*see* **Note 1**).
3. High copy number plasmid (e.g. pBluescript® II; Stratagene).

2.1.2 Buffers and Solutions

1. Luria-Bertani (LB) medium: dissolve 10 g NaCl, 10 g bacto tryptone and 5 g bacto-yeast extract in 1 L water, autoclave for 20 min at 121 °C.
2. LB Agar plates: add 15 g bacto agar to LB medium prior to autoclaving, let cool to about 50 °C, add appropriate antibiotic, pour plates and let them solidify.
3. Chloramphenicol: 12.5 mg/mL (1,000× stock) dissolved in 75 % ethanol, working concentration of 12.5 μg/mL (stored at –20 °C).
4. Resuspension buffer P1 (supplied in QIAGEN DNA Maxi Kit): 50 mM Tris–HCl (pH 8.0), 10 mM EDTA, 100 μg/mL RNase A [6.06 g Tris base, 3.72 g $Na_2EDTA{\cdot}H_2O$, adjust pH with HCl to 8.0].
5. Lysis buffer P2 (supplied in QIAGEN DNA Maxi Kit): 200 mM NaOH, 1 % SDS (w/v) [950 mL water and 8.0 g NaOH, 50 mL 20 % SDS (w/v)].
6. Neutralization buffer P3 (supplied in QIAGEN DNA Maxi Kit): 3.0 M potassium acetate, pH 5.0 [294.5 g potassium acetate in 500 mL water, adjust pH with glacial acetic acid (~110 mL) and adjust volume to 1 L with water].
7. 100 % and 70 % (v/v) ethanol.
8. 100 % isopropanol.
9. TE: 10 mM Tris–HCl, pH 8.0, 0.1 mM EDTA.

2.1.3 Equipment and Commercial Kits

1. An incubator set at 37 °C.
2. An incubator set at 37 °C and shaking at 220 rpm.
3. Electroporator (e.g., BIORAD Gene Pulser).
4. Electroporation Cuvettes 0.1 cm (BIORAD).
5. High speed centrifuge equipped with rotor and tubes (e.g., Sorvall RC6+, GSA rotor and tubes or equivalent).
6. DNA extraction kit (e.g., NucleoBond® Xtra Maxi, Macherey & Nagel or large construct kit, Qiagen).

2.2 Materials for Working with Eukaryotes

2.2.1 Cell Lines and Culture Media

1. Madin-Darby Bovine Kidney (MDBK) cells (ATCC).
2. DMEM (Dulbecco's Modified Eagle's Medium; Invitrogen).
3. DMEM supplemented with 1 % agarose (*see* **Note 2**).
4. OptiMEM (Invitrogen).

2.2.2 Buffers and Solutions

1. Phosphate buffered saline (PBS).
2. TNE: 50 mM Tris–HCl (pH 7.4), 100 mM NaCl, 0.1 mM EDTA.
3. 10 % sodium dodecyl sulfate (SDS) in TE.
4. 35 % sucrose (w/v) in TNE: dissolve 350 g/L sucrose in TNE and autoclave.
5. 3 M sodium acetate, pH 5.2.
6. 70 % (v/v) ethanol.

2.2.3 Equipment and Commercial Products

1. Ultraspeed centrifuge equipped with rotor and tubes (e.g., Sorvall WX Ultra 80 with AH629 rotor or equivalent).
2. Cell culture incubator set at 37 °C and 5 % CO_2 atmosphere.
3. Phenol/chloroform/isoamylalcohol (25:24:1) (e.g., Invitrogen).
4. Proteinase K (20 g/L, >600 mAU/mL (e.g., QIAGEN).
5. Agarose cell culture quality (e.g., Fermentas; TopVision LMGQ).
6. Lipofectamine™ 2000 (Invitrogen).
7. Cell scrapers (e.g., Corning or equivalent).
8. Liquid N_2.

3 Methods

3.1 Preparation of Transfer Plasmid and Virus DNA

The BAC cloning strategy described here is based on homologous recombination between transfected transfer plasmid DNA and viral genomic DNA in mammalian cells. The preparation of the transfer plasmid DNA is performed by simple plasmid isolation protocol for high copy plasmids (since the transfer plasmid contains a bacterial high copy origin of DNA replication). The viral genomic DNA is prepared from purified virions obtained from infected cell cultures.

3.1.1 Cloning and Preparation of the Transfer Plasmid

To construct the transfer plasmid, the sequences required for homologous recombination with the herpesvirus DNA of choice are introduced between the *lox*P sites of pBeloBacModified. A location on the viral genome is chosen that does not influence virus viability (*see* **Note 3**). It is convenient to amplify the homology arms as two separate amplimers by PCR of about 500 bp each and combine these in a high copy number plasmid as pBluescript® II KS+. The large fragment (the *Nsi*I fragment of pBeloBacModified) is ligated into the *Nsi*I or *Pst*I site of the high copy number plasmid containing the homologous arms resulting in the transfer plasmid ready for homologous recombination. A schematic presentation of the transfer plasmid is provided in Fig. 1a.

3.1.2 Extraction of Viral DNA

To prepare genomic viral DNA, use 150-cm² tissue culture flasks (*see* **Note 4**). The expected yield is 50–100 μg of viral DNA per 150-cm² tissue-culture flask. This is sufficient for several transfection experiments as well as for analyses by means of PCR and/or restriction enzymes.

1. Seed MDBK cells (or permissive cells for other alphaherpesvirus) into 150-cm² tissue-culture flasks and incubate overnight.
2. Wash the cell monolayer once with 20 mL PBS.
3. Infect cells with BoHV-1 (or other alphaherpesvirus) at a MOI of ~0.05 in 10 mL cell culture medium (0 % FCS), leave the inoculum for 2 h on the cells while incubating at 37 °C in 5 % CO_2 atmosphere.
4. Replace the inoculum with 20 mL DMEM supplemented with 2 % FCS.
5. Incubate the infected cells until the cultures show 90–100 % cytopathic effect.
6. Scrape infected cells into culture medium supernatant and transfer cell suspension of one 150-cm² tissue-culture flask to a fresh 50-mL conical tube.
7. Pellet the cell debris by centrifugation for 10 min at 4 °C and 2,000 × *g*.
8. Pour the supernatant into a fresh 50-mL conical tube leaving ~0.5 mL of liquid and the cell pellet.
9. Freeze the pellet in liquid N_2 and thaw it at RT; repeat this step twice.
10. Combine the supernatant from the low-speed centrifugation with the freeze/thawed pellet, mix and centrifuge for 10 min at 4 °C and 2,000 × *g*.

11. Transfer the cleared supernatant (~20 mL) into an ultracentrifuge tube and submerse the virus containing cell culture medium with 15 mL 35 % sucrose in TNE (*see* **Note 5**).
12. Centrifuge at 28,000 × *g* and 4 °C for 2 h.
13. Discard the supernatant, resuspend the pellet containing the purified virions in 348 μL TNE buffer, and transfer into an 1.5-mL Eppendorf tube.
14. Add 12 μL of Proteinase K solution (20 g/L; >600 AU/mL) to a final concentration of 0.6 g/L and 40 μL of a 10 % SDS solution to a final concentration of 1 %; incubate 6 h or overnight at 56 °C.
15. Extract the viral DNA with 400 μL phenol/chloroform/isoamylalcohol (25:24:1) and subsequently with 400 μL chloroform.
16. Add 40 μL of 3 M sodium acetate (pH 5.5) and 1 mL of ice-cold 100 % ethanol; precipitate DNA by centrifugation for 10 min at 15,000 × *g* and 4 °C.
17. Remove supernatant, wash the pellet with 500 μL ice-cold 70 % ethanol, air dry and resuspend in 100 μL of sterile H_2O (*see* **Note 6**).
18. Determine DNA concentration by UV spectroscopy at 260 nm wavelength.

3.2 Generation and Selection of Recombinant Viruses

3.2.1 Transfection

1. Seed cells at 50 % confluency in 6-cm tissue culture dishes and incubate overnight in a 37 °C and 5 % CO_2 atmosphere.
2. Prepare 250 μL OptiMEM and 10 μL Lipofectamine™ 2000 in one Eppendorf tube and 250 μL OptiMEM, 3 μg of viral DNA, and 0.5 μg of transfer plasmid in a second Eppendorf tube. Mix solutions by flicking (*see* **Note 7**).
3. Combine the contents of the two tubes, mix by flicking, and incubate for 30 min at RT.
4. Wash cells once with OptiMEM and add the DNA and lipid mix dropwise onto the cells. Add 0.8 mL OptiMEM to cover the cells with medium.
5. Incubate 6 h or overnight at 37 °C and 5 % CO_2.
6. Wash transfected cells three times with OptiMEM and add 3 mL DMEM supplemented with 2 % FCS. Incubate at 37 °C and 5 % CO_2 until plaques become visible (*see* **Note 8**).

3.2.2 Plaque Purification

Infection for the first round of plaque purification is done in six-well tissue culture plates with a serial dilution of transfection supernatant. Subsequent plaque purifications are performed in 10-cm plates, which are infected with the progeny of one picked plaque (*see* **Note 9**).

1. Seed MDBK cells to 80 % confluency into a six-well tissue culture plate and incubate overnight at 37 °C and 5 % CO_2 atmosphere.
2. Wash the cells once with PBS and infect with serial dilutions of supernatant from 10^0 to 10^{-6} for 2 h.
3. Remove inoculum and replace with 2 mL DMEM supplemented with 2 % FCS and 1 % agarose. Let overlay solidify for 15 min and incubate at 37 °C and 5 % CO_2 atmosphere (*see* **Note 10**).
4. Observe plates until plaques become visible (*see* **Note 11**).
5. Choose an infected well showing sufficiently separated plaques. Plaques are picked using a Pasteur pipette and transferred into cell culture medium for further rounds of plaque purification. Screening based on expression of EGFP is sufficient for screening of recombinant viruses (*see* **Note 12**).
6. Prepare five sterile 1.5-mL Eppendorf tubes containing 300 μL of DMEM.
7. Mark the plaques at the bottom of the plates by drawing circles around them using a lab marker.
8. Observe the plaques in the fluorescence microscope and examine if they contain GFP-expressing cells indicating recombinant viruses.
9. Pick the plaques under the laminar flow using a Pasteur pipette and transfer into 300 μL of cell culture medium (0 % FCS) in an Eppendorf tube (*see* **Note 13**).
10. Incubate the picked plaque overnight at 4 °C to release the viruses into the medium.
11. Seed cells to 80 % confluency into a 10-cm tissue culture plate and incubate overnight at 37 °C and 5 % CO_2 (*see* **Note 14**).
12. Wash cells once with PBS and infect with 150 μL of picked plaque suspension diluted in 5 mL DMEM for 2 h.
13. Remove inoculum and replace with 10 mL DMEM supplemented with 2 % FCS and 1 % agarose. Let agarose solidify for 15 min and then transfer the plates to 37 °C and 5 % CO_2 atmosphere.
14. Observe plates daily until plaques become visible.
15. Pick plaques as described above and repeat **steps 11–15** twice (*see* **Note 15**).

3.2.3 Propagation of Progeny Virus

1. Seed MDBK cells (or permissive cells for other alphaherpesviruses) into 10-cm tissue-culture dishes and incubate overnight.
2. Wash cell monolayer once with 10 mL PBS.

3. Infect cells with 150 μL of plaque-purified recombinant BoHV-1 (or other alphaherpesvirus) in 5 mL cell culture medium (0 % FCS), leave the inoculum for 2 h on the cells while incubating at 37 °C in 5 % CO_2 atmosphere.
4. Replace the inoculum with 10 mL DMEM supplemented with 2 % FCS.
5. Incubate the infected cells until the cultures show 90–100 % cytopathic effect.

3.3 Preparation of Infectious Herpesvirus BAC DNA

Recombinant virus DNA is isolated from infected eukaryotic cells and transferred to prokaryotic cells by means of transformation (*see* **Note 16**).

3.3.1 Transfer of BAC DNA from Eukaryotes to Prokaryotes

1. Seed cells to 80 % confluency into 6-cm tissue culture plates and incubate overnight.
2. Infect cells with BoHV-1 at a MOI of ~1 in 1.5 mL cell culture medium (0 % FCS), leave the inoculum for 2 h on the cells while incubating at 37 °C in a 5 % CO_2 atmosphere (*see* **Note 17**).
3. Wash cells once with 3 mL PBS and then with 3 mL TE.
4. Add 400 μL of TE and scrape cells from plates. Transfer to Eppendorf tube.
5. Add 5 μL of Proteinase K solution (20 g/L; >600 AU/mL) to a final concentration of 0.25 g/L and 24 μL 10 % SDS to a final concentration of 0.6 %.
6. Incubate for 30 min at RT.
7. Use 2 μL of DNA for electroporation into commercially available *E. coli* strain DH10B. Set the electroporation conditions to 1.8 kV, 200 Ω, 25 μF (*see* **Note 18**).
8. Immediately after administrating the electric pulse, transfer the bacteria in 500 μL SOC medium into a Falcon tube and incubate on a shaker for 1 h at 37 °C and 180 rpm.
9. Pellet bacteria for 10 min at 4 °C and 1,000 × *g*. Discard supernatant and plate the resuspended bacterial pellet using Drigalski spatula onto LB agar plates containing 12.5 μg/mL chloramphenicol.
10. Incubate overnight at 37 °C. Colonies resulting from electroporation are screened for the presence of herpesvirus BAC DNA. For this DNA is extracted using the Miniprep protocol as follows.
11. Inoculate 6 mL of LB medium containing 12.5 μg/mL chloramphenicol with a single colony and grow for 12–16 h at 37 °C with shaking at 220 rpm.
12. Place 700 μL aliquots of bacterial culture in cryogenic storage vials and add 300 μL of 50 % glycerol resulting in a final

concentration of 15 % glycerol. Mix well and freeze at −80 °C for long-term storage (up to several years).

13. Centrifuge the remaining overnight cultures for 5 min at 2,000 × *g* and 4 °C.
14. Decant medium and invert each tube on a paper towel in order to remove remaining LB-medium.
15. Place pellets into 250 μL resuspension buffer P1 by pipetting up and down and transfer to fresh Eppendorf tube.
16. Add 250 μL of lysis buffer P2 to each tube and mix gently by inverting the tubes four to six times. Incubate for 5 min at room temperature.
17. Add 250 μL of neutralization buffer P3 and mix immediately by inverting the tubes six times. Incubate for 10 min on ice.
18. Centrifuge for 15 min at 13,000 × *g* and 4 °C.
19. Transfer supernatant to a fresh Eppendorf tube and add 750 μL isopropanol (−20 °C).
20. Centrifuge for 10 min at 13,000 × *g* and 4 °C. Discard supernatant.
21. Wash pellet once with ice cold 70 % ethanol.
22. Dry DNA pellet and then resuspend in 30 μL H_2O. The miniprep DNA can be used for restriction enzyme analysis.

3.3.2 Bulk Prep of BAC DNA

1. Inoculate 6 mL of LB medium containing 12.5 μg/mL chloramphenicol with a single bacterial colony or with 50 μL of saved overnight culture (*see* Subheading 3.3.1, **step 12**) harboring the alphaherpesvirus-BAC DNA.
2. Grow for 8 h at 37 °C with shaking at 220 rpm.
3. The next day, inoculate 500 mL of LB medium containing 12.5 μg/mL chloramphenicol with 1 mL of the preculture.
4. Grow for 12–16 h at 37 °C with shaking at 220 rpm.
5. Proceed according to the kit manufacturer's instructions.
6. Determine the DNA concentration by UV spectrometry at 260 nm.

3.4 Characterization of alphaherpesvirus BAC DNA

3.4.1 Transfection and Recovery of Infectivity

The final proof for successful cloning of herpesvirus DNA as BAC is the test of infectivity of the cloned DNA. The transfection protocol described in Subheading 3.2.1 can be used also for the reconstitution of infectious virus from herpesvirus BAC DNA and the protocol described in Subheading 3.2.3 for the propagation of BACs-derived recombinant herpesvirus. Since all appearing plaques are derived from genetically identical herpesvirus BAC, there is no need for plaque purification. The progeny of the transfection can directly be used for infection and propagation of recombinant herpesvirus (*see* **Note 19**). However, when a plasmid encoding for *Cre*-recombinase is cotransfected with the alphaherpesvirus BAC DNA, plaque purification with screening for non-GFP expressing virus might be useful.

3.5 Characterization of Reconstituted Virus

Further assays to access the properties of the BAC cloned alphaherpesvirus genome are advised. In particular, the infectivity and genomic integrity may be important. The infectivity and properties of recombinant virus in cell culture can be characterized by single-step growth curves (*see* **Note 20**). The integrity of the viral DNA is demonstrated by the restriction enzyme analysis (*see* **Note 21**) and compared with DNA isolated from bacteria.

4 Notes

1. The cloning of pBeloBACModified was as follows: The EGFP coding sequence under the control of the CMV immediate early promoter was cloned between the *Hin*dIII and *Bam*HI sites of pBS246 (Invitrogen) resulting in p111. The 1.7-kb fragment of p111 excised with *Eco*RI and *Mlu*I was blunted and ligated into the blunted *Sal*I site of the cloned *Hin*dIII N-fragment of BoHV1 resulting in p112. The cos and *lox*P sites of pBeloBAC11 were deleted by collapsing the *Sac*I sites resulting in p114. The *Hin*dIII linearized p114 was cloned into the *Hin*dIII site of p112 resulting in p115. Two *Nsi*I sites outside the *lox*P sites were cloned into p115 resulting in pBeloBACModified. These *Nsi*I sites were used for exchange of BoHV-1 sequences with sequences from BoHV-5 or FeHV-1 in order to generate the BoHV-1 or FeHV-1 BACs, respectively and might be used for generation of transfer plasmids for other alphaherpesvirus BACs.
2. Prepare 6 mL 2 % agarose in H_2O per plate, autoclave for 20 min at 121 °C and keep it at 42 °C in a water bath. Mix 6 mL of 2× DMEM supplemented with 4 % FCS and prewarm to 42 °C. Mix the medium and agarose solution and keep at 42 °C until use.
3. Protein coding sequences should be avoided as site for insertion of heterologous DNA sequences. Furthermore, be aware of promoter sequences and signals for polyadenylation, as well as DNA motifs important for packaging and replication.
4. The volume of virus-containing cell culture supernatant is optimized for downstream processing: The culture supernatant from one 150-cm^2 tissue-culture flask fits onto one sucrose cushion in one SW28 tube. Subsequently, purified virions can

be processed in one Eppendorf tube for phenol/chloroform extraction and ethanol precipitation. The protocol may be scaled up to several flasks.

5. Draw 20 mL of sucrose solution into a syringe and mount a 70 mm long 20G needle on the syringe. Carefully move with the tip of the needle through the virus solution to the bottom of the ultracentrifuge tube and submerge the virus solution with sucrose. After releasing 13 mL of sucrose solution, slowly take the needle out of the tube.
6. Sterile DNA is required for transfection into eukaryotic cells. Washing with ethanol (70 % v/v), drying and resuspension of DNA should be performed under aseptic conditions in a laminar flow. The DNA is stable for several weeks at 4 °C. Neither freeze nor vortex the viral DNA, because of its large size it is fragile and can break easily.
7. MDBK cells are known to be very difficult to transfect. Several lipids for transfection of MDBK cells were evaluated. Lipofectamine™ 2000 gave the best results. Alternatively to lipofection, Ca-phosphate precipitation in BES buffer may be used for introduction of DNA into MDBK cells [20].
8. Successful cotransfection should yield several hundred plaques. Only with such a high number of plaques, the chance of obtaining the desired homologous recombination is given.
9. The yield of progeny virus upon transfection might vary significantly as it depends on factors such as transfection and recombination efficiency. It is convenient to start picking three to five plaques from passage 0 and then one or two plaques during the following rounds of purification. As such, ending up with approximately ten recombinant viruses is reasonable to attempt further steps of BAC cloning.
10. The agarose-containing overlay is prepared as follows: The freshly autoclaved liquid agarose solution (2 % in H_2O) is cooled in a water bath to 42 °C and then combined with 42 °C preheated 2× DMEM supplemented with 4 % FCS. The overlay medium is prepared as 12 mL aliquots per plate and kept liquid in a 42 °C water bath until applied to the tissue culture plates.
11. Plaques might be visible as opaque dots when observed from the bottom side of the plate. Circles can be drawn around the plaques and checked under the fluorescence microscope for the presence of a green fluorescent signal.
12. Homologous recombination was applied for years in order to generate recombinant herpesviruses [22]. However, most protocols involve an active selection for the presence or absence of the viral thymidine-kinase (TK) gene. When recombinant viruses are generated, which are expected to have a fitness

defect as compared to the wild-type (wt) virus, an active selection might be mandatory. However, the introduction of the BAC backbone does not seem to impair the fitness of the viruses and therefore, passive GFP screening is sufficient for the identification of recombinant viruses. Even the increased size of the viral DNA due to the introduction of heterologous sequences does not impair packaging of the genome.

13. The Pasteur pipette is equipped with a rubber pipette bulb, which is softly squeezed before piercing into the agarose overlay at the site of the plaques. By releasing the bulb, a small plug of agarose containing virus is sucked into the pipette, which then can be released into the prepared Eppendorf tube by pipetting the medium up and down.
14. To recognize plaques it is crucial to get a perfect monolayer of the cells. Inhomogeneous distribution of the cultured cells may appear as holes which are difficult to distinguish from viral plaques.
15. Plaque purification should be repeated three times in order to obtain pure recombinant progeny virus.
16. Herpesvirus DNA can be transferred into replication competent BAC DNA because of the fact that early in infection (2–3 hpi) the herpesvirus genome becomes circular. The circular form of DNA is required for the successful propagation in *E. coli*. Infection with high MOI yields enough circular DNA for transformation into bacteria.
17. The multiplicity of infection (MOI) is calculated by dividing the number of plaque forming units of virus (virus titer) by the number of cells in the culture vessel. The titer of the virus can be determined by infection with a serial dilution of the virus suspension under agarose overlay (*see* Subheading 3.2.2), counting of plaques and calculating the number of plaque forming units of virus per mL of virus suspension.
18. The time constant (τ) should be above 4.0 ms as a measure for the success of the electroporation.
19. In this stage, the herpesvirus BAC DNA can be altered by means of recombinant DNA technologies and recombinant herpesviruses can then be generated by transfection.
20. One-step growth curves are prepared by infection of permissive cells, progeny virus is harvested at different time points and titrated [20].
21. Restriction enzyme analysis (REA) might be useful for screening of BAC DNA minipreps or for confirming the integrity of BAC DNA bulk preps. As a rule of thumb 15 μL of miniprep DNA or 3 μg of bulk prep DNA or viral DNA, respectively, should be used for REA. The digested DNA separated in agarose gels might be further analyzed by Southern blot.

22. The transfer plasmid carries two compatible prokaryotic origins of replication. The high copy is derived from pBluescript® II (Genbank #X52327) and the low copy is derived from pBeloBAC11 (Genbank #U51113).

References

1. Domi A, Moss B (2002) Cloning the vaccinia virus genome as a bacterial artificial chromosome in Escherichia coli and recovery of infectious virus in mammalian cells. Proc Natl Acad Sci U S A 99:12415–12420
2. Adler H, Messerle M, Koszinowski UH (2003) Cloning of herpesviral genomes as bacterial artificial chromosomes. Rev Med Virol 13:111–121
3. Brune W, Messerle M, Koszinowski UH (2000) Forward with BACs – new tools for herpesvirus genomics. Trends Genet 16:254–259
4. Hall RN, Meers J, Fowler E, Mahony T (2012) Back to BAC: the use of infectious clone technologies for viral mutagenesis. Viruses 4:211–235
5. Davison AJ (2010) Herpesvirus systematics. Vet Microbiol 143:52–69
6. Davison AJ, Eberle R, Ehlers B, Hayward GS et al (2008) The order herpesvirales. Arch Virol 154:171–177
7. Curanovic D, Enquist LW (2009) Directional transneuronal spread of α-herpesvirus infection. Future Virol 4:591–603
8. Gaskell R, Dawson S, Radford A, Thiry E (2007) Feline herpesvirus. Vet Res 38:337–354
9. Thiry J, Keuser V, Muylkens B, Meurens F et al (2006) Ruminant alphaherpesviruses related to bovine herpesvirus 1. Vet Res 37:169–190
10. Messerle M, Crnkovic I, Hammerschmidt W, Ziegler H et al (1997) Cloning and mutagenesis of a herpesvirus genome as an infectious bacterial artificial chromosome. Proc Natl Acad Sci U S A 94:14759–14763
11. Mahony TJ, McCarthy FM, Gravel JL, West L et al (2002) Construction and manipulation of an infectious clone of the bovine herpesvirus 1 genome maintained as a bacterial artificial chromosome. J Virol 76:6660–6668
12. Chang WLW, Barry PA (2003) Cloning of the full-length rhesus cytomegalovirus genome as an infectious and self-excisable bacterial artificial chromosome for analysis of viral pathogenesis. J Virol 77:5073–5083
13. Zhou F-C, Zhang Y-J, Deng J-H, Wang X-P et al (2002) Efficient infection by a recombinant Kaposi's sarcoma-associated herpesvirus cloned in a bacterial artificial chromosome: application for genetic analysis. J Virol 76: 6185–6196
14. Smith GA, Enquist LW (2000) A self-recombining bacterial artificial chromosome and its application for analysis of herpesvirus pathogenesis. Proc Natl Acad Sci U S A 97: 4873–4878
15. Baigent SJ (2006) Herpesvirus of turkey reconstituted from bacterial artificial chromosome clones induces protection against Marek's disease. J Gen Virol 87:769–776
16. Dewals B (2006) Cloning of the genome of Alcelaphine herpesvirus 1 as an infectious and pathogenic bacterial artificial chromosome. J Gen Virol 87:509–517
17. Yu D, Smith GA, Enquist LW, Shenk T (2002) Construction of a self-excisable bacterial artificial chromosome containing the human cytomegalovirus genome and mutagenesis of the diploid TRL/IRL13 gene. J Virol 76: 2316–2328
18. Petherbridge L, Howes K, Baigent SJ, Sacco MA et al (2003) Replication-competent bacterial artificial chromosomes of Marek's disease virus: novel tools for generation of molecularly defined herpesvirus vaccines. J Virol 77:8712–8718
19. Estep RD, Powers MF, Yen BK, Li H et al (2007) Construction of an infectious rhesus rhadinovirus bacterial artificial chromosome for the analysis of Kaposi's sarcoma-associated herpesvirus-related disease development. J Virol 81:2957–2969
20. Gabev E, Fraefel C, Ackermann M, Tobler K (2009) Cloning of Bovine herpesvirus type 1 and type 5 as infectious bacterial artifical chromosomes. BMC Res Notes 2:209
21. Richter M, Schudel L, Tobler K, Matheis F et al (2009) Clinical, virological, and immunological parameters associated with superinfection of latently with FeHV-1 infected cats. Vet Microbiol 138:205–216
22. Mocarski ES, Post LE, Roizman B (1980) Molecular engineering of the herpes simplex virus genome: insertion of a second L-S junction into the genome causes additional genome inversions. Cell 22:243–255

Part IV

Applications of BACs in Model Organisms, Medical Genetics, and Drug Discovery

Chapter 11

A Recombineering-Based Gene Tagging System for Arabidopsis

Jose M. Alonso and Anna N. Stepanova

Abstract

Many of the experimental approaches aimed at studying gene function heavily rely on the ability to make precise modifications in the gene's DNA sequence. Homologous recombination (HR)-based strategies provide a convenient way to create such types of modifications. HR-based DNA sequence manipulations can be enormously facilitated by expressing in *E. coli* a small set of bacteriophage proteins that make the exchange of DNA between a linear donor and the target DNA molecules extremely efficient. These in vivo recombineering techniques have been incorporated as essential components of the molecular toolbox in many model organisms. In this chapter, we describe the experimental procedures involved in recombineering-based tagging of an Arabidopsis gene contained in a plant transformation-ready bacterial artificial chromosome (TAC).

Key words Recombineering, Arabidopsis, Fluorescent protein, Lambda RED

1 Introduction

The ability to precisely modify the DNA sequence of a gene in its chromosomal context represents one of the most informative experimental approaches to study gene function. Recombineering provides the opportunity to expand the benefits of such experimental strategies to organisms in which the direct manipulation of the chromosomal sequence is impractical [1, 2]. Recombineering relies on (1) the availability of bacterial vectors capable of carrying large fragments of genomic DNA (up to 150–200 kb) from a species of interest (e.g., bacterial artificial chromosomes (BAC) harboring Arabidopsis genomic DNA), and (2) an engineered *E. coli* strain with high rates of homologous recombination (reviewed in [3]).

Although several alternative recombineering strategies have been developed [4, 5], we will focus this discussion exclusively on the λ-RED system [5] and its implementation in Arabidopsis' gene functional studies using a transformation-ready end-sequenced JAtY library of bacterial artificial chromosomes [6]. The λ-RED

Kumaran Narayanan (ed.), *Bacterial Artificial Chromosomes*, Methods in Molecular Biology, vol. 1227, DOI 10.1007/978-1-4939-1652-8_11, © Springer Science+Business Media New York 2015

system utilized here was originally developed by Yu et al. [5], and later modified by Warming and co-workers [7]. The key components of this system are three λ phage genes, *exo, bet* and *gam*, integrated into the *E. coli* genome and expressed under the control of the temperature-regulated λ*PL* promoter. In addition to the λ-RED system, the strain used herein, SW105, also contains an L-Arabinose-inducible *flipase* (*Flp*) gene in its genome [7]. Thus, in this system the expression of the *exo, bet* and *gam* genes can be turned on and off by simply changing the growth temperature conditions, allowing for a rapid and transient regulation of the recombineering genes. On the other hand, the expression of *Flp* can be activated by adding arabinose to the growth media, leading to the removal of any DNA sequence flanked by direct repeats of a flipase recognition target sequence (FRT). The use of the SW105 recombineering *E. coli* [7] strain allows for the insertion at any position of the genomic DNA (harbored in a bacterial artificial chromosome (BAC) or a transformation-ready BAC (TAC)) of a PCR-amplified linear DNA fragment containing the desired tag (such as GFP) and an antibiotic resistance selectable marker (all together referred to as the recombineering cassette). By flanking the selectable marker in the recombineering cassette with compatible FRTs, it is possible to eliminate the marker sequence from the construct after the cassette has been inserted (Fig. 1), generating an in-frame fusion of the gene of interest with the desired tag [8].

Importantly, only 50 nt of the target-specific sequence at each end of the recombineering cassette needs to be added to the linear DNA fragment to precisely direct the insertion of the cassette into the desired location in the BAC clone. These relatively short targeting sequences can be easily incorporated during the synthesis of the primers used to amplify the recombineering cassette. Thus, the recombineering system described here (Fig. 1) represents a convenient experimental strategy to generate whole-gene translational fusions for almost any gene in the Arabidopsis genome [2].

2 Materials

2.1 PCR Amplification of the Recombineering Cassette

1. Recombineering cassette DNA template (*see* **Note 1**).
2. Recombineering forward primer. This gene-specific primer should contain, from the 5′ to the 3′, the sequence of the 50 nucleotides immediately upstream of the insertion site in the gene of interest followed by the sequence "5′-GGAGGT GGAGGTGGAGCT-3′" (*see* **Notes 1** and **2**).
3. Recombineering reverse primer. This gene-specific primer should contain, from the 5′ to the 3′, the reverse complement sequence corresponding to the 50 nucleotides immediately downstream of the insertion site in the gene of interest followed

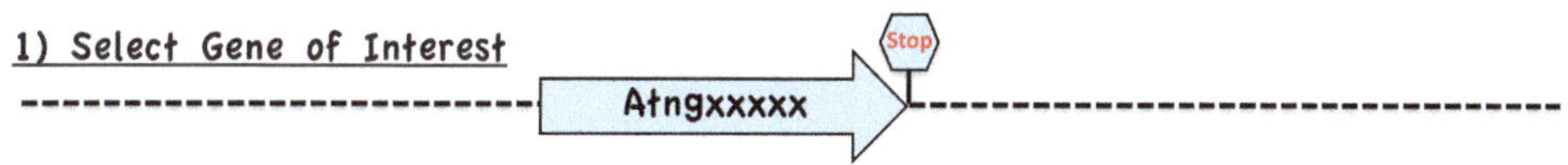

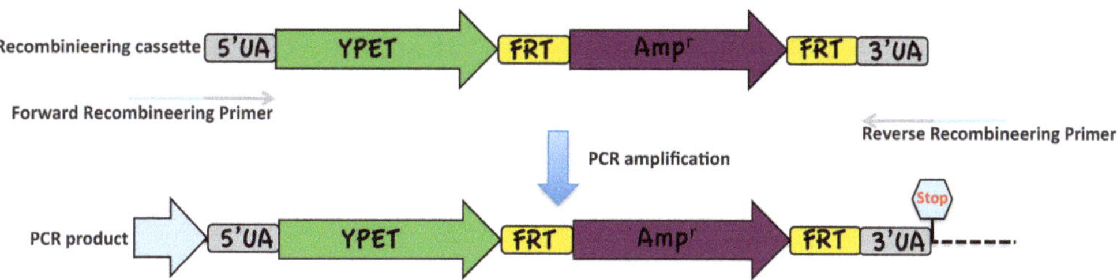

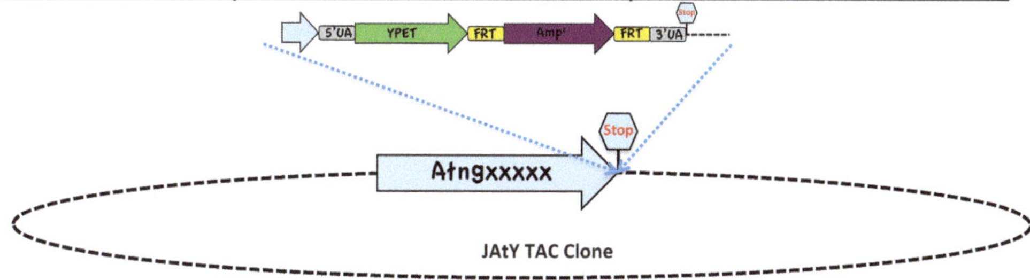

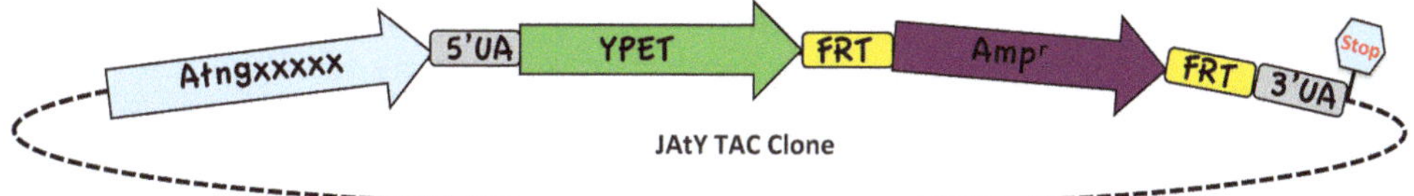

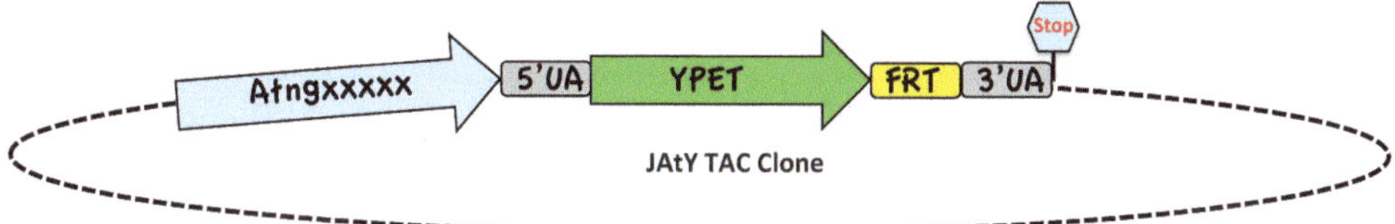

6) Sequence the junction sites and the coding region of Ypet

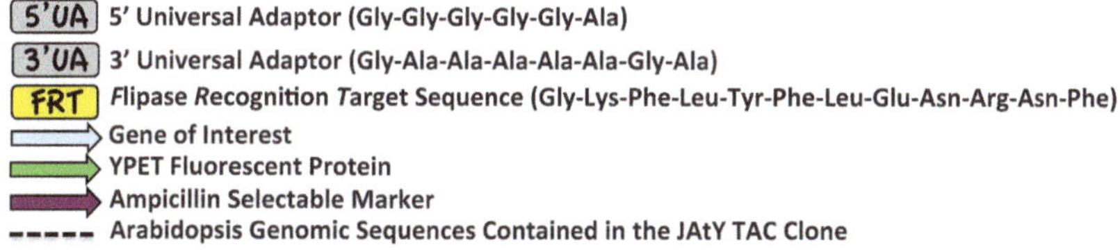

Fig. 1 Schematic representation of the experimental recombineering procedure. Genes are depicted as *arrows* while other sequence elements are shown as *boxes*. The label Atngxxxxx represents any Arabidopsis gene. The stop codon of the gene of interest is depicted with the shape of a traffic stop sign. All other abbreviations and symbols are explained in the figure

by the sequence "5′-GGCCCCAGCGGCCGCAGCAGCACC-3′" (*see* **Notes 1** and **2**).

4. Proofreading DNA polymerase, e.g., iProof High-Fidelity DNA Polymerase (Bio-Rad Laboratories) or equivalent.
5. dNTPs (2 mM each).
6. Thermocycler.
7. Agarose gel electrophoresis setup.
8. 10 mg/ml Ethidium Bromide.
9. Chloroform.
10. Ethanol: 95–100 % (v/v), and 70 % (v/v).
11. Table-top centrifuge with a rotor for microcentrifuge tubes.
12. Microcentrifuge plastic tubes (1.5 ml).
13. DNA template to generate the recombineering cassette (*see* **Note 1**).

2.2 Transformation of Recombineering-Ready Competent Cells

1. SW105 *E. coli* strain harboring a TAC clone with the gene of interest (*see* **Note 3**).
2. Luria-Bertani (LB) broth: 10 g/l tryptone, 5 g/l yeast extract, 10 g/l NaCl. Sterilize by autoclaving for 30 min.
3. Low-salt LB broth: 20 g/l tryptone, 5 g/l yeast extract, 0.5 g/l NaCl. Sterilize by autoclaving for 30 min.
4. LB-agar medium: LB broth supplemented with 15 g/l agar. Sterilize by autoclaving for 30 min.
5. 15-ml sterile round-bottom snap-cap plastic culture tubes.
6. Temperature-controlled shaker incubator.
7. Temperature-controlled water-bath shaker.
8. Antibiotics: 25 μg/ml Kanamycin (stock: 100 mg/ml stock in water); 100 μg/ml Ampicillin (stock: 100 mg/ml stock in water). Sterilize both antibiotics by filtering.
9. Sterile plastic Petri dishes (100 × 15 mm).
10. 250-ml Erlenmeyer flasks. Sterilize by autoclaving for 30 min.
11. 10 % v/v glycerol. Sterilize by autoclaving for 30 min.
12. Refrigerated centrifuge (Sorvall RC-5B) and rotor (Sorvall SS-34) or equivalents.
13. Transparent 50-ml Nalgene polypropylene tubes (3118-0050 Oak Ridge). Sterilize by autoclaving for 30 min.
14. 9-in. long glass Pasteur pipets. Sterilize by autoclaving for 30 min.
15. Faucet aspirator vacuum pump with a liquid trap.
16. Electroporator (Eppendorf 2510 or equivalent).
17. Electroporation cuvettes (2-mm gap).

2.3 Validating Recombinants

1. Test primers. A pair of regular 18–22 nucleotide long forward and reverse PCR primers flanking the insertion point and annealing between 100 and 150 base pairs upstream and downstream of the insertion point of the gene of interest, respectively.
2. *Taq* polymerase.
3. 50 % (v/v) glycerol. Sterilize by autoclaving for 30 min.

2.4 Removal of the Ampicillin Resistance Selectable Marker

1. L-Arabinose: 10 % (w/v). Sterilize by filtering.

3 Methods

3.1 PCR Amplification of the Recombineering Cassette

1. Prepare the template DNA to be used in PCR amplification in order to generate the recombineering cassette (*see* **Note 4**).
2. Amplify the DNA template corresponding to the recombineering cassette using the following PCR mixture: 29.5 μl of water, 10 μl of iProof 5× buffer GC, 1 μl of recombineering forward primer (5 μM), 1 μl of recombineering reverse primer (5 μM), 5 μl of dNTPs (2 mM), 2 μl of DNA template (1/10 of a mini-prep from 1.5 ml overnight culture), 1 μl of 10 % DMSO and 0.5 μl of iProof polymerase (*see* **Note 5**).
3. Perform the PCR amplification using the following PCR conditions:
   Step 1: 94 °C for 1 min, Step 2: 94 °C for 15 s, Step 3: 64 °C for 15 s and reducing 1 °C per cycle, Step 4: 72 °C for 60 s, Step 5: go to Step 2 10 times, Step 6: 94 °C for 15 s, Step 7: 54 °C for 15 s, Step 8: 72 °C for 60 s, Step 9 go to Step 6 20 times.
4. After the cycles are finished, transfer the entire 50 μl of amplification mixture to a 1.5-ml microfuge tube and extract the DNA with 50 μl of chloroform. Centrifuge for 10 min at 20,817 ×*g* in a table-top centrifuge at room temperature. Transfer the aqueous upper phase to a new 1.5-ml microfuge tube.
5. Precipitate the DNA by adding 125 μl of ethanol. Mix and centrifuge for 10 min at 20,817 ×*g* in a table-top centrifuge at room temperature. Remove the supernatant by aspiration with the faucet aspirator and add 1 ml of 70 % ethanol. After another 10 min centrifugation at 20,817 ×*g*, remove all of the 70 % ethanol and let the pellet dry for 5–10 min.
6. Resuspend the pellet in 15 μl of distilled water. Run 2 μl on an agarose gel. A single band of about 1.8 kb should be obtained (*see* **Note 5**).

3.2 Transformation of Recombineering-Ready Competent Cells

1. Streak the bacterial stock of the SW105 strain harboring the JAtY clone that contains your gene of interest (*see* **Note 6**) on LB-agar plates supplemented with 25 μg/ml of kanamycin and incubate at 32 °C for 48 h.
2. Pick a freshly grown colony and inoculate 3 ml of liquid LB media supplemented with kanamycin (25 μg/ml) in a 15-ml sterile plastic tube. Grow overnight at 32 °C with constant shaking.
3. Inoculate 50 ml of liquid low-salt LB supplemented with kanamycin (25 μg/ml) with 1 ml of the overnight culture in a 250-ml Erlenmeyer flask. Grow at 32 °C with constant shaking for about 2.5–3 h until the culture reaches OD 600–0.6.
4. Place the 50 ml culture in the water bath shaker preheated to 42 °C for exactly 15 min with constant shaking (*see* **Note 7**).
5. Place the flask with the culture on ice for at least 10–15 min, but not more than 1–2 h.
6. Transfer the culture to a prechilled 50-ml Nalgene centrifuge tube.
7. Pellet the cells by centrifugation using a prechilled Sorvall SS-34 rotor (or equivalent) at 4 °C for 5 min at 3,000 × *g*.
8. Decant the media by inverting the tube and quickly place the tube on ice. Resuspend the cells in the remaining media or glycerol by gently stirring the tube in a slurry of ice and water. Carefully fill the tube with ice-cold 10 % glycerol solution.
9. Pellet the cells by centrifugation using a prechilled Sorvall SS-34 rotor (or equivalent) at 4 °C for 10 min at 6,700 × *g*.
10. Repeat **steps 8** and **9**.
11. Immediately after centrifugation, place the tube on ice (*see* **Note 8**) and then remove as much of the residual 10 % glycerol as possible by aspiration using the glass Pasteur pipet connected to a vacuum aspirator. Resuspend the cells in the residual drop of glycerol (typically, about 100–200 μl are left in the tube's bottom and walls) by gently stirring the tube in a slurry of water and ice.
12. Place 7 μl of the amplified purified recombineering cassette DNA (from Subheading 3.1, **step 6**) in a 1.5-ml microfuge tube and chill the sample on ice.
13. Gently pipette 40 μl of the competent cells (from Subheading 3.2, **step 11**) and add them to the prechilled tube with the DNA.
14. Gently pipette out the mix of cells and DNA and place it in a prechilled 2-mm-gap electroporation cuvette.
15. Take the cuvette out of the ice, quickly dry the outside surface and electroporate at 1,800 V (*see* **Note 9**).

16. Remove the cells from the electroporation cuvette by adding 1 ml of LB media (without antibiotics) and transferring the mixture to a new 15-ml sterile plastic tube. Recover the cells for 1.5 h at 32 °C with constant shaking.
17. Transfer the cells to a sterile 1.5-ml microfuge tube and collect all of the cells by spinning at maximum speed in a table-top centrifuge for 1 min at room temperature.
18. Decant most of the media by inverting the tube, leaving about 25–50 μl to resuspend the cells by pipetting.
19. Spread all of the cells on a LB-agar plate supplemented with 25 μg/ml of kanamycin and 100 μg/ml of ampicillin.
20. Incubate the plates upside down at 32 °C. Nice colonies should be observed after 48 h.

3.3 Validating Recombinants

1. Perform colony PCR from three independent actively growing colonies obtained after the recombination steps described above. After picking the cells from one isolated colony, resuspend them in 20 μl of sterile distilled water. Use 2 μl of the cell suspension as a template in a regular 20 μl PCR reaction (*see* **Note 10**).
2. Re-streak PCR-positive colonies on a fresh LB-agar plate supplemented with kanamycin (25 μg/ml) and ampicillin (100 μg/ml) and incubate the plate upside down at 32 °C (*see* **Note 11**). Test once again one colony from each of the selected three original recombinants per construct.
3. Make glycerol stocks of three independent recombinants per construct (*see* **Note 12**) by mixing 700 μl of a freshly grown liquid culture with 300 μl of sterile 50 % glycerol in a 1.5-ml microfuge tube and freezing/storing the samples at −80 °C.

3.4 Removal of the Ampicillin Resistance Selectable Marker

This step is necessary to minimize the amount of unnecessary DNA sequences inserted into the gene of interest. If not removed, the ampicillin marker sequences would cause problems when making amino-terminal in-frame fusions, and could trigger nonsense mRNA decay when making carboxyl-terminal fusions.

1. Streak on a fresh LB-agar plate supplemented with kanamycin (25 μg/ml) and ampicillin (100 μg/ml) one of the positive recombinants from the step above and incubate at 32 °C for 48 h.
2. Inoculate 3 ml of liquid LB media supplemented with kanamycin (25 μg/ml) and ampicillin (100 μg/ml) in a 15-ml sterile plastic culture tube with a single colony. Grow overnight at 32 °C with constant shaking.
3. Inoculate 1 ml of liquid LB supplemented with kanamycin (25 μg/ml) (*see* **Note 13**) in a 15-ml sterile plastic culture tube

with 20 μl of the overnight culture. Grow for 2–3 h (to OD 600–0.6) at 32 °C with constant shaking.

4. Add 10 μl of 10 % filter-sterilized L-Arabinose and incubate for another 3 h at 32 °C with constant shaking.
5. Immerse the tip of a sterile toothpick (or equivalent) into the cell culture and streak the culture for single colonies on a fresh LB-agar plate supplemented with kanamycin (25 μg/ml).
6. After 48 h at 32 °C, test two isolated colonies by colony PCR exactly as described in Subheading 3.3 (*see* **Note 14**).
7. Re-streak one flipase-positive colony on a fresh LB-agar plate supplemented with kanamycin (25 μg/ml). After 48 h and 32 °C, pick a single colony and perform colony PCR as indicated in Subheading 3.3.
8. Sequence the PCR product using the internal primers YpetF "5′-TGATACTCTCGTTAACAGGATCG-3′" and YpetR "5′-CTTATCAGCAGTGATGTACACG-3′."
9. Make glycerol stock of the strain containing the sequence-verified final construct by mixing 700 μl of a freshly grown liquid culture with 300 μl of sterile 50 % glycerol in a 1.5-ml microfuge tube and freezing/storing the samples at −80 °C.

4 Notes

1. A recombineering cassette with the desired antibiotic resistance and fluorescent protein sequences can be generated using classical molecular biology approaches or directly synthesized. The DNA template for the recombineering cassette used as an example herein (*Ypet-FRT-Amp-FRT*) can be obtained from the Arabidopsis Stock Center (stock number CD3-1726). This cassette is contained in the pYLTAC17 vector and consists of the 5′ universal adapter (GGAGGTGGAGGTGGAGCT), the codon optimized sequence for the fluorescent protein gene Ypet [2], the ampicillin-resistance gene flanked by direct repeats of a flipase recognition target sequence (*FRT*) [8], followed by the 3′ universal adapter sequence (GGCCCCAGC GGCCGCAGCAGCACC). These adapter sequences provide a flexible bridge of poly-glycine and poly-alanine, respectively, between the protein of interest and the fluorescent tag.
2. The quality of the long recombineering primers is very important, as very different efficiencies can be obtained dependent on the brand of primers utilized. We routinely use primers from Integrated DNA Technologies Inc. without any additional purification steps.
3. The SW105 strain can be obtained from the Frederick National Laboratories for Cancer Research, *see*: http://ncifrederick.cancer.gov/research/brb/recombineeringInformation.aspx.

4. When using the *Ypet-FRT-Amp-FRT* cassette or any other cassette contained in a TAC clone as a template, the quality of the DNA is important to ensure good amplification using the long recombineering primers. A regular alkaline lysis miniprep where a phenol/chloroform purification step is incorporated prior to the alcohol precipitation produces sufficiently pure DNA.
5. Some weak additional bands may sometimes appear upon PCR amplification that have no detrimental effect on the recombineering efficiency. On the other hand, if poor amplification of the correct-size band and/or strong additional bands are observed in the PCR product, gel purification of the correct band or a new PCR amplification is advisable. Reducing the concentration of the recombineering primers, using a new DNA polymerase and/or preparing a new aliquot of the template recombineering cassette DNA typically solves the problem. Ideally, between 0.1 and 1 μg of PCR product should be used for electroporation. Nevertheless, acceptable recombination efficiencies can be achieved even when using lower or higher than advisable amounts of the PCR product as long as the purity of the DNA used is good.
6. The JAtY clone containing the gene of your interest can be identified using that gene's ID as a query at http://Arabidopsislocalizome.org. The individual JAtY clone containing the gene of interest can be ordered from http://www.genome-enterprise.com. The procedure to transfer the JAtY clone of interest from DH10B to SW105 is described in [2]. This process is the same as that described in Subheading 3.2 with few modifications. Thus, for the transfer of the JAtY clones to SW105, 7 μl of freshly prepared TAC DNA (1/3 of that obtained from an alkaline DNA preparation from a 1.5 ml overnight culture) is used instead of the recombineering cassette. The optimal electroporation conditions for TAC clones are: 2.5 kV, 100 Ω and 25 μF [9].
7. It is important that the whole culture is submerged below the water level, but the amount of water in the bath needs to be kept relatively low so that during the vigorous shaking the water does not reach the top of the Erlenmeyer flask.
8. After the second wash, the pellet of cells on tube's bottom is usually very loose. To avoid losing too many cells, right after the last centrifugation the tube should be placed on ice at a 45° angle, with the pellet positioned in the lower part of the tube during glycerol aspiration.
9. For the bacterial transformation to work, the time constant readout after the electroporation should be between 4.5 and 6 ms. Lower reads usually means dirty (salt-contaminated) sample.

10. The typical conditions used for colony PCR are: 14 μl of distilled water, 2 μl of 10× PCR buffer, 0.5 μl of dNTPs (2 mM), 0.5 μl of forward test primer (20 μM), 0.5 μl of reverse test primer (20 μM), 0.5 μl of *Taq* DNA polymerase, and 2 μl of cell suspension. The recommended PCR cycle is as follows: Step 1: 94 °C 15 s, Step 2: 54 °C 15 s, Step 3: 72 °C 1.5 min, Step 4: go to Step 1 25 times.
11. In most cases, two bands will be observed when analyzing the results of the colony PCR. The bigger band corresponds to true recombination event and should be around 1.8–2.0 kb, while the smaller band will have the size expected when no recombination has occurred and is probably due to the presence of a background of dead or not growing cells that did not undergo recombination but are still present in the plate. The lower band should disappear after re-streaking the primary positives and selecting a new single colony.
12. It is advisable to save three independent recombinants, as we commonly observe mutations in about 5–10 % of samples due to errors in the long recombineering primers (this issue is the most common and is caused by the mistakes during primer synthesis; please, beware that some primer brands give an even higher percentage of synthesis-related mistakes than Integrated DNA Technologies Inc.) or due to mistakes introduced by the DNA proofreading polymerase (less common, but also dependent on the enzyme brand). However, we never observe the same mutation in two independent primary recombinants and, therefore, it is more than sufficient to keep three clones per construct to ensure the identification of at least one error-free construct.
13. Please note that ampicillin is omitted from this step on.
14. It is advisable to run in parallel a control colony PCR from the same strain before the flipase step. A successful flipase reaction will be manifested by a reduction in the size of the band by about 950 base pairs. The flipase reaction is extremely efficient and 100 % of the colonies tested will usually show the expected reduction in size. On some occasions (in less than 10 % of the cases), two bands corresponding to the flipped and nonflipped DNA can be observed in a single colony. A single colony showing only the smaller flipped band should be identified after re-streaking this colony on a fresh LB-agar plate supplemented with kanamycin (25 μg/ml).

Acknowledgements

Work in the Alonso-Stepanova lab is supported by NSF-MCB0923727 grant to JMA and ANS and NSF-MCB 1158181 grant to JMA.

References

1. Sarov M, Murray JI, Schanze K, Pozniakovski A, Niu W, Angermann K, Hasse S, Rupprecht M, Vinis E, Tinney M, Preston E, Zinke A, Enst S, Teichgraber T, Janette J, Reis K, Janosch S, Schloissnig S, Ejsmont RK, Slightam C, Xu X, Kim SK, Reinke V, Stewart AF, Snyder M, Waterston RH, Hyman AA (2012) A genome-scale resource for in vivo tag-based protein function exploration in C. elegans. Cell 150:855–866
2. Zhou R, Benavente LM, Stepanova AN, Alonso JM (2011) A recombineering-based gene tagging system for Arabidopsis. Plant J 66:712–723
3. Court DL, Sawitzke JA, Thomason LC (2002) Genetic engineering using homologous recombination. Annu Rev Genet 36:361–388
4. Zhang Y, Buchholz F, Muyrers JP, Stewart AF (1998) A new logic for DNA engineering using recombination in Escherichia coli. Nat Genet 20:123–128
5. Yu D, Ellis HM, Lee EC, Jenkins NA, Copeland NG, Court DL (2000) An efficient recombination system for chromosome engineering in Escherichia coli. Proc Natl Acad Sci U S A 97: 5978–5983
6. Liu YG, Shirano Y, Fukaki H, Yanai Y, Tasaka M, Tabata S, Shibata D (1999) Complementation of plant mutants with large genomic DNA fragments by a transformation-competent artificial chromosome vector accelerates positional cloning. Proc Natl Acad Sci U S A 96: 6535–6540
7. Warming S, Costantino N, Court DL, Jenkins NA, Copeland NG (2005) Simple and highly efficient BAC recombineering using galK selection. Nucleic Acids Res 33:e36
8. Tursun B, Cochella L, Carrera I, Hobert O (2009) A toolkit and robust pipeline for the generation of fosmid-based reporter genes in C. elegans. PLoS One 4:e4625
9. Alonso JM, Stepanova AN (2014) Arabidopsis transformation with large bacterial chromosomes. Methods Mol Biol 1062:271–283

Chapter 12

BAC Transgenic Zebrafish for Transcriptional Promoter and Enhancer Studies

Petra Kraus, Cecilia L. Winata, and Thomas Lufkin

Abstract

With the advent of BAC recombineering techniques, transcriptional promoter and enhancer isolation studies have become much more feasible in zebrafish than in mouse given the easy access to large numbers of fertilized zebrafish eggs and offspring in general, the easy to follow ex-utero development of zebrafish, an overall less skill demand and a more cost-effective technique. Here we provide guidelines for the generation of BAC recombineering-based transgenic zebrafish for DNA transcriptional promoter and enhancer identification studies as well as protocols for their analysis, which have been successfully applied in our laboratories many times. BAC recombineering in zebrafish allows for economical functional genomics studies, for example by integrating developmental biology with comparative genomics approaches to validate potential enhancer elements of vertebrate transcription factors.

Key words BAC recombination technology, Enhancer, *lacZ*, EGFP, Zebrafish, Paraffin sectioning

1 Introduction

Bacterial artificial chromosomes (BAC) are stable large capacity cloning vehicles for genomic DNA [1]. BAC recombination technology has revolutionized the field of mouse transgenesis significantly in recent years [2–5], yet pronuclear injection for the production of transgenic mice, embryonic stem (ES) cell manipulation and ES cell microinjection for targeted mutagenesis, despite many advances in the field [6–8], still remains very costly and time-consuming, given the in-utero development and relatively small litter size of mice. While gene targeting is still fairly inefficient in zebrafish (*Danio rerio*), zebrafish has emerged as a cost-effective alternative model system for identifying gene regulatory elements such as promoters and enhancers [9, 10] as well as gene trap experiments [11, 12].

Not long ago BAC recombineering found its way into the fish tank. The ability to clone up to 300 kb DNA fragments dramatically improved the efficiency for searching and identifying gene

Kumaran Narayanan (ed.), *Bacterial Artificial Chromosomes*, Methods in Molecular Biology, vol. 1227,
DOI 10.1007/978-1-4939-1652-8_12,

regulatory elements and the recent development of *tol2*-mediated transgenesis facilitated the delivery of a single copy modified BAC transgene with the option to identify and characterize the integration site [13–15]. These achievements in combination with a large number of easily accessible offspring and ex-utero development [16] makes the zebrafish an ever increasing valuable asset to study vertebrate development in a statistically significant and cost-effective manner [17–19].

There are two major approaches undertaken when generating transgenic zebrafish, namely "transient" expression without selecting for the integration of the transgene into the zebrafish genome or the generation of "stable transgenic lines" aiming for integration. The choice depends on the objective of the experiment, whether the path of transient transgenics is sufficient or the establishment of a stable genome integrated germ-line transmitting transgenic line is necessary. Transient transgenics can be generated quickly and in very high number, which can be statistically preferable as the rapid development of zebrafish embryos generally leads to mosaic expression in founder animals, but can be averaged out statistically by analyzing hundreds of animals. In stable lines, the integration of the transgene into the zebrafish genome will occur at random and the epigenetic modification of the integration site can impact on the expression pattern. The typically mosaic founder fish need to be tested for germline transmission should stable lines be desired.

Assaying a statistically sufficient number of zebrafish ($\geq$100) per transgene construct and comparison to available RNA in situ expression data is important for the identification of DNA regulatory elements. Stable lines will have the benefit to allow for the characterization of the integration site and a non-mosaic expression pattern given all cells in the animal will carry the transgene after successful germ line transmission, as well as facilitating the analysis of transgene expression in adult tissues. The generation of stable transgenic lines, however, is more labor-intensive and costly to establish and maintain. We will focus here on transient transgenics, the more common initial approach in zebrafish enhancer identification studies. Protocols to generate stable lines [20] and for transposon-mediated generation of BAC transgenics [14] have been described previously and can be referred to.

For enhancer identification studies, a BAC carrying the maximal amount of flanking genomic sequence (presumably containing enhancer regulatory elements) is modified in an *E. coli* host via homologous recombination (*see* Fig. 1). To insert a suitable reporter gene immediately upstream of the translation start site (TLSS), 50 bp oligonucleotides of sequence flanking the insertion site are often sufficient to serve as 5′ and 3′ homology arms [9]. Generally for in vivo monitoring, EGFP [21] or a related fluorescent reporter is preferable. Other reporter genes like *lacZ* are more

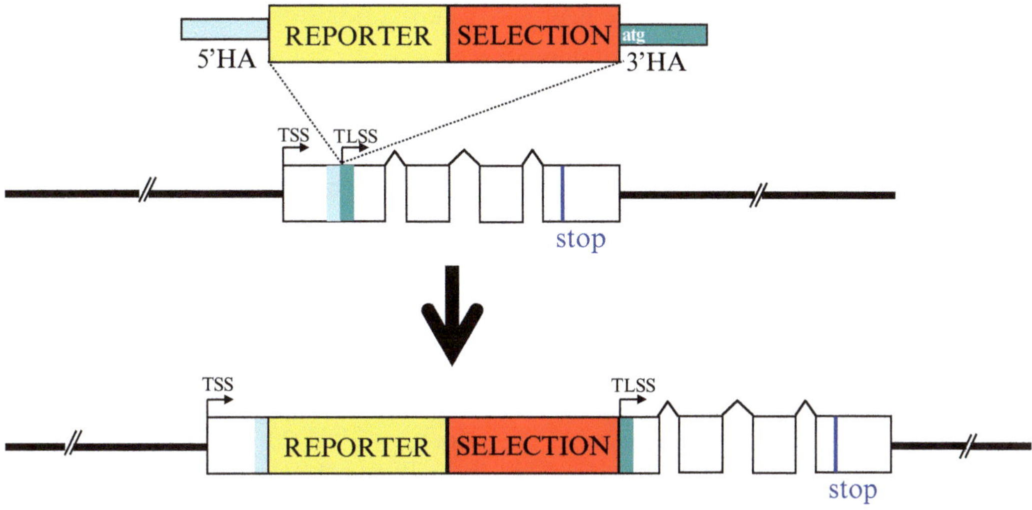

Fig. 1 Schematic of a construct for enhancer trap studies in zebrafish. A reporter gene (*yellow rectangle*) and a gene for selection (*red rectangle*) are flanked by 50 bp oligonucleotides immediately adjacent to the translation start site (TLSS) providing the 5′ homology arm (*light green rectangle*, 5′HA) and the 3′ homology arm (*green rectangle*, 3′HA). After homologous recombination with a BAC carrying the gene of interest the reporter gene will be expressed according to the enhancer elements present in the BAC

tedious to work with because they require additional steps, including euthanization and fixation of the zebrafish. However given the accumulation of the beta-galactosidase product over the course of the staining reaction, *lacZ* might be a suitable reporter for studying genes which are expressed at low transcription rates [22].

In zebrafish, postfertilization development occurs more rapidly than in mouse, hence there is a high possibility of generating mosaic fish following DNA injection [23, 24]. Therefore, a larger number of embryos will have to be injected and analyzed to determine the role of any given gene regulatory element. However, the ease and simplicity of the zebrafish microinjection technology compared to mouse pronuclear injection and the abundance of available fertilized zebrafish 1-cell embryos makes this a very feasible approach.

2 Materials

2.1 Zebrafish BAC Libraries

Suitable BACs and sources thereof like zebrafish BAC libraries from SourceBioscience Life Sciences, BACPAC Resources Center at the Children's Hospital Oakland Research Institute (CHORI), and others can be identified via The Zebrafish Model Organism Database at http://zfin.org. Additional helpful information is available via The Wellcome Trust Sanger Institute browser (http://www.sanger.ac.uk/resources/zebrafish/) as well as from the University of California, Santa Cruz (UCSC) genome browser (http://genome.ucsc.edu/).

2.2 Buffers and Kits

2.2.1 BAC Modification and Purification

1. Epicentre BAC DNA Purification Kit.
2. UV transilluminator.
3. Gene Bridges Quick and Easy BAC Modification kit.
4. Gene Bridges Quick and Easy BAC Subcloning kit.
5. Ethidium bromide stock (10 mg/ml) [final concentration (0.5 μg/ml)].
6. 1× Tris/EDTA (TE): 10 mM Tris–HCl, pH 8, 1 mM EDTA.
7. 0.2 % Phenol Red in diH_2O.

2.2.2 Histology

1. X-Gal staining solution: 2 mM $MgCl_2$, 0.01 % deoxycholic acid, 0.02 % Igepal CA-630, 0.1 % X-Gal in dimethylformamide (4 % stock), 5 mM $K_4[Fe(CN)_6]\cdot 3H_2O$, 5 mM $K_3[Fe(CN)_6$ in 1× PBS, pH 8.
2. 1× Phosphate buffered saline (PBS): 13.7 mM NaCl, 0.27 mM KCl, 0.81 mM Na_2HPO_4, 0.147 mM KH_2PO_4.
3. PBST: 1× PBS with 0.1 % Tween20.
4. 3 % H_2O_2 in 1× PBS.
5. 2 % Boehringer Blocking Reagent (Roche).
6. DAB (3,3′ Diaminobenzidine) solution: 13 μl of DAB, 5 μl of DMSO, 5 μl of 20 % Triton X-100 in 1 ml 1× PBS.
7. Anti-GFP antibody (polyclonal), anti-rabbit IgG HRP-linked antibody.
8. 1 % NaN_3 in 1× PBS.
9. 50 % Glycerol in 1× PBS.
10. 4 % Paraformaldehyde (PFA) in 1× PBS.
11. 0.5 % Eosin in 95 % Ethanol (EtOH).
12. 100 % Ethanol (EtOH) and dilutions thereof in water to 95 % (v/v), 90 % (v/v), 70 % (v/v), 50 %(v/v) and 30 % (v/v).
13. Histoclear or Histochoice® (Amresco).
14. Paraffin.
15. Paramount.

2.3 Zebrafish (See Note 1)

Members of the *Danio rerio* species can be easily purchased from local pet stores, or from the Zebrafish International Resource Center (ZIRC), University of Oregon, USA (http://zebrafish.org).

2.3.1 Zebrafish Aquatic Equipment

1. Zebrafish housing system (from, e.g., Aquatic Habitats or Tecniplast), usually consisting of:

   Aluminum racks.
   Tanks with cover.
   Large cardboard box to cover mating tanks.
   Overflow racks.

UV sterilizer.
Biofilters.
Air pump.
Reservoir tanks with circulation pump.
Heating element.

2. Tank water: 60 mg/l Instant Ocean® sea salt in diH_2O pH 6.8–7.5.
3. Test kits for pH, NO_2, NO_3, and NH_4.
4. Crossing tanks with mesh insert and divider.
5. Fish food (e.g., TetraMin flakes and/or brine shrimp).

2.3.2 Zebrafish Injection Supply

1. Egg water: 60 μg/ml Instant Ocean®sea salt in diH_2O.
2. Injector (Harvard Apparatus PLI-100).
3. Compressed nitrogen gas cylinder.
4. Dissecting scope (Leica).
5. Puller for injection needles (Sutter).
6. Glass capillaries (World Prec. Inst. Inc. Glass w/fil OD: 1 mm, ID: 0.75 mm, 4 in.).
7. 60 and 100 mm culture dishes.
8. Kitchen sauce strainer.

2.3.3 Transgenic Zebrafish Monitoring and Histology Equipment

1. Dissecting-type microscope (Leica).
2. Octopus light source (Leica).
3. Embedding station (Leica).
4. Rotary microtome (Leica).
5. Water bath (Leica).
6. Heating plate (Leica).
7. Polylysine slides (Thermo Scientific).
8. Coverslips 24 × 60 mm.
9. Multiwell glass dish (EMS).
10. Incubator at 28 °C (Thermo Scientific).

3 Methods

3.1 BAC Engineering

For enhancer isolation order one or more BACs that include the TLSS and cover as much of the adjacent upstream and downstream regions as possible to ensure all regulatory elements are present. If necessary, plan to co-inject overlapping or adjacent BACs. Design primers for screening, modification and subcloning of the chosen BACs by using the NCBI primer-BLAST function (http://www.ncbi.nlm.nih.gov/tools/primer-blast) or the Primer 3 (v. 0.4.0) software (http://frodo.wi.mit.edu/primer3/). Use the Gene

Bridges Quick and Easy BAC Modification kit and the Gene Bridges Quick and Easy BAC Subcloning kit following the manufacturer's protocol for genetic modifications of the BAC applying the Red/ET recombineering technology.

3.2 BAC DNA Purification and Preparation for Injection

Isolate BAC DNA from an *E. coli* host with one of the commercially available BAC DNA isolation kits. There is no need to linearize the BAC prior to injection [9, 10]. Measure the BAC DNA concentration with a Nanodrop, or under UV fluorescence after mixing a 1 μl aliquot with ethidium bromide on Saran wrap in comparison to a similarly prepared sample of known DNA concentration. Assay for BAC DNA quality by electrophoresis of an undigested or linearized aliquot on a 0.8 % agarose gel, any smear will indicate degradation of the BAC DNA.

3.3 Pulling Injection Needles

Prepare glass needles for zebrafish injections using a P97 needle puller from Sutter.

Unfortunately, the program varies with the surrounding conditions of the puller and has to be established individually for best results. The following settings can be used as a guideline with the capillaries indicated under Subheading 2.3.2 when using a box filament: Pressure 200/Heat 525/Pull 120/Velocity 100/Time 175.

3.4 Setting up a Small-Scale Zebrafish Unit

Please take note that it is mandatory in most countries to have an approved IACUC protocol in place before embarking on any work involving live animals. Maintain zebrafish in 45-l tanks with a density of 25 fish per tank in a 14 h light/10 h dark cycle. Maintain water temperature at 28.5 °C. Clean tanks daily by siphoning up debris from the bottom of the tank and replacing $^1/_3$ of the tank water. Alternatively, use filters within a circulating system in combination with replacing ½ of tank water once a week.

For larger scale studies over longer periods of time, set up a small-scale zebrafish unit from several established companies such as Aquatic Habitats (www.aquatichabitats.com). A small-scale fish unit generally consists of several shelves of a zebrafish housing system, each holding several tanks with interconnected water flow with a controlled light and dark cycle. Clean tanks once a week and monitor pH and salt content closely. For guidelines on these parameters see the ZFIN webpage (https://wiki.zfin.org/display/prot/Fish+Pathology+and+Health).

Generally, fish will thrive in tap water aged for a day to release chlorine, or distilled water with added sea salts (60 mg/l of Instant Ocean® Sea Salt). Feed fish twice or three times a day with a protein-rich diet, such as live brine shrimp or dry flake food such as TetraMin flakes. Adult male (smaller and more yellowish in color) and female zebrafish (bigger belly) are maintained in separate holding tanks and rotated for the use of breeding according to breeding records to avoid exhaustion of individual fish yet at the same time ensure a good quality of eggs. Ideally, mate fish once a week.

3.5 Setting Up Matings

1. Feed the zebrafish at around 3.30 pm in the afternoon on the day preceding the actual injection day. Prepare a typical tank for matings consisting of a holding tank with a mesh insert and a divider. The mesh insert should be wide enough to allow for eggs to fall through to the bottom of the tank. Approximately 30 min after the feeding, mate the fish as following: for one round of microinjection set up 5–10 mating tanks with 2–3 females on one side of the divider and 2–3 males on the other side of the divider.
2. Cover the tank with a plate so that the fish cannot jump out of the tank.
3. Cover the set-up with cardboard boxes to simulate the dark cycle.
4. The next morning, lift the cardboard box, one tank at a time, then open the divider and allow the fish to mate for approximately 20 min (*see* **Note 2**).
5. Lift the mesh insert and place it at an angle so the female fish are more likely to rub their belly against the mesh. This will stimulate the laying of eggs. Adding of plastic sea grass imitations or similar stimuli can increase the number of eggs.

3.6 Harvesting Fertilized Eggs

1. Move fish to a fresh tank then collect the eggs by filtering the water of the mating tank through a small kitchen strainer. Separate males and females later.
2. Collect eggs in a 100 mm culture dish by rinsing the strainer with egg water. Turn the strainer inside out to remove all eggs from the strainer and dip still attached eggs into the dish with egg water.
3. Immediately proceed with the injection as fish embryos develop fast. If necessary, slow down the fish development by placing the culture dish with eggs and egg water on ice or speed it up in a 28 °C incubator (*see* **Note 3**).

3.7 Injection of Zebrafish Embryos

1. Prepare the BAC DNA to a concentration of 15–30 ng/μl. Add Phenol Red to a final concentration of 0.1 % that allows visualization of the injection and the ensuing DNA uptake.
2. Add the DNA dilution to the needle, then place the needle in a needle holder and break the needle tip at an angle such that the size of the opening will not cause physical damage to the embryo.
3. Prepare the injection apparatus by adjusting pressure and injection time to release approximately 1–2 nl per injection (*see* Fig. 2).
4. Use a plastic Pasteur pipet to place about 50–100 embryos on a glass slide (*see* Fig. 3a) (*see* **Note 4**).

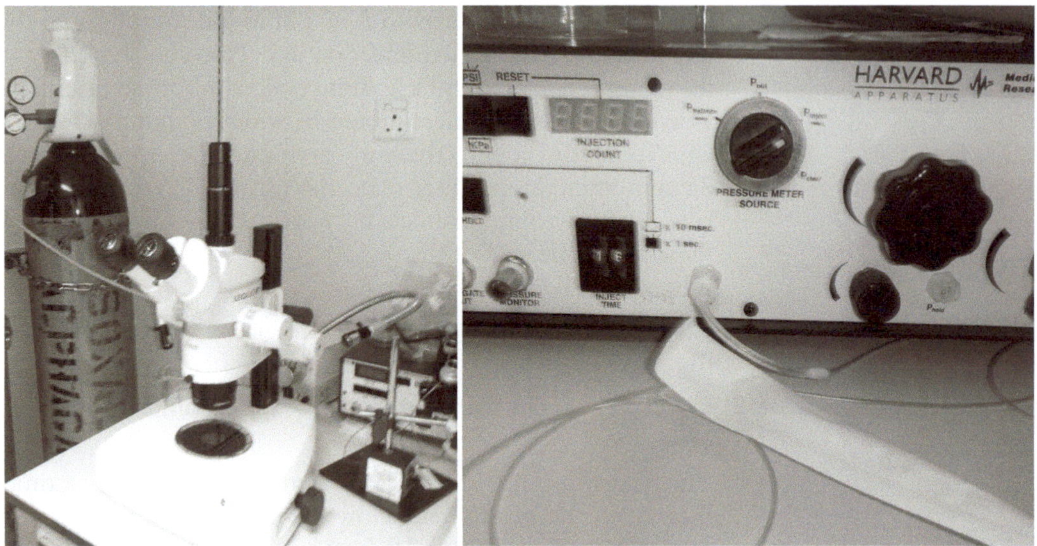

Fig. 2 Typical set-up for zebrafish injections with an injection apparatus, dissecting microscope, octopus light source, needle holder, and N_2 tank

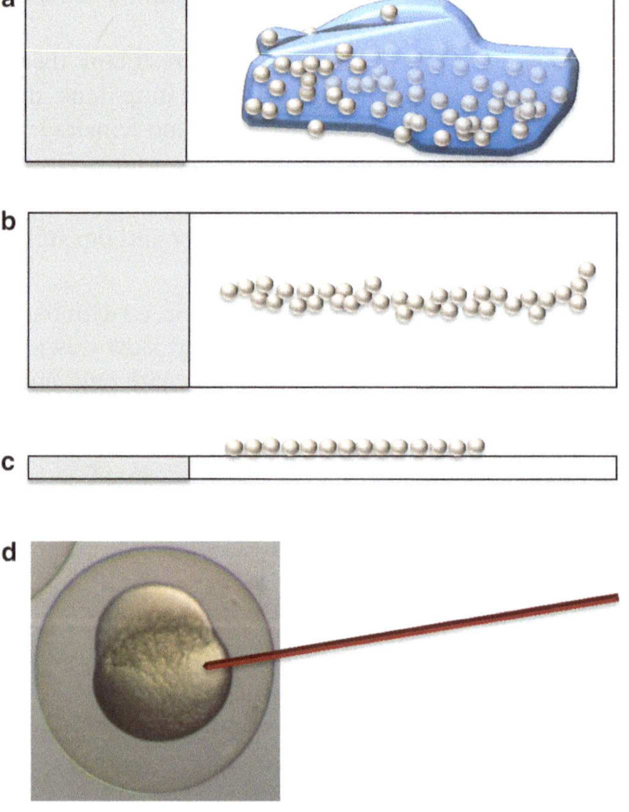

Fig. 3 Schematics of zebrafish eggs in egg water placed on a glass slide (**a**) and a *top* (**b**) and side (**c**) view after the egg water has been removed with a Kim wipe and the eggs have been aligned in a monolayer for injection. The DNA/Phenol Red dilution is injected into the yolk just beneath the blastomere (**d**)

5. Use a Kim Wipe to gently remove as much fluid as possible (*see* **Note 5**).
6. Fill a second 100 mm culture dish with egg water to collect the injected embryos from each batch.
7. Use a razor blade to gently align the embryos in a monolayer on the slide (*see* Fig. 3b, c), then place the slide with the aligned embryos on an upside down bottom of a 60 mm culture dish and proceed with the injection (*see* **Note 6**).
8. Inject the DNA dilution into the yolk just beneath each healthy looking one-cell embryo. Once all embryos on the slide are injected, wash them gently with egg water into a 100 mm culture dish containing egg water.
9. Return injected and control embryos to the 28 °C incubator and change egg water the evening of the injection day and then daily (*see* **Note 7**).
10. Observe and document survival rate/phenotype and expression pattern at several relevant time-points indicating hpf (hours past fertilization) or dpf (days past fertilization). Suggested time points for photo documentation are: 6–12, 24, and 36 hpf. To study gene expression in the skeleton or jaw 24 and 3 dpf, respectively, are important time points to analyze (*see* **Notes 8–10**).

3.8 Observation of Transgenic Zebrafish In Vivo

If the reporter is EGFP or any other fluorescent reporter, screen the developing zebrafish embryos for reporter gene expression in vivo using a stereomicroscope with a GFP2 filter or equivalent.

3.9 Whole Mount Antibody Staining for EGFP

1. Dechorionate embryos by gently tearing the chorion with fine forceps or a 27G needle.
2. Pick up the dechorionated embryos using a glass Pasteur pipet and transfer them into a microcentrifuge tube containing 1 ml of Histochoice and store overnight at 4 °C.
3. Replace ½ of the Histochoice solution with 4 % PFA in 1× PBS and incubate for 15 min on a shaker or nutator at room temperature.
4. Replace ½ of the solution volume with PBST and incubate for 15 min on a shaker or nutator at room temperature.
5. Repeat **step 4**.
6. Replace the solution with 1× PBS and incubate for 15 min on a shaker or nutator at room temperature.
7. Replace the solution with a solution of 3 % H_2O_2 in 1× PBS and incubate for 15 min on a shaker or nutator at room temperature.
8. Wash with PBST three times for 5 min each on a shaker or nutator at room temperature.

9. Block in 2 % Boehringer Blocking Reagent dissolved in 1× PBS for 1 h on a shaker or nutator at room temperature.
10. Add 1:500 diluted anti-GFP antibody (polyclonal) in 1× PBS solution and incubate for 6 h at room temperature or overnight at 4 °C.
11. Wash with PBST four times for 30 min each on a nutator at room temperature.
12. Add anti-rabbit IgG, HRP-linked antibody diluted to 1:1,000 in PBST and incubate for at least 2 h at room temperature or overnight at 4 °C.
13. Wash four times in PBST for 30 min each on a nutator at room temperature.
14. Pre-soak with DAB solution for 5 min.
15. Add 2.5 μl of 3 % H_2O_2 to develop and constantly monitor the staining. Usually the signal develops within 1–5 min.
16. When the desired signal intensity is reached, stop the staining progression by replacing the solution with 1%NaN_3 in 1× PBS.
17. To clear the yolk for imaging purposes store the embryos in 50 % glycerol in 1× PBS overnight prior to imaging. Alternatively, for long-term storage, keep embryos in 4 % PFA.

3.10 lacZ Staining of Transgenic Zebrafish

1. Fix zebrafish for 10 min in 4 % fresh PFA on ice.
2. Wash zebrafish 5× for 10 min in 1× PBS on ice.
3. Incubate in *lacZ* staining solution in the dark at 4 °C until signal becomes visible.
4. Wash stained zebrafish 3× for 5 min with cold 1× PBS.
5. Post-fix stained zebrafish for 30 min in 4 % PFA at 4 °C.
6. Wash post-fixed zebrafish 3× for 5 min with cold 1× PBS.
7. Photograph and document the staining pattern (*see* Fig. 4) then proceed to sectioning.

3.11 Microtome Sectioning of 48 hpf lacZ Stained Zebrafish

1. Prepare a set of 35 mm Petri dishes with the following EtOH concentrations for gradual dehydration of the zebrafish: 1× 30 %/2× 50 %/2× 70 %/3× 95 % (*see* Fig. 5a).
2. Gently move them through the EtOH gradient using a disposable plastic Pasteur pipet with cut off tip, once in the last 95 % EtOH sort out the *lacZ* stained fish and proceed. Keep the others in 95 % EtOH in an Eppendorf tube at 4 °C if desired, or discard.
3. Prepare another set of 35 mm culture dishes: One with 0.5 % Eosin in 95 % EtOH and six or more dishes with 100 % EtOH (*see* Fig. 5b) (*see* **Note 11**).

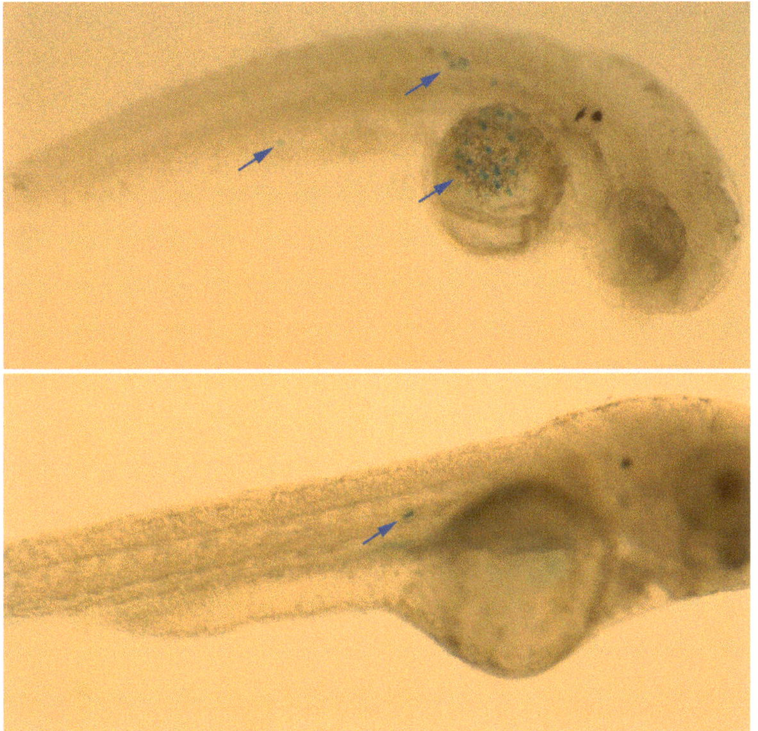

Fig. 4 Blue arrows point to *lacZ* stained cells of a BAC transgenic fish

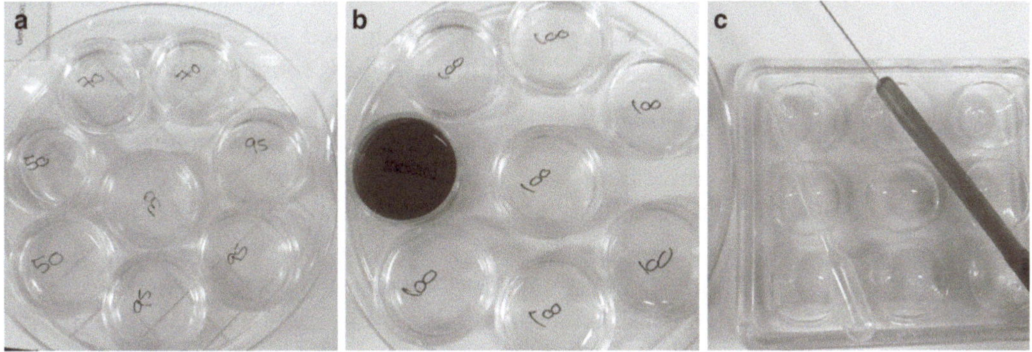

Fig. 5 Typical set up of 35 mm culture dishes with an EtOH gradient for the dehydration of zebrafish 24–48 hpf (**a**), and Eosin staining for better visibility during the embedding process (**b**). Suggested materials for Histoclear treatment prior to paraffin embedding of 24–48 hpf zebrafish (**c**)

4. Stain the zebrafish in the Eosin staining solution, then wash them by passing them through the 100 % EtOH dishes until the ethanol remains clear. The stain will not penetrate much into the fish but will label the fish enough to visualize it during the embedding procedure.

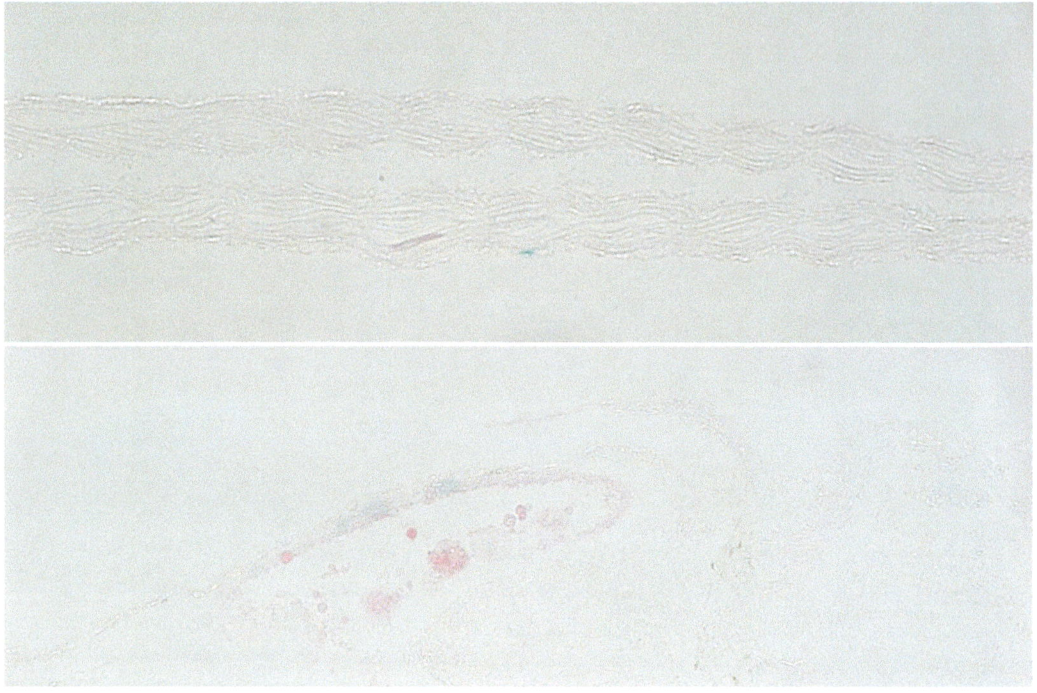

Fig. 6 10 μm sections through a *lacZ* stained 48 hpf zebrafish cut on a rotary microtome after the described method of dehydration and paraffin embedding

5. Prepare several wells in a multi-well glass dish with Histoclear (*see* Fig. 5c) and gently passage the zebrafish through them (*see* **Notes 12–14**).
6. Fill 2–3 molds with liquid paraffin on the hot plate of a tissue embedding station, then use a dissection needle to transfer each fish carefully over and wash 2× in liquid paraffin before embedding them (*see* **Note 15**).
7. Store the paraffin block overnight at −20 °C, then section the zebrafish on a rotary microtome and if necessary counterstain with Eosin (*see* Fig. 6).

4 Notes

1. The equipment listed is solely based on what we have been using, yet there are many other brands that might work equally well. This is more considered to be a guideline for what is needed in general not in particular.
2. Fertilized fish eggs develop fast, do not open all tanks at once.
3. No CO_2 necessary.

4. For each embryo batch (embryos from same mating tank) some should be left behind un-injected as control to estimate the percentage of healthy developing fish.
5. Too much fluid will make the embryos turn upon injection.
6. With little practice it is quite feasible to inject zebrafish embryos by hand, which is faster than using a device to hold and lower the needle.
7. Methylene Blue can be added to the egg water to prevent fungal growth.
8. For better photo documentation the fish might be gently mechanically dechorionated with fine Dumont#5 forceps or dissection needles.
9. For the ease of photo documentation after 16 hpf, the movement of embryos or fish might be slowed down on ice or they might be temporarily immobilized by adding Tricaine (3-amino benzoic acidethylester) at 100–200 mg/l to the egg water. Tricaine is an anesthetic drug and over dosage could be lethal to the animal.
10. After 20 hpf and before 24 hpf upcoming pigmentation might be suppressed by adding 0.15–0.2 mM 1-phenyl-2-thiourea (PTU) to the egg water. PTU should not be added prior to 15 hfp as it might negatively impact on embryo development and survival.
11. If *lacZ* staining is very limited it will be very difficult to visualize the zebrafish in the paraffin. Staining the zebrafish with Eosin is not absolutely necessary but it will help to see the zebrafish during the embedding step.
12. Avoid the use of plastic here, it will dissolve in the Histoclear.
13. For the first transfer from the EtOH into Histoclear a plastic pipet can still be used, but for the following transfers it is better to use a dissection needle as the plastic starts to dissolve in the Histoclear, including the one from pipet aid tips.
14. By the second or third wash the solution may suddenly get milky and the fish will shrink and look "plastic like," yet they will be good enough for sectioning.
15. Having the fish stand up in the mold is hard but possible with lots of patience. Careful the fish are very fragile.

Acknowledgements

The authors are grateful to Serene Lee and Song Jie as well as Drs Igor Kondrychyn, V Sivakamasundari, Zhen Li, Sumantra Chatterjee, Vladimir Korzh, and especially Mathavan Sinnakaruppan for invaluable advice, patience and access to the zebrafish equipment.

References

1. Shizuya H, Birren B, Kim UJ et al (1992) Cloning and stable maintenance of 300-kilobase-pair fragments of human DNA in Escherichia coli using an F-factor-based vector. Proc Natl Acad Sci U S A 89:8794–8797
2. Testa G, Zhang Y, Vintersten K et al (2003) Engineering the mouse genome with bacterial artificial chromosomes to create multipurpose alleles. Nat Biotechnol 21:443–447
3. Lee EC, Yu D, Martinez De Velasco J et al (2001) A highly efficient Escherichia coli-based chromosome engineering system adapted for recombinogenic targeting and subcloning of BAC DNA. Genomics 73:56–65
4. Muyrers JP, Zhang Y, Testa G et al (1999) Rapid modification of bacterial artificial chromosomes by ET-recombination. Nucleic Acids Res 27:1555–1557
5. Zhang Y, Buchholz F, Muyrers JP et al (1998) A new logic for DNA engineering using recombination in Escherichia coli. Nat Genet 20: 123–128
6. Dechiara TM, Poueymirou WT, Auerbach W et al (2009) VelociMouse: fully ES cell-derived F0-generation mice obtained from the injection of ES cells into eight-cell-stage embryos. Methods Mol Biol 530:311–324
7. Dechiara TM, Poueymirou WT, Auerbach W et al (2010) Producing fully ES cell-derived mice from eight-cell stage embryo injections. Methods Enzymol 476:285–294
8. Kraus P, Leong G, Tan V et al (2010) A more cost effective and rapid high percentage germline transmitting chimeric mouse generation procedure via microinjection of 2-cell, 4-cell, and 8-cell embryos with ES and iPS cells. Genesis 48:394–399
9. Chatterjee S, Bourque G, Lufkin T (2011) Conserved and non-conserved enhancers direct tissue specific transcription in ancient germ layer specific developmental control genes. BMC Dev Biol 11:63
10. Chatterjee S, Lufkin T (2011) Fishing for function: zebrafish BAC transgenics for functional genomics. Mol Biosyst 7:2345–2351
11. Asakawa K, Abe G, Kawakami K (2013) Cellular dissection of the spinal cord motor column by BAC transgenesis and gene trapping in zebrafish. Front Neural Circuits 7:100
12. Shakes LA, Du H, Wolf HM et al (2012) Using BAC transgenesis in zebrafish to identify regulatory sequences of the amyloid precursor protein gene in humans. BMC Genomics 13:451
13. Clark KJ, Urban MD, Skuster KJ et al (2011) Transgenic zebrafish using transposable elements. Methods Cell Biol 104:137–149
14. Suster ML, Abe G, Schouw A et al (2011) Transposon-mediated BAC transgenesis in zebrafish. Nat Protoc 6:1998–2021
15. Suster ML, Kikuta H, Urasaki A et al (2009) Transgenesis in zebrafish with the tol2 transposon system. Methods Mol Biol 561:41–63
16. Wixon J (2000) Featured organism: Danio rerio, the zebrafish. Yeast 17:225–231
17. Brittijn SA, Duivesteijn SJ, Belmamoune M et al (2009) Zebrafish development and regeneration: new tools for biomedical research. Int J Dev Biol 53:835–850
18. Grunwald DJ, Eisen JS (2002) Headwaters of the zebrafish – emergence of a new model vertebrate. Nat Rev Genet 3:717–724
19. Seabra R, Bhogal N (2010) In vivo research using early life stage models. In Vivo 24: 457–462
20. Gong Z, Ju B, Wang X et al (2002) Green fluorescent protein expression in germ-line transmitted transgenic zebrafish under a stratified epithelial promoter from keratin8. Dev Dyn 223:204–215
21. Zhang G, Gurtu V, Kain SR (1996) An enhanced green fluorescent protein allows sensitive detection of gene transfer in mammalian cells. Biochem Biophys Res Commun 227: 707–711
22. Culp P, Nusslein-Volhard C, Hopkins N (1991) High-frequency germ-line transmission of plasmid DNA sequences injected into fertilized zebrafish eggs. Proc Natl Acad Sci U S A 88:7953–7957
23. Kimmel CB, Ballard WW, Kimmel SR et al (1995) Stages of embryonic development of the zebrafish. Dev Dyn 203:253–310
24. Stuart GW, Mcmurray JV, Westerfield M (1988) Replication, integration and stable germ-line transmission of foreign sequences injected into early zebrafish embryos. Development 103: 403–412

Chapter 13

BACs-on-Beads™ (BoBs™) Assay for the Genetic Evaluation of Prenatal Samples and Products of Conception

Francesca Romana Grati, François Vialard, and Susan Gross

Abstract

BACs-on-Beads™ (BoBs™) is a new emerging technology, a modification of comparative genomic hybridization that can be used to detect DNA copy number gains and losses. Here, we describe the application of two different types of BoBs™ assays: (1) Prenatal BoBs (CE-IVD) to detect the most frequent syndromes associated with chromosome microdeletions, as well as the trisomy 13, 18 and 21, and (2) KaryoLite BoBs (RUO) which can detect aneuploidy in all chromosomes by quantifying proximal and terminal regions of each chromosomal arm. The interpretation of the results by BoBsoft™ software is also described. Although BoBs™ may not have the breadth and scope to replace chromosomal microarrays (array comparative genomic hybridization and single nucleotide polymorphism array) in the prenatal setting, particularly when a fetal anomaly has been detected, it is a well suited alternative for FISH or QF-PCR because BoBs™ is comparable, if not superior in terms of cost, turnaround time (TAT) and throughput and accuracy. BoBs™ also has the ability to detect significant fetal mosaicism (≥30 % with Prenatal BoBs and ≥50 % with KaryoLite BoBs). However, perhaps the greatest strength of this new technology is the fact that unlike FISH or QF-PCR, it has the ability to detect common microdeletion syndromes or additional aneuploidies, both of which may be easily missed despite excellent prenatal sonography. Thus, when BoBs™ is applied in the correct clinical setting and run and analyzed in appropriate laboratories this technique can improve and augment best practices with a personalization of prenatal care.

Key words Prenatal BACs on beads, KaryoLite BACs on beads, Prenatal diagnosis, Product of conception, Karyotype, Microdeletion, Microduplication, Aneuploidies

1 Introduction

BACs-on-Beads™ (BoBs™) is a microsphere-based suspension array technology supported by a Luminex® xMAP™ system. Briefly, the Luminex xMAP system is designed using a color encoding strategy: it incorporates 5.6 µm polystyrene microspheres that are internally dyed with two spectrally distinct fluorochromes. Mixing precise proportions of ten different amounts of each fluorochrome (red, emission wavelength, $\lambda_{em} > 650$ nm; infrared, $\lambda_{em} \sim 585$ nm), a set

Kumaran Narayanan (ed.), *Bacterial Artificial Chromosomes*, Methods in Molecular Biology, vol. 1227, DOI 10.1007/978-1-4939-1652-8_13, © Springer Science+Business Media New York 2015

(array) of 100 different microspheres with specific spectral addresses is created. Microsphere sets can then be distinguished by their spectral addresses so they can be used simultaneously allowing the multiplex measurement of up to 100 different analytes in a single experiment. Using the same strategy, the addition of four supplementary layers will theoretically permit the analysis of 500 different analytes in the future. Different analytes can be quantified since different capture molecules can be conjugated on the beads' surface allowing for multiple uses in many clinical laboratory areas: genotyping, cytokine and thyroid level measurements, cystic-fibrosis screening, genetic human lymphocyte antigen (HLA) typing, kinase testing, and allergy testing [1–13].

A third green fluorescent reporter molecule (R-phycoerythrin; λ_{em} ~ 530 nm) is used to quantify the target analyte [14]. Microspheres pass into the flow-cytometry Luminex analyzer and are individually assayed by two lasers: a 635-nm laser excites the two fluorochromes contained in each microsphere and performs the "identification" of analyte; a 532-nm laser excites the reporter fluorochrome and performs "quantification" of the analyte conjugated to each bead. High-speed digital signal processing makes a dual-classification of each microsphere thereby merging in only a few seconds the data resulting from the analysis of thousands of microspheres per sample. As compared to planar microarrays, suspension arrays have the benefits of ease of use, low cost, statistical superiority, faster hybridization kinetics and more flexibility in array preparation [14]. In BoBs™ assays, probes derived from bacterial artificial chromosomes (BACs) mapping in different chromosome regions have been immobilized on xMAP Luminex® beads. BACs-on-Beads is, therefore, a molecular cytogenetic technology comparable to fluorescence in situ hybridization (FISH) in a liquid format.

For prenatal diagnostic purposes, two types of BoBs™ assays are available: Prenatal BoBs (CE-IVD) and KaryoLite BoBs (RUO) (PerkinElmer, Wallac, Turku, Finland). The Prenatal BACs-on-Beads™ assay has been developed to detect gains and losses of DNA in chromosomal critical regions associated with nine microdeletion syndromes (Wolf-Hirschhorn, 4p16.3; Cri du Chat, 5p15.3-p15.2; Williams-Beuren, 7q11.2; Langer-Giedion, 8q23-q24; Prader-Willi/Angelman, 15q11-q12; Miller-Dieker, 17p13.3; Smith-Magenis, 17p11.2; DiGeorge-1 and 2, 22q11.2 and 10p14) and recurrent aneuploidies of chromosomes 13, 18, 21, X, and Y [15–19]. Each bead contains multiple copies of the same BAC that are immobilized on its surface. Five independent BACs-on-Beads probes are included for chromosomes 13, 18, 21, X and Y, and four to eight probes for each of the microdeletion critical regions [15–19].

KaryoLite™ BoBs™ assay provides dosage information about the proximal and terminal regions of each chromosome arm. Each

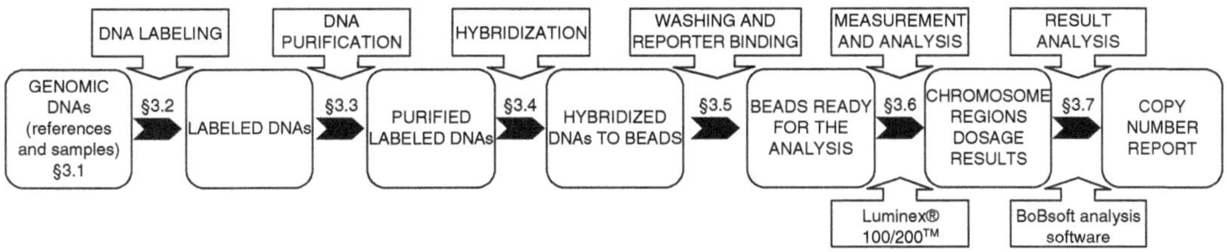

Fig. 1 BACs-ON-Beads workflow

bead is a composite of three neighboring BACs in equal amount according to the Human Genome Build GRCh37.2/hg19 Assembly [18]. The results represent the sum of the three BACs intensity. A total of two beads per arm are used providing dosage information on terminal and pericentromeric regions. For acrocentric chromosomes, a total of three beads provide dosage information on pericentromeric, interstitial and terminal regions of the q-arms.

BACs-on-Beads products are manufactured under or encompassed by one or more patent applications owned by PerkinElmer Inc. Figure 1 depicts the scheme of the BoBs experiment: genomic DNA is labeled (Labeling of genomic DNA), purified (Purification of labeled DNA), hybridized to the BACs-on-Beads probes (DNA hybridization), washed and bound to the reporter molecule and then washed again (DNA washing and Reporter binding). Thereafter, the fluorescent signals are measured and the results analyzed (Measurement and analysis).

BoBs assays comprise the analysis of the test sample and of a female and a male reference DNAs throughout the hybridization to their complementary BACs-on-Beads probes. Test sample analysis is performed in singlicate and reference samples are analyzed in duplicate. After hybridization, the signal intensities are read using the Luminex®100/200™ instrument system with either IS 2.3 or xPONENT®3.1 software versions. The mean green fluorescence intensity of the sample DNA hybridized to a specific BAC probe is compared to the mean green fluorescence intensity of reference DNA binding to that same specific BACs-on-Beads probe. Ratios of sample versus male reference and sample versus female reference are calculated for each probe.

A sample is defined as normal disomic when the ratio between the green fluorescence intensities in the test and in male/female reference is approximately 1.0 for all analyzed loci. A sample is defined as duplicated/deleted in a chromosome locus when the fluorescence in the test is higher/lower than that in the reference: single copy gains and losses generate ratios ranging from 1.3 to 1.4 and from 0.6 to 0.8, respectively. A more detailed description of the statistical analytic approach is described below under "Result Analysis."

The BoBsoft™ software (PerkinElmer Wallac, Turku, Finland) generates a "Results tab" with a numeric and graphic representation of probe and group ratios against female (F—red line/dots) and male (M—blue line/dots) references. A normal disomic pattern for autosomes is defined when the both red and blue lines/dots are included in the normal expected ratio threshold (Fig. 2a, b). Deletion or duplication is defined when both lines/dots are outside the normal expected range in the haploinsufficiency (left) or overdosage (right) window of the scheme, respectively (Fig. 2c–h). A female probe pattern is defined when X and Y probe ratios are included in the expected range for a female sample (red line/dots inside and blue line/dots outside the normal expected X/Y range) (Fig. 2a); a male pattern is defined by a reverse pattern (blue line/dots inside and red line/dots outside the normal expected X/Y range) (Fig. 2b); sex chromosomes aneuploidies are depicted as deviations from normal patterns and consequently interpreted (Fig. 2i, l).

2 Materials

2.1 Reagents

2.1.1 Reference DNAs

1. Human female genomic DNA (Promega Corporation).
2. Human male genomic DNA (Promega Corporation).

2.1.2 BoBs Package P1 (Perkin Elmer)

The reagents in this package should be stored at −30 to −16 °C.

1. Random Primer Solution (two vials): oligonucleotides are resuspended in Tris–HCl buffered solution with $MgCl_2$ and <0.2 % 2-mercaptoethanol (pH 7.2) in a volume of 980 μL/vial.
2. Biotin-dNTP Mix (one vial): deoxyribonucleotide phosphates are resuspended in TE (Tris-EDTA) buffer (pH 7.0) in 590 μL.
3. Polymerase (one vial): DNA polymerase is resuspended in potassium phosphate solution (pH 7.0) in 120 μL.
4. Hybridization Buffer (one vial): a ready-to-use SSC (saline sodium citrate) buffer with blocking DNA, and dextran sulfate in a 1,100 μL volume. This buffer contains 50 % formamide.

2.1.3 BoBs Package P2 (Perkin Elmer)

The reagents in this package should be stored at +2 to +8 °C.

1. Sample Diluent (two vials): ultra-filtered autoclaved water, 1,000 μL/vial.
2. BACs-on-Beads Mix (one vial): contains fluorescently coded Luminex® beads in TE buffer with DNA probes immobilized onto the beads, in 110 μL (store protected from light).
3. Reporter Concentrate (one vial): is a 250-fold concentrate of streptavidin phycoerythrin, in 45 μL (store protected from light).

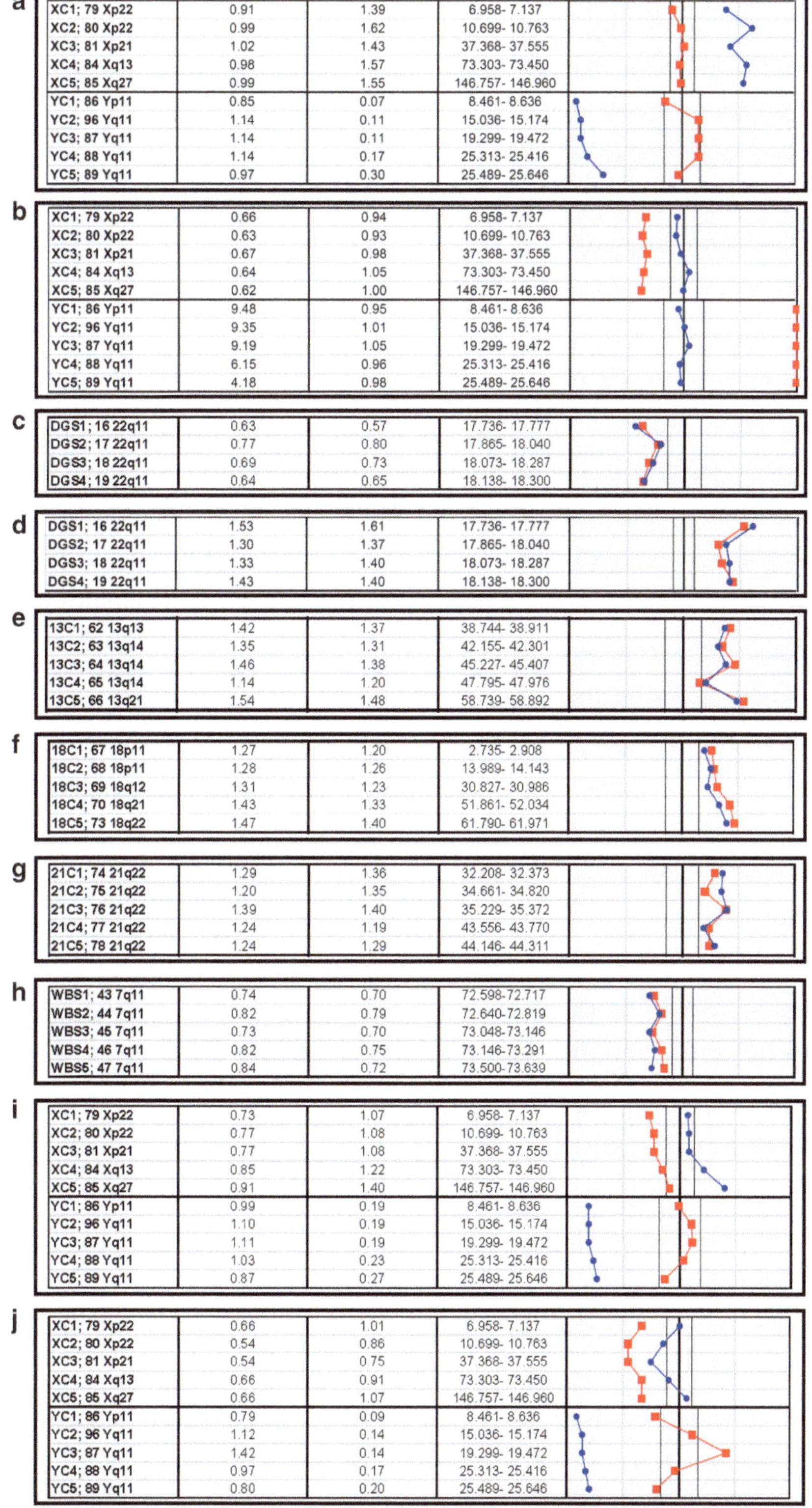

a

XC1; 79 Xp22	0.91	1.39	6.958- 7.137
XC2; 80 Xp22	0.99	1.62	10.699- 10.763
XC3; 81 Xp21	1.02	1.43	37.368- 37.555
XC4; 84 Xq13	0.98	1.57	73.303- 73.450
XC5; 85 Xq27	0.99	1.55	146.757- 146.960
YC1; 86 Yp11	0.85	0.07	8.461- 8.636
YC2; 96 Yq11	1.14	0.11	15.036- 15.174
YC3; 87 Yq11	1.14	0.11	19.299- 19.472
YC4; 88 Yq11	1.14	0.17	25.313- 25.416
YC5; 89 Yq11	0.97	0.30	25.489- 25.646

b

XC1; 79 Xp22	0.66	0.94	6.958- 7.137
XC2; 80 Xp22	0.63	0.93	10.699- 10.763
XC3; 81 Xp21	0.67	0.98	37.368- 37.555
XC4; 84 Xq13	0.64	1.05	73.303- 73.450
XC5; 85 Xq27	0.62	1.00	146.757- 146.960
YC1; 86 Yp11	9.48	0.95	8.461- 8.636
YC2; 96 Yq11	9.35	1.01	15.036- 15.174
YC3; 87 Yq11	9.19	1.05	19.299- 19.472
YC4; 88 Yq11	6.15	0.96	25.313- 25.416
YC5; 89 Yq11	4.18	0.98	25.489- 25.646

c

DGS1; 16 22q11	0.63	0.57	17.736- 17.777
DGS2; 17 22q11	0.77	0.80	17.865- 18.040
DGS3; 18 22q11	0.69	0.73	18.073- 18.287
DGS4; 19 22q11	0.64	0.65	18.138- 18.300

d

DGS1; 16 22q11	1.53	1.61	17.736- 17.777
DGS2; 17 22q11	1.30	1.37	17.865- 18.040
DGS3; 18 22q11	1.33	1.40	18.073- 18.287
DGS4; 19 22q11	1.43	1.40	18.138- 18.300

e

13C1; 62 13q13	1.42	1.37	38.744- 38.911
13C2; 63 13q14	1.35	1.31	42.155- 42.301
13C3; 64 13q14	1.46	1.38	45.227- 45.407
13C4; 65 13q14	1.14	1.20	47.795- 47.976
13C5; 66 13q21	1.54	1.48	58.739- 58.892

f

18C1; 67 18p11	1.27	1.20	2.735- 2.908
18C2; 68 18p11	1.28	1.26	13.989- 14.143
18C3; 69 18q12	1.31	1.23	30.827- 30.986
18C4; 70 18q21	1.43	1.33	51.861- 52.034
18C5; 73 18q22	1.47	1.40	61.790- 61.971

g

21C1; 74 21q22	1.29	1.36	32.208- 32.373
21C2; 75 21q22	1.20	1.35	34.661- 34.820
21C3; 76 21q22	1.39	1.40	35.229- 35.372
21C4; 77 21q22	1.24	1.19	43.556- 43.770
21C5; 78 21q22	1.24	1.29	44.146- 44.311

h

WBS1; 43 7q11	0.74	0.70	72.598-72.717
WBS2; 44 7q11	0.82	0.79	72.640-72.819
WBS3; 45 7q11	0.73	0.70	73.048-73.146
WBS4; 46 7q11	0.82	0.75	73.146-73.291
WBS5; 47 7q11	0.84	0.72	73.500-73.639

i

XC1; 79 Xp22	0.73	1.07	6.958- 7.137
XC2; 80 Xp22	0.77	1.08	10.699- 10.763
XC3; 81 Xp21	0.77	1.08	37.368- 37.555
XC4; 84 Xq13	0.85	1.22	73.303- 73.450
XC5; 85 Xq27	0.91	1.40	146.757- 146.960
YC1; 86 Yp11	0.99	0.19	8.461- 8.636
YC2; 96 Yq11	1.10	0.19	15.036- 15.174
YC3; 87 Yq11	1.11	0.19	19.299- 19.472
YC4; 88 Yq11	1.03	0.23	25.313- 25.416
YC5; 89 Yq11	0.87	0.27	25.489- 25.646

j

XC1; 79 Xp22	0.66	1.01	6.958- 7.137
XC2; 80 Xp22	0.54	0.86	10.699- 10.763
XC3; 81 Xp21	0.54	0.75	37.368- 37.555
XC4; 84 Xq13	0.66	0.91	73.303- 73.450
XC5; 85 Xq27	0.66	1.07	146.757- 146.960
YC1; 86 Yp11	0.79	0.09	8.461- 8.636
YC2; 96 Yq11	1.12	0.14	15.036- 15.174
YC3; 87 Yq11	1.42	0.14	19.299- 19.472
YC4; 88 Yq11	0.97	0.17	25.313- 25.416
YC5; 89 Yq11	0.80	0.20	25.489- 25.646

Fig. 2 Examples of different BoBs™ profiles. (**a** and **b**) Normal male and female profiles, respectively; (**c**–**h**) Abnormal autosome profiles: (**c** and **d**) microdeletion and microduplication of 22q11 region; (**e**–**g**) trisomies of chromosomes 13, 18 and 21, respectively; (**h**) microdeletion of 7q11 region; (**i** and **l**) Abnormal sex chromosomes profiles: (**i**) mos46,XY/45,X; (**l**) 45,X

4. Reporter Diluent (one bottle): is a ready-to-use 11.2 mL solution of phosphate buffered saline with BSA and Surfact-Amps 20 (a registered trademark of Pierce Biotechnology, Inc.).
5. Wash Buffer 1 (one bottle): is 12.5 mL of ready-to-use SSC buffer containing 50 % formamide.
6. Wash Buffer 2 (one bottle): is 46 mL of ready-to-use phosphate buffered saline with Surfact-Amps 20.

2.1.4 Purification Kit for Labeled DNA

1. Macherey Nagel NucleoFast10 96 PCR clean-up Kit (Macherey-Nagel) with TE buffer (10 mM Tris–HCl, pH 8.0, 1 mM EDTA) for washing and elution. For an alternative kit, *see* **Note 1**.

2.1.5 Luminex Reagents

1. xMAP® Sheath Fluid (Luminex Corporation).
2. xMAP® Calibration Microspheres Kit (Luminex Corporation).
3. xMAP® Verification Microspheres Kit (Luminex Corporation).

2.1.6 Disposables

1. Thin wall PCR tubes (250 μL) or strips with caps.
2. Low-retention tips (200 μL volume).
3. Polypropylene tubes (15 mL).
4. Hard shell skirted 96-well PCR plate suitable for the PCR thermal cycler and incubator/shaker.
5. MultiScreenHTS HV (Millipore Corporation), 0.45 μm, clear 96-well filter plate suitable for the incubator/shaker.
6. Microseal F Film PCR Sealer (Bio-Rad Laboratories, Inc.) aluminum foil sealer for microplates, PCR grade.
7. 4titude FrameStar 96 skirted PCR plates and Strips of Flat Optical Caps (4titude Ltd.) for DNA hybridization (4titude Ltd).
8. Clear plastic adhesive plate sealer.
9. Liquid reagent reservoirs (25 mL volume) (optional).
10. Paper towels.
11. Aluminum foil.

2.2 Equipment

1. Microliter UV spectrophotometer (e.g., Thermo Scientific NanoDrop, Thermo Electron).
2. Filter plate vacuum apparatus (*see* **Note 2**) with a pump and a waste collection vessel (e.g., MultiScreen$_{HTS}$ Vacuum Manifold, Vacuum Filtering Flask 1 L, and No. 8 stopper).
3. 96-well PCR thermal cycler with a heated lid accommodating full skirted hard shell 96-well PCR plates.
4. Microcentrifuge capable of reaching 12,000 × *g*.
5. Minicentrifuge for PCR strips (e.g., Spectrafuge Mini with strip tube rotor, Labnet International, Inc.).

6. Multichannel pipette (volume range 5–300 μL).
7. Plate sealing roller (optional).
8. Incubator/shaker(s) capable of maintaining 37 (±1)°C, 50 (±1)°C, 52 (±1)°C, and 1,200 rpm with a 2–3 mm diameter orbit (3×*g*-force). The device(s) must accommodate to a 96-well PCR plate preserving the seal, and to a 96-well filter plate, e.g., TriNEST™ Microplate Incubator Shakers (PerkinElmer, Inc.) or equivalent.
9. Luminex® 100/200™ instrument system (Luminex Corp.).
10. BoBsoft analysis software (PerkinElmer, Inc.)
11. Nanodrop DNA quantifier.

2.3 Equipment and Supplies Needed for Each Phase

2.3.1 Labeling of Genomic DNA

1. MilliQ Water.
2. Pipettes and low-retention tips (volume range 0.5–1,000 μL).
3. 250 μL thin wall PCR tubes or strips with caps (if performing the labeling reaction in PCR tubes).
4. Hard shell skirted 96-well PCR plate fitting the PCR thermal cycler in use (if performing the labeling reaction in a 96-well plate).
5. Minicentrifuge for PCR strips (if performing the labeling reaction in PCR tube strips).
6. Aluminum foil sealer for microplates.
7. PCR grade or Strips of eight Flat Optical Caps for PCR plates.
8. 1.5 mL microcentrifuge tubes.
9. 96-well PCR thermal cycler with a heated lid, preprogrammed for step 1: 98 °C for 5 min, then 37 °C steady state.
10. Vortexer.
11. Microcentrifuge capable of reaching 12,000×*g*.

2.3.2 Purification of Labeled DNA Using Macherey-Nagel NucleoFast 96 PCR Plate

1. Macherey-Nagel NucleoFast 96 PCR plate.
2. Pipettes and low-retention tips (volume range 0.5–1,000 μL).
3. Filter plate vacuum apparatus with a pump and a waste collection vessel.
4. 96-well PCR plate.
5. Multichannel pipette (5–50 μL).
6. 25 mL liquid reagent reservoir.

2.3.3 Purification of Labeled DNA Using PureLink PCR Purification Kit (See **Note 3**)

1. Invitrogen PureLink PCR Purification Kit.
2. Extra bottle of PureLink Wash Buffer for PureLink PCR Purification Kit.
3. Microcentrifuge capable of reaching 12,000×*g*.
4. Microliter UV spectrophotometer.

5. Pipettes and low-retention tips (volume range 0.5–1,000 μL).
6. 100 % Isopropanol (as specified in PureLink PCR Purification Kit User Manual).
7. 96–100 % EtOH (as specified in PureLink PCR Purification Kit User Manual).
8. Timekeeper.
9. 96-well PCR plate.

2.3.4 Measurement of DNA Concentration

1. Microliter UV spectrophotometer.
2. 1× TE buffer (10 mM Tris–HCl, pH 8.0, 1 mM EDTA) or Elution Buffer (E1).
3. Pipette and low-retention tips (volume range 0.5–10 μL).

2.3.5 DNA Hybridization

1. Pipettes and low-retention tips (volume range 0.5–1,000 μL).
2. Incubator/shaker.
3. Preheated to 52 °C.
4. Hard shell skirted 96-well PCR plate fitting the PCR thermal cycler and incubator/shaker in use (e.g., 4titude FrameStar 96 skirted PCR plates).
5. Aluminum foil sealer for microplates.
6. PCR grade or Strips of eight Flat Optical Caps for 4titude FrameStar 96 skirted PCR plates.
7. 96-well PCR thermal cycler with a heated lid preprogrammed for Subheading 3.5, **step 13** (85 °C for 5 min, then 52 °C steady state).
8. 1× TE buffer (10 mM Tris–HCl, pH 8.0, 1 mM EDTA) or Elution Buffer (E1).
9. Multichannel pipette (volume range 5–300 μL).
10. Vortexer.
11. 1.5 mL microcentrifuge tubes.
12. Low-retention tips.
13. Plate sealing roller (if using the aluminum foil sealer).

2.3.6 DNA Washing and Reporter Binding

1. Luminex® 100/200™ instrument system, with product-specific template or protocol file uploaded (available at the PerkinElmer website).
2. Incubator/shaker(s) (temperatures needed 50 and 37 °C): a device capable of cooling or two separate incubator/shakers are recommended.
3. Multichannel pipette (volume range 30–300 μL).
4. Pipettes and low-retention tips (volume range 0.5–1,000 μL).
5. Aluminum foil.

6. MultiScreen$_{HTS}$ HV 0.45 μm, clear 96-well filter plate fitting the incubator/shaker in use.
7. Filter plate vacuum apparatus with a pump and a waste collection vessel.
8. Clear plastic adhesive plate sealer.
9. 15 mL polypropylene tubes.
10. Paper towels.
11. 25 mL liquid reagent reservoirs.

3 Methods

3.1 Preparation of Reagents and DNAs

1. Remove the following reagents from the freezer and refrigerator and place them in room temperature 15–30 min prior to use to facilitate maximum solubility and mixing:
   - Genomic DNAs: test sample/s, female reference and male reference
   - Random Primer Solution
   - Biotin-dNTP Mix
   - Polymerase
   - Sample Diluent

   Make sure that the reagents are thawed before use.
2. Prior to use, vigorously vortex and then spin down all sample DNAs, reference DNAs and reagents, except for the Polymerase.
3. Quantify DNA using a Nanodrop.
4. Dilute M/F Reference DNAs with sample diluent to obtain a concentration of 2 ng/μL; use as input DNA 24 μL of this solution (total ~50 ng).
5. Dilute sample DNA with sample diluent to obtain a concentration of 10 ng/μL; use as input DNA 24 μL of this solution (total 240 ng) (*see* **Note 4**).

3.2 Labeling of Genomic DNA

1. Prior to use, vigorously vortex and then spin down all sample DNAs, reference DNAs and reagents, except for the Polymerase.
2. Prepare two labeling reactions for each reference DNA. Transfer 50 ng of female and male reference DNA into the corresponding tubes/wells, respectively. Add Sample Diluent to bring the volume in all tubes to 24 μL.
3. Prepare one labeling reaction for each sample DNA. Transfer 240 ng (minimum 50 ng) of sample DNA into the corresponding tubes/wells (*see* **Note 5**).

Table 1
Volumes needed to prepare the Labeling Mix

	Vol/sample (μL)	Vol × 6 reactions	Vol × 8 reactions	Vol × 12 reactions	Vol × 20 reactions	Vol × *n* reactions
Biotin-dNTP mix	5.5	33	44	66	110	5.5 × *n*
Polymerase	1.1	6.6	8.8	13.2	22	1.1 × *n*

4. Vortex and spin down Random Primer Solution and add 20 μL into each tube/well, pipetting each one up and down five to ten times to mix, or vortex the tubes well and spin down.
5. Cap the tubes/seal the wells and place into a PCR thermal cycler.
6. Denature at 98 °C for 5 min, then cool to 37 °C. Continue with the next step as soon as the PCR machine has reached 37 °C. Leave the thermal cycler in 37 °C steady state to continue the incubation in Subheading 3.2, **step 9**.
7. Prepare Labeling Mix in a 1.5 mL centrifuge tube. Use Table 1 to calculate the volumes needed to prepare the Labeling Mix. Vortex the Biotin-dNTP Mix (but not the Polymerase) and spin both down prior to dispensing. The extra volumes needed for pipetting are already included in the volumes shown in Table 1. Mix well by pipetting five to ten times up and down and spin down the Labeling Mix. Pipette the Polymerase carefully to avoid extra droplets outside the tip.
8. Pipette 6 μL of the Labeling Mix into each reaction tube/well. Mix the whole reaction volume well by pipetting up and down at least five times and spin down. Do not create bubbles.
9. Cap the tubes/seal the wells tightly and incubate for 60 min in the PCR thermal cycler at 37 °C. Replace unused labeling reagents to their recommended storage conditions.
10. While waiting for the reaction to finish, proceed to set up for the next step (Purification of labeled DNA). Set the shaker-incubator at 37 °C. When the labeling program of the thermal cycler is complete, remove tubes from the thermal cycler. If purification cannot be performed immediately after the labeling, labeled samples/references can be stored at 4 °C for few hours or at −20 °C overnight.

3.3 Purification of Labeled DNA Using NucleoFast 96 PCR Plate

1. Place the Hybridization Buffer at room temperature.
2. Pipette the labeled DNA up and down five times.
3. Transfer 50 μL of each labeled DNA sample into the NucleoFast 96 PCR purification plate (for an alternative DNA purification kit *see* **Note 3**) well with a multichannel pipette. Slowly dispense the sample directly onto the membrane; avoid dispensing of the samples to the inner walls of the wells.

4. Place the purification plate on top of the vacuum manifold. Apply a vacuum and maintain at approximately 190 mmHg until no liquid is left (normally about 10 min). Switch off vacuum source.
5. Add 50 μL of 1× TE buffer to each well. Apply a vacuum and maintain at approximately 190 mmHg until no liquid is left (normally about 10 min). Switch off vacuum source.
6. Dry the bottom of the filter plate by pressing it firmly on a stack of paper towels.
7. Add 25 μL of 1× TE buffer to each sample well.
8. Place the plate in a Microplate Incubator Shakers and shake at room temperature, 3×*g* (1,200 rpm) for 5–10 min.
9. Transfer 20 μL of each labeled DNA to a fresh PCR plate for dilution as needed.
10. Measure and record the DNA concentration with a microliter UV spectrophotometer. Use 1.5 μL of each eluted DNA sample for the measurement of DNA concentration. For DNA concentration below 200 ng/μL and for storage of DNA *see* **Note 6**.
11. If the concentration is 150–200 ng/μL, proceed without making an adjustment. After purification and concentration measurement the volume of the purified DNA per well should be 18.5 μL (20.0–1.5 μL) if purified with the Nucleofast kit (*see* **Note 7** for dilution with the alternative kit). If the initial concentration is >200 ng/μL, add suitable TE buffer (E1 or TE) to the sample well to lower the DNA concentration to 200 ng/μL. TE Volume to add = ([measured concentration/200] × 18.5) − 18.5 μL.

3.4 DNA Hybridization

1. Remove the Hybridization Buffer from freezer and BACs-on-Beads Mix from refrigerator.
2. Set microplate incubator/shaker to 52 °C.
3. Create a plate map for hybridizations so that the correspondence of each sample to the plate well is recorded. An example plate map is shown in Table 2. The reference DNAs (F and M) are labeled in duplicate and these two labeling reactions are hybridized in their own wells.
4. Use Table 3 to calculate the volumes needed to prepare the Hybridization Bead Mix for the hybridizations. The extra volumes needed for pipetting are already included in the volumes shown in Table 3.
5. Make sure the Hybridization Buffer has reached ambient temperature. Vortex the Hybridization Buffer vigorously three times for 10 s.

Table 2
Example of plate map

	1	2	3	4	5	6	7	8	9	10	11	12
A	F	S5										
B	F	S6										
C	M	S7										
D	M	S8										
E	S1	S9										
F	S2	S10										
G	S3	S11										
H	S4	S12										

F female reference, *M* male reference, *S* sample

Table 3
Volumes needed to prepare the Hybridization bead Mix

	Vol/sample (μL)	Vol × 8 reactions	Vol × 16 reactions	Vol × 24 reactions	Vol × 32 reactions	Vol × *n* reactions
Hyb Buffer	11	88	176	264	352	$11 \times n$
BACs on beads	1	8	8	24	32	$1 \times n$

6. Pipette slowly, due to the high viscosity of the Hybridization Buffer, the required volume of Hybridization Buffer into a microcentrifuge tube of appropriate size to achieve the total volume of the Hybridization Bead Mix listed in Table 3.
7. Vortex BACs-on-Beads Mix vigorously three times for 10 s immediately before use and pipette the required volume into the Hybridization Bead Mix tube. Protect beads from extended light exposure. Beads are sensitive to photobleaching. To ensure homogenous suspension of BACs-on-Beads Mix, pipette the required volume immediately after vortexing and visually inspect that the solution is opaque.
8. Quick spin the Hybridization Bead Mix and vortex three times for 10 s.
9. Transfer 11 μL of Hybridization Bead Mix to each plate well according to plate map in Table 2. Pipette slowly due to the high viscosity of the Hybridization Bead Mix. Use the low-retention tips. The beads will remain in suspension for approximately 15 min. If subsequent pipetting steps take longer than this, repeat the quick spin and vortex as needed.

10. Using a multichannel pipette with low-retention tips transfer 5 μL of labeled sample DNA to the bottom of each well containing Hybridization Bead Mix according to Table 2 plate layout. If there are many samples, do not hesitate to vortex again.
11. Seal the hybridization plate with PCR foil plate sealer or with 4titude optical caps (if using the 4titude plate). Use a roller or finger pressure to ensure tight sealing of the plate sealer.
12. Transfer the sealed hybridization plate into Microplate Incubator Shakers at 52 °C and shake for 5 min at 2×*g* (800 rpm).
13. Place the sealed hybridization plate into a thermal cycler and denature at 85 °C for 5 min, then cool to 52 °C (preset program).
14. When the temperature has reached 52 °C remove the hybridization plate from the thermal cycler and ensure tight sealing with fingers or a roller. Plate may still be hot after denaturation (*see* **Note 8**).
15. Place the plate immediately into the preprogrammed incubator/shaker at 52 °C and 2×*g* (800 rpm) for overnight hybridization (16–20 h).
16. If using only one Microplate Incubator Shakers, place a cool block in 2–8 °C or in freezer to be used later in Subheading 3.5, **step 16**.

3.5 DNA Washing and Reporter Binding

1. Place the Wash Buffer 1, Wash 2, Reporter Concentrate, Reporter Diluent at room temperature.
2. Calculate the amount of Wash Buffer 1 needed (*see* Table 4). A total of 125 μL is needed per well (includes extra volume needed for multichannel pipetting).
3. Remove the sealed hybridization plate from the 52 °C incubator/shaker and immediately set the temperature of the device to 50 °C for the subsequent wash step. Check visually that volume is approximately same in each well. If a well has a lower volume, the results might be affected. Write down the well number for data interpretation.
4. Add the required volume of Wash Buffer 1 to a reagent reservoir. Carefully remove the plate sealer, and use a multichannel

Table 4
Calculus the amount of Wash Buffer 1

Vol per well	Vol × 8 reactions	Vol × 16 reactions	Vol × 24 reactions	Vol × 32 reactions	Vol × *n* reactions
125 μL	1 mL	2 mL	3 mL	4 mL	125 × *n* μL

Table 5
Volumes needed to prepare the Reporter Mix

	Vol/sample (μL)	Vol × 10 reactions (μL)	Vol × 14 reactions (μL)	Vol × 16 reactions (μL)	Vol × 21 reactions (μL)	Vol × *n* reactions
Reporter diluent	112.5	1,125	1,575	1,800	2,362.5	112.5 × *n*
Reporter concentrate	0.45	4.5	6.3	7.2	9.45	0.45 × *n*

pipette to add 100 μL of Wash Buffer 1 to each well. Ensure thorough mixing by pipetting up and down for ten times rapidly (avoid bubbles formation). Wash Buffer 1 contains formamide, therefore avoid inhalation and/or contact with the solution.

5. Wait until incubator/shaker has reached 50 °C, and incubate the plate for 20 min in the incubator/shaker with 1,200 rpm.
6. Set the other incubator/shaker to 37 °C if there is one.
7. While incubating prepare a Reporter Mix in a 15 mL polypropylene tube: use Table 5 to calculate the volumes needed to prepare the Reporter Mix. The extra volumes needed for pipetting are already included in the volumes shown in Table 5.
8. Vortex the Reporter Concentrate for 5 s, spin down, and add into the Reporter Mix tube according to the calculation (Table 5). Immediately wrap the tube in foil to protect from photobleaching. Pipette the Reporter Concentrate carefully to avoid extra droplets outside the tip.
9. Vortex the Reporter Mix tube for 5 s.
10. Use a 0.45 μm filter plate for washing and reporter binding and create a plate map before transferring hybridized beads onto the filter plate, or follow the same plate map used for hybridization.
11. Seal the unused wells of the filter plate with adhesive plastic sealer to ensure that adequate vacuum is applied to the wells in use.
12. Calculate the amount of Wash Buffer 2 needed. A volume of 0.47 mL is needed per well (includes extra volume needed for multichannel pipetting). Add needed volume of Wash Buffer 2 into a reagent reservoir.
13. Use a multichannel pipette to add 30 μL of Wash Buffer 2 to each well to be used on the filter plate to wet the filter membrane. Be careful not to touch the filter membrane not to damage it.

14. Remove the hybridization plate from the 50 °C incubator/shaker. Using a multi-channel pipette with low-retention tips, immediately transfer the contents of the well to the prewetted filter plate. Do not touch the filter membrane.
15. If there is not another Microplate Incubator Shakers, to cool it at 37 °C rapidly, add the block previously cooled (*see* Subheading 3.4, **step 16**). Check the plate map and orientation of the filter plate.
16. Switch on the laser of the Luminex®100/200™ instrument system.
17. Use vacuum manifold at 125 mmHg for approximately 5 s to empty the filter plate wells as quickly as possible after the transfer to the filter plate, avoiding drying filters. Apply vacuum manifold sufficient to remove liquid from the filter plate wells in approximately 5 s. Manifold vacuum gauges may not be accurate and reliable.
18. Add 100 μL of Wash Buffer 2 into each well. Be careful not to touch the filter membrane. Repeat the Subheading 3.5, **step 17**.
19. Dry the bottom of the filter plate by pressing it firmly on a stack of paper towels.
20. Add 100 μL of Reporter Mix into each filter plate well in use. Be careful not to touch the filter membrane. Do not cover the wells in use. Make sure the incubator/shaker has reached 37 °C. Wait no longer than 15 min keeping the filter plate at room temperature protected from light prior to incubation at 37 °C.
21. Incubate the filter plate in the incubator/shaker for 30 min at 37 °C and 3 × *g* (1,200 rpm) shaking.
22. After the Reporter Mix incubation, empty the filter plate wells using the vacuum apparatus as in Subheading 3.6, **step 17**.
23. Add 100 μL of Wash Buffer 2 into each well. Be careful not to touch the filter membrane.
24. Dry the bottom of the plate by pressing firmly down on a stack of paper towels.
25. Add 100 μL of Wash Buffer 2 into each well. Be careful not to touch the filter membrane.
26. Do not empty the wells. Resuspended beads are read with the Luminex®100/200™ instrument system in next step (Assay measurement). If the measurement cannot be started within 15 min, protect the plate from light until the instrument is ready.

3.6 Assay Measurement

1. Make sure that the daily Start-up maintenance of Luminex® 100/200™ instrument system has been performed, that calibration and verification are current, and that the probe height is properly adjusted for use with the filter plate.

Table 6
Values for the acceptance of a BoBsTM assay after the performance verification evaluations

Parameter	Quality criteria
Reference average mean fluorescent intensity (MFI)	≥180
Reference autosomal CV	≤6.0 %
Reference total × separation	≥0.6
Sample autosomal CV	≤8.0 %

2. Open the product-specific protocol file or template.
3. Name output folder as appropriate.
4. Enter the sample name to the well map (F, M, and sample identifiers). This step can be done in advance, e.g., during incubation, but must be done before reading the plate. Make sure that the well map and the plate map from Subheading 3.5, **step 10** are identical. The references should be named always as "M" for the male reference and "F" for the female reference in order to the BoBsoft analysis software to recognize the references.
5. Load the filter plate into the Luminex® 100/200™ instrument. Follow the instrument instructions, using the appropriate product-specific template or protocol file to start reading.
6. After reading, the unused filter plate wells can be reused. Mark the used columns, use vacuum manifold to empty the filter plate wells, and seal the used, emptied wells.
7. Discard Wash Buffers remaining in reservoirs. Wash Buffer 1 contains formamide, discard according to local regulations.

3.7 Result Analysis

1. Analyze and visualize the results with BoBsoft (PerkinElmer Wallac, Turku, Finland). Details of the software is provided in **Note 9**.
2. Look at quality control, expected values: typical values for the acceptance of a BoBs™ assay after the performance verification evaluations are described in Table 6. Samples not meeting the criteria should be reported as "failed."
3. Interpret the Results (*see* **Note 10**).

4 Notes

1. Alternative kit for labeled DNA purification (Subheading 2.1.4): Invitrogen PureLink PCR Purification Kit (Invitrogen Corporation) with 100 % Isopropanol and 96–100 % EtOH (as specified in PureLink PCR Purification Kit User Manual).

2. Equipment (Subheading 2.2): Filter plate vacuum apparatus must accommodate Millipore HTS series filter plates and MACHEREY-NAGEL NucleoFast 96 PCR clean-up plates and it must be able to produce a controlled vacuum for the installed filter plate.
3. Preparation of DNAs (Subheading 3.1): Twenty-four microliters of each sample is required for the labeling reactions. Samples with concentrations greater than 10 ng/μL need to be diluted to 10 ng/μL with Sample Diluent. Samples between 2.1 ng/μL and 10 ng/μL can be used as such. Samples lower than 2.1 ng/μL have insufficient DNA for the assay.
4. Labeling can be carried out in PCR plates to aid in streamlining the assay. If using PCR tubes, use permanent marker to identify 250 μL PCR tubes or strips for female and male reference and samples.
5. An alternative kit for DNA purification is the PureLink PCR Purification Kit (Life Technologies) (Subheading 3.3):
   (a) Place the Hybridization Buffer at room temperature.
   (b) Prepare the reagents as instructed in PureLink User Manual, Binding Buffer (B2) and Elution Buffer (E1):
   (c) Mark purification columns and elution tubes to match the labeling reactions. Use a permanent marker.
   (d) • If labeling is performed in tubes, add 200 μL of Binding Buffer (B2) into each labeling reaction tube from Subheading 3.2, **step 9**. Pipette up and down five times and transfer the whole content of each tube into the corresponding purification column.
   • If labeling is performed in wells, add 100 μL of Binding Buffer (B2) into each reaction well from Subheading 3.2, **step 9**. Pipette up and down five times and transfer the whole content of each well into the corresponding purification column and add additional 100 μL of Binding Buffer (B2) into each purification column.
   (e) Incubate columns for 5 min at room temperature.
   (f) Centrifuge incubated columns 12,000 × *g* for 3 min. Discard the flow-through and place each column back to the same collection tube.
   (g) Add 650 μL of PureLink Wash Buffer into the columns and centrifuge columns 12,000 × *g* for 1 min, discard the flow-through.
   (h) Repeat Subheading 3.3, **step 7**.

(i) Centrifuge columns 12,000×*g* for 3 min to remove remaining residual PureLink Wash Buffer.

(j) Place each column into a corresponding elution tube marked with permanent lab marker. Collection tubes with the flow-through can be discarded.

(k) Add 15 μL of Elution Buffer (E1) onto the filter in the center of column. Do not touch the filter. Incubate for 10 min at room temperature.

(l) Centrifuge columns 12,000×*g* for 5 min to elute purified labeled DNA.

(m) Transfer the labeled DNA to a fresh PCR plate for dilution as needed.

(n) Continue to Subheading 3.4.

6. Typical yield is between 200 and 400 ng/μL of labeled DNA. Yield depends on the amount and quality of the sample DNA. DNA concentration of 150 ng/μL is still adequate to proceed for DNA hybridization but lower concentration of DNA will not work. Labeled DNA can be stored overnight at 2–8 °C before use with no detrimental effect on assay performance. If stored overnight, briefly mix the plate in the incubator/shaker before use (*see* Subheading 3.4).

7. The dilution should be made with 1× TE buffer (if purified with the Nucleofast 96 PCR Purification kit) or with E1 buffer (if purified with the Purelink PCR Purification Kit). E1 buffer Volume to add = ([measured concentration/200] × 12.5) − 12.5 μL.

8. Place the incubator/shaker under the ventilation to make sure any gases arising from formamide are directed away from the laboratory room.

9. BoBsoft 2.0 analyzes the output file of the Luminex® 100/200™. It includes subtraction of the background of the negative controls, calculation of overall assay QC metrics from the results of the male and female references, calculation of assay QC metrics for each sample, production of graphical ratio line plots and bar graph for each sample. BoBsoft configures itself to the panel defined in the Luminex® 100/200™ instrument system output, which is in turn defined by a template or a protocol file loaded into the Luminex® software.

10. Quality control, expected values, and interpretation of results: the male and female reference samples included in all runs are used as internal quality controls (QC) and as controls against which patient data are scaled. The reference samples can be utilized to assess the quality of the runs in two ways. When male and female reference samples are scaled against each

other, the CV shown by the autosomal probes and the total separation shown by the sex chromosome probes represent assay success. Patient results should only be reported if both patient and control results of the assay meet the laboratory's criteria for acceptability. To determine whether a DNA copy number change had occurred in a sample a ratio-based cut-off is used. A gain or loss of a region was called if three or more probes within a given target region exceeded the cut-off. The ratio cut-off was set at ±2× trimmed standard deviation (SD) of the autosomal probes. The trimmed SD is defined as the SD of autosomal probes showing ratios within the range of ±2× SDs of the autosomal ratios.

Acknowledgments

We would like to thank PerkinElmer, Wallac, Turku, Finland, for providing support and reagents for Prenatal BoBs and KaryoLite BoBs validations.

References

1. Prabhakar U, Eirikis E, Davis HM (2002) Simultaneous quantification of proinflammatory cytokines in human plasma using the LabMAP (TM) assay. J Immunol Methods 260:207–218
2. Martins TB (2002) Development of internal controls for the Luminex instrument as part of a multiplex seven-analyte viral respiratory antibody profile. Clin Diagn Lab Immunol 9:41–45
3. Ye F, Li MS et al (2001) Fluorescent microsphere-based readout technology for multiplexed human single nucleotide polymorphism analysis and bacterial identification. Hum Mutat 17:305–316
4. Taylor JD, Briley D et al (2001) Flow cytometric platform for highthroughput single nucleotide polymorphism analysis. Biotechniques 30: 661–675
5. Yang L, Tran DK, Wang X (2001) BADGE, BeadsArray for the detection of gene expression, a high-throughput diagnostic bioassay. Genome Res 11:1888–1898
6. Dunbar SA, Jacobson JW (2000) Application of the Luminex LabMAP in rapid screening for mutations in the cystic fibrosis transmembrane conductance regulator gene: a pilot study. Clin Chem 46:1498–1500
7. Vignali DAA (2000) Multiplexed particle-based flow cytometric assays. J Immunol Methods 243:243–255
8. Carson RT, Vignali DAA (1999) Simultaneous quantitation of 15 cytokines using a multiplexed flow cytometric assay. J Immunol Methods 227:41–52
9. Gordon RF, McDade RL (1997) Multiplexed quantification of human IgG, IgA, and IgM with the FlowMetrix(TM) system. Clin Chem 43:1799–1801
10. WalkerPeach CR, Smith PL et al (1997) A novel rapid multiplexed assay for herpes simplex virus DNA using the FlowMetrix(TM) cytometric microsphere technology. Clin Chem 43:21
11. Smith PL, WalkerPeach CR et al (1998) A rapid, sensitive, multiplexed assay for detection of viral nucleic acids using the FlowMetrix system. Clin Chem 44:2054–2056
12. Oliver KG, Kettman JR, Fulton RJ (1998) Multiplexed analysis of human cytokines by use of the FlowMetrix system. Clin Chem 44: 2057–2060
13. Bellisario R, Colinas RJ, Pass KA (2000) Simultaneous measurement of thyroxine and thyrotropin from newborn dried blood-spot specimens using a multiplexed fluorescent

microsphere immunoassay. Clin Chem 46: 1422–1424

14. Dunbar SA (2006) Applications of Luminex xMAP technology for rapid, high-throughput multiplexed nucleic acid detection. Clin Chim Acta 363:71–82
15. Vialard F, Simoni G et al (2011) Prenatal BACs-on-Beads™: a new technology for rapid detection of aneuploidies and microdeletions in prenatal diagnosis. Prenat Diagn 31: 500–508
16. Popowski T, Molina-Gomes D et al (2012) Prenatal diagnosis of the duplication 17p11.2 associated with Potocki-Lupski syndrome in a foetus presenting with mildly dysmorphic features. Eur J Med Genet 55:723–726
17. Gross SJ, Bajaj K et al (2011) Rapid and novel prenatal molecular assay for detecting aneuploidies and microdeletion syndromes. Prenat Diagn 31:259–266
18. Grati FR, Gomes DM et al (2013) Application of a new molecular technique for the genetic evaluation of products of conception. Prenat Diagn 33:32–41
19. Vialard F, Simoni G et al (2012) Prenatal BACs-on-Beads™: the prospective experience of five prenatal diagnosis laboratories. Prenat Diagn 32:329–335

Chapter 14

New BAC Probe Set to Narrow Down Chromosomal Breakpoints in Small and Large Derivative Chromosomes, Especially Suited for Mosaic Conditions

Ahmed B. Hamid, Xiaobo Fan, Nadezda Kosyakova, Gopakumar Radhakrishnan, Thomas Liehr, and Tatyana Karamysheva

Abstract

Fluorescence in situ hybridization (FISH) and/or array-comparative genomic hybridization (aCGH) performed after initial banding cytogenetics is still the gold standard for detection of chromosomal rearrangements. Although aCGH provides a higher resolution, FISH has two main advantages over the array-based approaches: (1) it can be applied to characterize balanced as well as unbalanced rearrangements, whereas aCGH is restricted to unbalanced ones, and (2) chromosomal aberrations present in low level or complex mosaics can be characterized by FISH without any problems, while aCGH requires presence of over 50 % of aberrant cells in the sample for detection. Recently, a new FISH-based probe set was presented: the so-called pericentric-ladder-FISH (PCL-FISH) that enables characterization of chromosomal breakpoints especially in mosaic small supernumerary marker chromosomes (sSMC). It can also be applied on large inborn or acquired derivative chromosomes. The main feature of this set is that the probes are applied in a chromosome-specific manner and they align along the chromosome in average intervals of ten megabasepairs. Hence PCL-FISH provides denser coverage and a more precise anchorage on the human DNA-sequence than most other FISH-banding approaches.

Key words Fluorescence in situ hybridization (FISH), Pericentric-ladder-FISH (PCL-FISH), Small supernumerary marker chromosomes (sSMC), Chromosomal breakpoints, Bacterial artificial chromosome (BAC)

1 Introduction

Chromosomal rearrangements detected in banding cytogenetic diagnostics can be characterized by fluorescence in situ hybridization (FISH) and/or array-comparative genomic hybridization (aCGH) [1, 2]. aCGH provides higher resolution, however, FISH-based methods surpass array-based approaches in two specifics [2, 3]. First, aCGH only detects unbalanced rearrangements, while FISH can also resolve balanced ones. Second, and the most important, chromosomal aberrations present in low mosaic level can be

Kumaran Narayanan (ed.), *Bacterial Artificial Chromosomes*, Methods in Molecular Biology, vol. 1227,
DOI 10.1007/978-1-4939-1652-8_14, © Springer Science+Business Media New York 2015

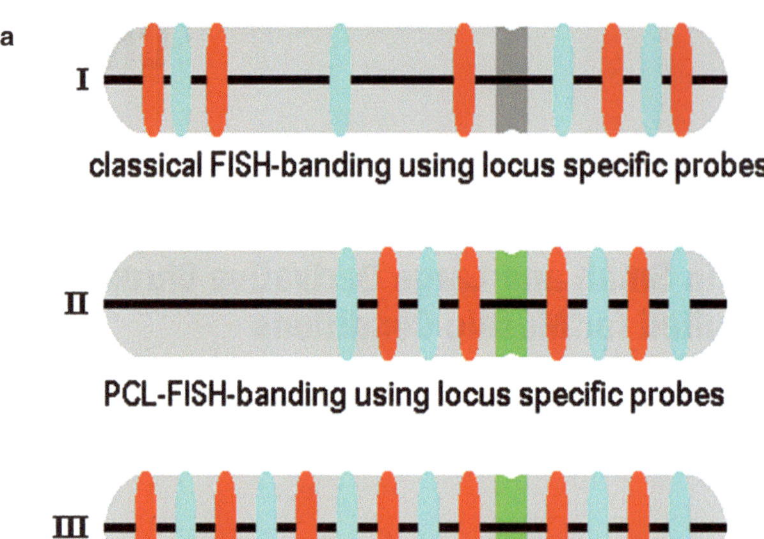

b

BAC-probe	cytoband	position (hg18)	Distance to centromere [Mb] / label
RP11-192H11	6p22.1	29,962,436-30,073,097	28.33 / blue
RP11-100B10	6p21.2	38,054,811-38,217,746	20.18 / red
RP11-334H12	6p12.3	48,824,692-48,956,050	9.44 / blue
RP11-421P21	6p11.2	57,228,514-57,292,599	1.11 / red
RP11-349P19	6q12	65,158,548-65,208,779	1.76 / red
RP11-256L9	6q13	73,180,924-73,217,550	9.78 / blue
RP11-25O6	6q14.1	83,407,494-83,562,457	20.01 / red
RP11-538A16	6q16.1	93,629,676-93,823,839	30.23 / blue

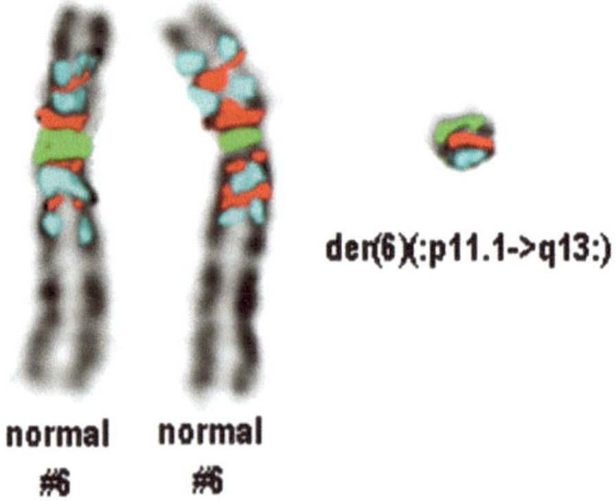

Fig. 1 Title: PCL-FISH depicted using chromosome 6 specific probe set. (**a**) Schematic depiction of three chromosomes after FISH-banding using locus-specific probes in different manners. Chromosome I shows the classical FISH-banding based on locus-specific probes, also called chromosome bar code. Here just an approximate coverage of the chromosome in question was the goal, irrespective of the individual distances of the used probes. Chromosome II shows the principle of PCL-FISH, i.e., that the pericentric region of the chromosome is covered with locus-specific probes having individual distances of 10 Mb. Also a centromeric probe is added (*green color*).

characterized by FISH, as this approach is single cell directed [4, 5]. In a research setting aCGH is reported to be able to detect low mosaic levels too [5], however, from our experience with routine diagnostics it is well established that presence of >50–70 % of aberrant cells is necessary for detection, thereby limiting its sensitivity [1, 3].

For exact breakpoint determination of derivative chromosomes various FISH-probe sets have been developed in the last few decades, particularly FISH-banding approaches [6]. An initial idea on how to implement FISH-banding was chromosome-bar coding, using well-mapped locus-specific probes [7]. However, no genome-wide probe set based on locus-specific probes was ever finalized for human chromosomes; all comparable cosmid, yeast artificial chromosome- (YAC-) or BAC-based approaches were established for single chromosomes and were not applicable for routine diagnostics. Furthermore, none of them was intended to have well-defined distances between the probes [8] (Fig. 1a).

Recently, a FISH probe set based on 174 bacterial artificial chromosome (BAC) probes called pericentric-ladder-FISH (PCL-FISH) was established to enable chromosome-specific characterization of breakpoints with a resolution of ten megabasepairs (Mb) (Fig. 1a). Its usefulness was proven in small and large inborn or acquired derivative chromosomes [9]. As its primary goal was to characterize mosaic small supernumerary marker chromosomes (sSMC) [10, 11] it was directed mainly toward the pericentric regions (Fig. 1b). Nevertheless, it may easily be implemented and enlarged also for whole chromosomes (Fig. 1a) as demonstrated for the characterization of large derivative chromosomes [9].

2 Materials

2.1 Chromosome Preparation from Peripheral Blood

1. Carnoy's fixative: methanol/glacial acetic acid 3:1, freshly prepared, at 4 °C.
2. Cell culture medium: RPMI 1640 medium with glutamine, 20 % fetal calf serum, 1 U/ml penicillin, 1 mg/ml streptomycin, 0.1/ml phytohemagglutinine.
3. Colcemide.
4. Hypotonic solution: 0.075 M KCl, freshly prepared.

Fig. 1 (continued) In Chromosome III, the PCL-FISH-principle is expanded to the whole length of the chromosome. (**b**) BACs used for PCL-probe set of chromosome 6 are listed in the table, including BAC-names, their cytoband position, their molecular position (according to hg18) and their centromeric distance in megabasepairs (MB) and the applied label. Below the table two normal chromosomes 6 and a small supernumerary marker chromosome (sSMC) is depicted. Before application of PCL-FISH it was determined that the sSMC contained only euchromatic material derived from the long arm material. Thus, the derivative could be characterized having one break in the long arm between 73.2 and 83.4 Mb. Finally, as the derivative had an intact centromere the second breakpoint was determined to be in 6p11.1, leading to a final karyotype of der (6)(:p11.1->q13:)

2.2 Slide and Probe Pretreatment

1. 24 × 50 and 24 × 24 mm coverslips.
2. 24 × 50 mm microscopic slides, precleaned.
3. 1 M $MgCl_2$ solution.
4. 3 M Sodium-acetate solution.
5. Ethanol, 100 %.
6. Formalin-buffer (1 %): 5 ml formaldehyde solution (Merck), 4.5 ml 1× PBS, 500 μl $MgCl_2$-solution.
7. PBS (1×): PBS Dulbecco w/o Ca, w/o Mg.
8. Pepsin stock solution: 1.0 g pepsin (Sigma), 50 ml distilled water, aliquot and store at −20 °C.
9. Pepsin-solution: 95 ml distilled water (37 °C) in a Coplin jar, 5 ml of 0.2 N HCl, 0.5 ml pepsin stock solution (make fresh when required).
10. Denaturation buffer: 70 ml formamide, 10 ml 20× SSC, 30 ml distilled water (adjust at pH 7–7.5).
11. Dextransulfate-solution: 2 g dextransulfate, 2 ml 20× SSC, add distilled water to obtain a final volume of 10 ml.
12. Hybridization buffer: 10 ml dextransulfate-solution, 10 ml formamide.
13. PCL-FISH-probe set:
    (a) Selection (*see* **Note 1**) and combination of suitable BAC-probes (*see* **Note 2**).
    (b) Labeling of BAC-probes: labeling can be done either by PCR methods [9], or more easily using Nick Translation (performed according to the manufacturer's protocol, Roche); the latter is recommended.
14. C_0t-1 DNA (Roche), 1 μg/μl.

2.3 FISH-Procedure and Washing

1. Rubber cement, Fixogum.
2. 4× SSC/0.2 % Tween solution: 100 ml 20× SSC, 400 ml ddH_2O, 250 μl Tween 20, adjusted to pH 7–7.5.
3. 1× SSC-solution, prepared from 20× SSC.

2.4 Microscopic Evaluation

1. Fluorescence microscope (e.g., Zeiss) with appropriate number of fluorescence filters equipped with a CCD camera. Evaluation assistant software may be used as well.

3 Methods

PCL-FISH is intended to be performed only on metaphase spreads (*see* **Note 3**). Here we describe how to prepare chromosome from different tissues [12], metaphase spreads (and interphase nuclei) from peripheral blood, pretreated slides with spread metaphases (and interphase nuclei), the PCL-FISH probe set before hybridization, and how to perform the FISH-procedure itself and evaluate the results.

3.1 Chromosome Preparation from Peripheral Blood

1. Add 1 ml of heparinized blood to 9 ml of cell culture medium, mix carefully, and incubate for 72 h at 37 °C/5 % CO_2 (*see* **Note 4**).
2. 30 min before harvesting the cells, add 1 mg of colcemide, mix gently, and incubate at 37 °C/5 % CO_2 (*see* **Note 4**).
3. Transfer fluid into a 15 ml tube (*see* **Note 5**).
4. Centrifuge the solution at room temperature for 8 min at 1,000 × *g*, and discard the supernatant by sucking it off carefully with a glass pipette (*see* **Note 6**).
5. For hypotonic treatment, the pellet is resuspended in 1 ml hypotonic solution (37 °C) and incubated at 37 °C for 20 min.
6. Slowly add 0.6 ml of Carnoy's fixative (4 °C), mix the solution carefully and repeat **step 4**.
7. Resuspend the pellet in 10 ml of fixative (4 °C), incubate at 4 °C for 20 min and repeat **step 4**.
8. Resuspend the pellet in 5 ml of fixative (4 °C) and repeat **step 4**.
9. Repeat **step 8** twice.
10. Depending on the density of the suspension, the pellet is finally resuspended in 0.3–1 ml of fixative (*see* **Note 7**).
11. Drop 1–2 drops of the suspension onto a clean and humid slide using a glass pipette and let the slide dry at room temperature.
12. After incubation overnight at room temperature, the slides can be subjected FISH, stored dust-free at room temperature for several weeks, or frozen at –20 °C for several months.

3.2 Probe Preparation for 24 × 24 Coverslip

1. Add 200 ng of each labeled BAC-DNAs (*see* **Note 8**) to 1 μg of C_0t-1-DNA per BAC by adding 5 μl sodium-acetate (3 M) and 90 μl 100 % ethanol and mix.
2. Incubate for 12–24 h at –20 °C or for 30–60 min at –80 °C.
3. Centrifuge for 20 min at 4 °C and 21,500 × *g*.
4. Remove supernatant and dry pellet in a vacuum centrifuge for 10 min.
5. Dissolve in 10 μl of hybridization buffer; probe-set solution is ready for use or may be stored at –20 °C for at least 1 year (*see* **Note 9**).

3.3 Slide and Probe Pretreatment

1. Incubate slides with the metaphase spreads in 100 ml 1× PBS at room temperature for 2 min on a shaking platform.
2. Put slides for 5–10 min in pepsin-solution at 37 °C in a Coplin jar (*see* **Note 10**).
3. Repeat **step 1** twice.
4. Postfix metaphase spreads on the slide surface by replacing 1× PBS with 100 ml of formalin-buffer for 10 min (room temperature, with gentle agitation).
5. Repeat **step 1** twice.
6. Dehydrate slides in an ethanol series (70 %, 90 %, 100 %, for 1 min each) and air-dry.
7. Add 100 μl of denaturation buffer to each slide and cover with a 24×50 mm coverslip.
8. Incubate slides on a hot plate for 2–3 min at 75 °C (*see* **Note 11**).
9. Remove the coverslip immediately and place slides in a Coplin jar filled with 70 % ethanol (4 °C) to conserve target DNA as single strands.
10. Dehydrate slide in ethanol (70 %, 90 %, 100 %, 4 °C, 1 min each) and air-dry.
11. Denature the probe set from Subheading 3.2, **step 5** in a thermocycler at 75 °C for 5 min, pre-anneal at 37 °C for 30 min and cool down to 4 °C.

3.4 FISH-Procedure, Washing, and Detection

1. Add probe-solution onto denatured slide from Subheading 3.3, **step 11**, cover with an appropriate coverslip and seal with rubber cement (*see* **Notes 9** and **12**).
2. Incubate slides overnight at 37 °C in a humid chamber.
3. Take the slides out of 37 °C humid chamber, remove rubber cement with forceps, and the coverslips by letting them swim off in 4× SSC/0.2 % Tween (room temperature, 100 ml Coplin jar). During FISH washing steps prevent the slide surface drying out, as otherwise background problems may arise.
4. Wash the slides 1×5 min in 1× SSC-solution (62–65 °C) with gentle agitation followed by 3×5 min in 4× SSC/0.2 % Tween (room temperature, with gentle agitation).
5. Dehydrate slide in ethanol (70, 90, 100 %, 4 °C, 3 min each) and air-dry.
6. Counter-stain the slides with DAPI-solution (100 ml in a Coplin jar, room temperature) for 8 min.
7. Wash slides three times in water for a few seconds and air-dry.
8. Add 15 μl of antifade, cover with coverslip and observe the results under a fluorescence microscope.

3.5 Evaluation

Image capturing is performed with a digital camera device and suitable software. The analysis can be done (1) directly by eyes using a fluorescence microscope, or (2) by analyzing after image acquisition with corresponding FISH-evaluation software. At least 10–20 metaphase spreads need to be evaluated.

4 Notes

1. The 174 BAC probes used for PCL-FISH probe set have been reported previously [9]. Before PCL-FISH, the subcentromere-specific multicolor FISH (subcenM-FISH) probe sets were published, which constituted of probes specific to the centromere-near regions; these BAC-probes served now as "starting points" of the probes used in PCL-FISH on each of the chromosome-arms [11]. Then probes were selected from published human DNA-sequence having an average distance of 10 Mb (acc. to NCBI36/hg18). BAC-probes are available via BACPAC Resources Center, Children's Hospital Oakland Research Institute (CHORI), Oakland, CA, USA (https://bacpac.chori.org/). As BACs are normally sold as plasmids in *Escherichia coli* with a specific antibiotic resistance, usually an overnight culture, followed by plasmid isolation (e.g., with Qiagen "QIAprep Spin Miniprep Kit" performed according to the manufacturer's protocol), PCR amplification [9] and labeling should give a useful locus-specific FISH probe.
2. BACs can be combined in a chromosome-specific manner. According to the question studied, two or more chromosomes may be targeted simultaneously, too. The number of probes combined in a single hybridization step depends on the availability of varied fluorescence microscope filter sets. An example of three-color PCL-FISH approach is shown in Fig. 1b.
3. Metaphase FISH is dependent on well spread chromosomes. Tumor cytogenetics chromosomes in particular are often not well spread, overlapping, and short. Thus, it may be difficult to study and evaluate the presence and order of the BAC probes tested on the chromosomes in question. Here, careful result interpretation is necessary. The "four-eye-principle" should be applied especially in doubtful cases, i.e., two experienced cytogeneticists should evaluate the results independently, and compare and discuss them if necessary.
4. Must be performed under sterile conditions to avoid contamination.
5. Sterile conditions no longer need to be observed from now on.

6. Between 1 and 1.5 ml of supernatant is left in the tube to avoid loss of material.
7. Remove as much of the suspension as necessary, i.e., after resuspension the suspension should be slightly gray to white.
8. For labeling of BACs either use a Nick-translation commercial kit (e.g., Roche) or use DOP-PCR (degenerate oligonucleotid primed—polymerase chain reaction); therefore you need following primer-mix: 5′ CCG ACT CGA GNN NNN NAT GTG G 3′, usual PCR-mix, and dNTPs, which are complemented with a dUTP carrying the fluorochrome or other hapten of your choice. As PCR-program use (a) 95 °C for 3 min, (b) 94 °C for 1 min, (c) 56 °C for 1 min, (d) 72 °C for 2 min—repeat steps b to d 20 times.
9. Coverslips larger than 24 × 24 mm can also be used. The ratio of "probe:solution" has to be adapted according to the coverslip size. In case the amount of hybridization buffer applied on the coverslip is not enough, it may damage the metaphases on the slide.
10. This pepsin pretreatment is performed in our laboratory as standard in every FISH approach including a pepsin treatment followed by postfixation with formalin-buffer with the aim to reduce the background. Each laboratory has to adapt an optimum pepsin treatment time, i.e., a balance between chromosome preservation and digestion must be found. Beginners: start with target samples that are not limited in availability, thus, enough backup material is on-hand in case of loss of metaphases due to too strong and/or too long pepsination.
11. Denaturation time of 2–3 min only is suggested for the maintenance of available metaphase chromosomes; similarly here (*see* **Note 9**) a balance between chromosome preservation and denaturation must be found.
12. For directly labeled fluorescence probes avoid direct light exposure; it can lead to fluorochrome fading and weakening of the FISH signals. Additional steps that may help reduce fading include wrapping probe-set solutions tubes in foil and dimming the light in the laboratory work area.

Acknowledgments

This work was supported in parts by the DAAD and the China Scholarship Council as well as the DFG (LI 820/22-1).

References

1. Manolakos E, Vetro A, Kefalas K et al (2010) The use of array-CGH in a cohort of Greek children with developmental delay. Mol Cytogenet 3:22
2. Weimer J, Heidemann S, von Kaisenberg CS et al (2011) Isolated trisomy 7q21.2-31.31 resulting from a complex familial rearrangement involving chromosomes 7, 9 and 10. Mol Cytogenet 4:28
3. Tsuchiya KD, Opheim KE, Hannibal MC et al (2008) Unexpected structural complexity of supernumerary marker chromosomes characterized by microarray comparative genomic hybridization. Mol Cytogenet 1:7
4. Iourov IY, Vorsanova SG, Yurov YB (2008) Chromosomal mosaicism goes global. Mol Cytogenet 1:26
5. van der Veken LT, Dieleman MMJ, Douben H et al (2010) Low grade mosaic for a complex supernumerary ring chromosome 18 in an adult patient with multiple congenital anomalies. Mol Cytogenet 3:13
6. Liehr T, Heller A, Starke H et al (2002) FISH banding methods: applications in research and diagnostics. Exp Rev Mol Diagn 2:217–225
7. Lengauer C, Green ED, Cremer T (1992) Fluorescence in situ hybridization of YAC clones after Alu-PCR amplification. Genomics 13:826–828
8. Liehr T, Starke H, Heller A et al (2006) Multicolor fluorescence in situ hybridization (FISH) applied to FISH-banding. Cytogenet Genome Res 114:240–244
9. Hamid AB, Kreskowski K, Weise A et al (2012) How to narrow down chromosomal breakpoints in small and large derivative chromosomes – a new probe set. J Appl Genet 53:259–269
10. Liehr T, Claussen U, Starke H (2004) Small supernumerary marker chromosomes (sSMC) in humans. Cytogenet Genome Res 107:55–67
11. Liehr T, Mrasek K, Weise A et al (2006) Small supernumerary marker chromosomes-progress towards a genotype-phenotype correlation. Cytogenet Genome Res 112:23–34
12. Weise A, Liehr T (2009) Pre- and postnatal diagnostics and research on peripheral blood, chorion, amniocytes, and fibroblasts. In: Liehr T (ed) Fluorescence in situ hybridization (FISH) – application guide. Springer, Berlin, pp 113–122

Chapter 15

BAC-Probes Applied for Characterization of Fragile Sites (FS)

Kristin Mrasek, Kathleen Wilhelm, Luciana G. Quintana, Luise Theuss, Thomas Liehr, Andreja Leskovac, Jelena Filipovic, Gordana Joksic, Ivana Joksic, and Anja Weise

Abstract

Genomic instability tends to occur at specific genomic regions known as common fragile sites (FS). FS are evolutionarily conserved and generally involve late replicating regions with AT-rich sequences. The possible correlation between some FS and cancer-related breakpoints emphasizes on the importance of understanding the mechanisms of chromosomal instability at these sites.

Although about 230 FS have already been mapped cytogenetically, only a few of them have been characterized on a molecular level. In this chapter, we provide a protocol for mapping of common FS using bacterial artificial chromosome (BAC) probes in fluorescence in situ hybridization (FISH) and suggest the usage of lymphocytes from Fanconi anemia patients as a model system. In the latter, rare FS are expressed much more frequently than in, for example, aphidicolin-induced blood lymphocyte preparations. Knowing the exact location of FS enables the molecular comparison of their location and breakpoints that appear during evolution, cancer development and inherited disorders.

Key words Fluorescence in situ hybridization (FISH), Bacterial artificial chromosome (BAC), Fanconi anemia, Fragile sites (FS)

1 Introduction

Human fragile sites (FS) were first observed in cytogenetics more than 40 years ago [1] and the term "fragile site" was introduced thereafter [2]. The initially identified FS were induced by cell culture media that had a different composition than the media used routinely nowadays. Subsequently, various chemical compounds, e.g. aphidicolin (Fig. 1a), bromodeoxyuridine (BrdU) and 5-azacytidine (5-azaC) were found to have the same effect [3]. Nowadays, FS are understood as specific loci inclined toward formation of gaps and breaks on metaphase chromosomes caused by partial inhibition of DNA synthesis. FS are classified into two types: the rare FS—present in a small proportion of individuals

Kumaran Narayanan (ed.), *Bacterial Artificial Chromosomes*, Methods in Molecular Biology, vol. 1227, DOI 10.1007/978-1-4939-1652-8_15,

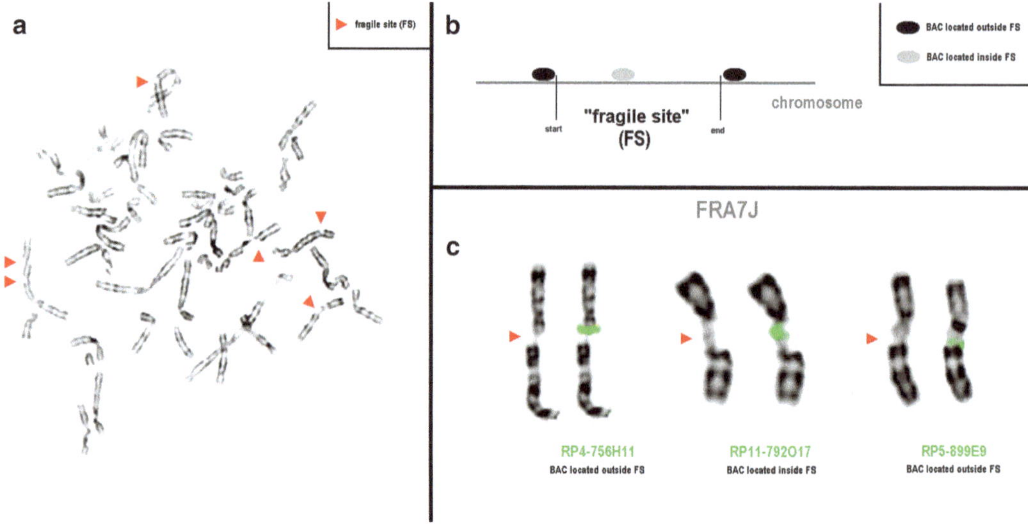

Fig. 1 Title: Fragile site (FS) mapping. (**a**) A typical metaphase prepared from aphidicolin-treated peripheral blood lymphocytes: Fragile sites (FS) are indicated by *arrowheads*. (**b**) Schematic depiction of a fragile site (FS), which may span several megabasepairs in size. Thus, BAC-probes may be located outside (near the border) or inside the FS. (**c**) BACs located at the proximal (RP4-756H11) and distal border of the FS (RP5-899E9) or inside the FS FRA7J (RP11-792017)

follow Mendelian patterns of inheritance and can be associated with specific disorders, and the common FS—much more frequently observed and inducible in any individual by the aforementioned substances [4]. Common FS are considered to be a normal part of the mammalian chromosome structure, spanning up to several mega basepairs of DNA (Fig. 1b, c). Common FS have a frequency of <5 % in normal population and they are usually not associated with disease. However, a possible correlation of some FS with cancer-related breakpoints [5, 6] or a molecular cytogenetic co-localization with breakpoints in Fanconi anemia patients was reported [7].

Recently, we carried out the largest ever banding cytogenetic study to track all available FS present in the human genome. We studied 25,000 aphidicolin-induced metaphase spreads and finally identified 230 common FS loci [3]. However, as of now the majority of the tracked common FS were just mapped cytogenetically and not on a molecular level. This is mainly due to the fact that most of these FS are present in extremely low frequencies of 0.01–0.3 % in aphidicolin-induced lymphocytes. It was also recently discovered that Fanconi anemia patients exhibit relatively high frequencies of breaks precisely within those extremely rare FS [7].

Consequently, provided here is a protocol for molecular mapping of FS using chromosome preparations from Fanconi anemia patients, enabling the detection of FS which normally occur infrequently in aphidicolin-induced cells, but are found frequently in Fanconi anemia-derived chromosome spreads. This makes Fanconi anemia-derived chromosome spreads a model system for the fine mapping of infrequent FS [7]. Table 1 shows 33 common FS, their yet unreported mapping results, and also the appropriate BACs that may be applied for their systematic studies.

Table 1
Common FS, their locations in cytobands and the BACs by which they may be tracked (yet unpublished data apart from FRA10F [5]

FS designation acc. to [3]	BACs located proximal and distal of the FS	Cytobands
FRA2H	2q32.1 2q32.2	RP11-234L7 RP11-114B11
FRA2J	2q37.1 2q37.3	RP11-534J17 RP11-27M15
FRA4C	4q31.21 4q31.22	RP11-54P19 RP11-324J17
FRA7J	7q11.22 7q11.23	RP11-358M3 RP11-419A13
FRA10F	10q26.13 10q26.2	RP11-12J10 RP11-310M21
FS designation acc. to [3]	**BACs located within the FS (exact borders not determined yet)**	**Cytobands**
FRA1E	1p21.3 1p21.3	RP11-14O19 RP11-272L13
FRA1F	1q21.1 1q21.2	RP11-301M17 RP11-54A4
FRA1K	1q31.2 1q31.3	RP11-258M18 RP11-166A4
FRA2A	2q11.2 2q11.2	RP11-478P16 RP11-353P23
FRA2B	2q13	RP11-1429F20
FRA2E	2p14	RP11-547F18
FRA2L	2p11.2 2q11.2	RP11-421K23 RP11-478P16
FRA2F	2q21.2 2q22.1	RP11-809C23 RP11-432O12
FRA2R	2p11.2	RP11-401C13
FRA4D	4p15.1 4p15.1	RP11-355H11 RP13-486L17

(continued)

Table 1 (continued)

FS designation acc. to [3]	BACs located proximal and distal of the FS	Cytobands
FRA4F	4q22.1 4q22.2	RP11-529H2 RP11-428L21
FRA4I	4q21.22 4q21.23	RP11-163O17 RP11-8N8
FRA5E	5p15.1 5p14.1	RP11-35A11 RP11-184E9
FRA5D	5q15 5q21.1	RP11-33A7 RP11-384D8
FRA5G	5q35.1 5q35.1	RP11-20O22 CTB-54I1
FRA6I	6p11.2 6p11	RP11-325M4 RP11-278J20
FRA6G	6q16.1	RP11-596A13
FRA7B	7p22.2 7p22.1	RP11-348A21 RP11-172O13
FRA7C	7p14.1 7p14.1	RP11-623K16 RP11-64I2
FRA7F+K	7q22.3 7q31.1	CTB-20D2 RP11-274I17
FRA7I	7q36.1 7q36.2	RP11-208G20 RP11-422E4
FRA7M	7q34 7q35	RP11-265A15 RP11-374N8
FRA9D	9q21.33 9q21.33	RP11-202I11 RP11-213G2
FRA10G	10q11.21 10q21.22	RP11-175I17 RP11-556L1
FRA13C	13q21.2 13q21.32	RP11-307O11 RP11-129M14
FRA14B	14q23.2 14q24.1	RP11-676P5 RP11-350H11
FRA16C	16q21 16q22.1	RP11-106J23 RP11-140I24
FRA18B	18q21.32 18q21.32	RP11-396N11 RP11-15C15

2 Materials

2.1 Chromosome Preparation from Lymphocyte Cell Lines

1. Carnoy's fixative: methanol/glacial acetic acid 3:1, freshly prepared, at 4 °C.
2. Colcemide.
3. Hypotonic solution: 0.075 M KCl, freshly prepared.

2.2 Slide and Probe Pretreatment

1. 24 × 50 and 24 × 24 mm coverslips.
2. 24 × 50 mm microscopic slides, precleaned.
3. 1 M $MgCl_2$-solution.
4. 3 M Sodium-acetate solution.
5. 100 % Ethanol.
6. 1 % Formalin-buffer: 5 ml formaldehyde solution, 4.5 ml of 1× PBS, 500 μl of 1 M $MgCl_2$-solution.
7. PBS (1×): PBS Dulbecco w/o Ca, w/o Mg.
8. Pepsin stock solution: 1.0 g pepsin, 50 ml distilled water, aliquot and store at −20 °C.
9. Pepsin-solution: 95 ml distilled water (37 °C) in a Coplin jar, 5 ml of 0.2 N HCl, 0.5 ml pepsin stock solution (make fresh when required).
10. Denaturation buffer: 70 ml formamide (Merck), 10 ml 20× SSC, 30 ml distilled water (adjust to pH 7–7.5).
11. Dextransulfate-solution: 2 g dextransulfate, 2 ml 20× SSC, add distilled water to obtain a final volume of 10 ml.
12. Hybridization buffer: 10 ml dextransulfate-solution, 10 ml formamide.
13. BAC-probes suited for FS mapping:
    (a) Selection (*see* **Note 1**) and combination of suitable BAC-probes (*see* **Note 2**).
    (b) Labeling of BAC-probes: labeling can be done either by PCR methods as described [7], or more easily using Nick Translation (performed according to the manufacturer's protocol, Roche); the latter is recommended.
14. 1 μg/μl C_0t-1 DNA.

2.3 FISH-Procedure and Washing

1. Rubber cement, Fixogum.
2. 4× SSC/0.2%Tween solution: 100 ml 20× SSC, 400 ml ddH_2O, 250 μl Tween 20, adjusted to pH 7–7.5.
3. 1× SSC-solution, prepared from 20× SSC.

2.4 Microscopic Evaluation

1. Fluorescence microscope (e.g. Zeiss) with appropriate number of fluorescence filters equipped with a CCD camera. Evaluation assistant software may be used, as well.

3 Methods

FISH mapping is intended to be performed only on metaphase spreads (*see* **Note 3**). Here we describe how to prepare chromosomes for FISH analysis [8], pretreated slides with spread metaphases, the FISH probes before hybridization, and how to perform the FISH-procedure itself and evaluate the results. For FS mapping Fanconi anemia cell lines can be used as described elsewhere [7] and below.

3.1 Chromosome Preparation from Lymphocyte Cell Lines

1. 30 min before harvesting the cell line, add 1 mg of colcemide, mix gently, and incubate at 37 °C/5 % CO_2 (*see* **Note 4**).
2. Transfer fluid into a 15 ml tube (*see* **Note 5**).
3. Centrifuge the solution at room temperature for 8 min at 1,000 × *g*, and discard the supernatant by sucking it off carefully with a glass pipette (*see* **Note 6**).
4. For hypotonic treatment, the pellet is resuspended in 1 ml hypotonic solution (37 °C) and incubated at 37 °C for 20 min.
5. Slowly add 0.6 ml of Carnoy's fixative (4 °C), mix the solution carefully and repeat **step 4**.
6. Resuspend the pellet in 10 ml of fixative (4 °C), incubate at 4 °C for 20 min and repeat **step 4**.
7. Resuspend the pellet in 5 ml of fixative (4 °C) and repeat **step 4**.
8. Repeat **step 7** twice.
9. Depending on the density of the suspension, the pellet is finally resuspended in 0.3–1 ml of fixative (*see* **Note 7**).
10. Drop 1–2 drops of the suspension onto a clean and humid slide using a glass pipette and let the slide dry at .
11. After incubation overnight at room temperature, the slides can be subjected FISH, stored dust-free at room temperature for several weeks, or frozen at −20 °C for several months.

3.2 Probe Preparation for 24 × 24 Coverslip

1. Add 200 ng of each labeled BAC-DNAs (*see* **Note 8**) to 1 μg of C_0t-1-DNA per BAC by adding 5 μl of 3 M sodium-acetate and 90 μl of 100 % ethanol and mix.
2. Incubate for 12–24 h at −20 °C or for 30–60 min at −80 °C.
3. Centrifuge for 20 min at 4 °C and 15,000 × *g*.

4. Remove supernatant and dry pellet in a vacuum centrifuge for 10 min.
5. Dissolve in 10 μl of hybridization buffer; probe-set solution is ready for use or may be stored at −20 °C for at least 1 year now (*see* **Note 9**).

3.3 Slide and Probe Pretreatment

1. Incubate slides with metaphase spreads in 100 ml 1× PBS at room temperature for 2 min on a shaking platform.
2. Put slides for 5–10 min in pepsin-solution at 37 °C in a Coplin jar (*see* **Note 10**).
3. Repeat **step 1** twice.
4. Postfix metaphase spreads on the slide surface by replacing 1× PBS with 100 ml of formalin-buffer for 10 min (room temperature, with gentle agitation).
5. Repeat **step 1** twice.
6. Dehydrate slides in an ethanol series (70 %, 90 %, 100 %, for 1 min each) and air-dry.
7. Add 100 μl of denaturation buffer to each slide and cover with a 24 × 50 mm coverslip.
8. Incubate slides on a hot plate for 2–3 min at 75 °C (*see* **Note 11**).
9. Remove the coverslip immediately and place slides in a Coplin jar filled with 70 % ethanol (4 °C) to conserve target DNA as single strands.
10. Dehydrate slide in ethanol (70 %, 90 %, 100 %, 4 °C, 1 min each) and air-dry.
11. Denature the probe set from Subheading 3.2, **step 5** in a thermocycler at 75 °C for 5 min, pre-anneal at 37 °C for 30 min and cool down to 4 °C.

3.4 FISH-Procedure, Washing, and Detection

1. Add probe-solution onto denatured slide from Subheading 3.3, **step 11**, cover with an appropriate coverslip and seal with rubber cement (*see* **Notes 9** and **12**).
2. Incubate slides overnight at 37 °C in a humid chamber.
3. Take the slides out of 37 °C humid chamber, remove rubber cement with forceps, and the coverslips by letting them swim off in 4× SSC/0.2 % Tween (room temperature, 100 ml Coplin jar). During FISH washing steps prevent the slide surface drying out, as otherwise background problems may arise.
4. Wash the slides 1 × 5 min in 1× SSC-solution (62–65 °C) with gentle agitation followed by 3 × 5 min in 4× SSC/0.2 % Tween (room temperature, with gentle agitation).
5. Dehydrate slide in ethanol (70, 90, 100 %, 4 °C, 3 min each) and air-dry.

6. Counter-stain the slides with DAPI-solution (100 ml in a Coplin jar, room temperature) for 8 min.
7. Wash slides three times in water for a few seconds and air-dry.
8. Add 15 µl of antifade, cover with coverslip, and observe the results under a fluorescence microscope.

3.5 Evaluation

Image capturing is performed with a digital camera device and suitable software. The analysis can be done (1) directly by eyes using a fluorescence microscope, or (2) by analyzing after image acquisition with corresponding FISH-evaluation software. At least 10–20 metaphase spreads need to be evaluated.

4 Notes

1. BAC-probes used for FISH mapping of a specific common FS are selected in the following way: the cytogenetic position of the FS of interest is determined by GTG-banding. Then the publicly available Genome browsers are used to select the BAC-probes for the putative FS-region—from our experience initially a region of approximately 10–20 Mb should be covered. After narrowing down the region of interest more BACs for regions of ~5 Mb can be selected [3, 5–7]. BAC-probes are available via BACPAC Resources Center, Children's Hospital Oakland Research Institute (CHORI), Oakland, CA, USA (https://bacpac.chori.org/). Since BACs are normally sold as plasmids in *Escherichia coli* with a specific antibiotic resistance, usually an overnight culture, followed by plasmid isolation (e.g. with Qiagen "QIAprep Spin Miniprep Kit" performed according to manufacturers protocol), PCR amplification [6] and labeling should give a useful locus-specific FISH probe.
2. For mapping BACs can be combined in two-color FISH experiments and labeling of the probes depends on the available fluorescence microscope filter sets. After an FS has been mapped exactly FS-spanning or FS-flanking BACs may be used for large-scale studies of interest. An example of FS mapping is shown in Fig. 1c.
3. Metaphase FISH is dependent on well spread chromosomes. Thus, for mapping of FS in cytogenetic preparations of Fanconi anemia patients we recommend the following: after dropping slides and aging for 1 week at room temperature the metaphases there on are stained with DAPI-counterstain and analyzed under a microscope. All metaphases and their corresponding break-events are recorded. According to the frequency of breaks in putative FS regions only such slides with two or more breaks within the regions of interest are used for

FISH mapping thereafter. It is also recommended to apply FISH on the same slides and metaphases of interest several times using different probes. In our hands at least 3–4 subsequent FISH experiments can be done per slide; however, denaturation time has to be extended up to 5 or 8 min.

4. Must be performed under sterile conditions to avoid contamination.
5. Sterile conditions no longer need to be observed from now on.
6. Between 1 and 1.5 ml of supernatant is left in the tube to avoid loss of material.
7. Remove as much of the suspension as necessary, i.e. after resuspension the suspension should be slightly gray to white.
8. For labeling of BACs either use a Nick-translation commercial kit (e.g. Roche) or use DOP-PCR (degenerate oligonucleotid primed—polymerase chain reaction); therefore you need following primer-mix: 5′ CCG ACT CGA GNN NNN NAT GTG G 3′, usual PCR-mix, and dNTPs, which are complemented with a dUTP carrying the fluorochrome or other hapten of your choice. As PCR-program use (a) 95 °C for 3 min, (b) 94 °C for 1 min, (c) 56 °C for 1 min, (d) 72 °C for 2 min—repeat steps b to d 20 times.
9. Coverslips larger than 24 × 24 mm can also be used. The ratio of "probe:solution" has to be adapted according to the coverslip size. In case the amount of hybridization buffer applied on the coverslip is not enough, it may damage the metaphases on the slide.
10. This pepsin pretreatment is performed in our laboratory as a standard in every FISH approach. It involves pepsin treatment followed by postfixation with formalin-buffer in order to reduce the background. Each laboratory has to adapt an optimum pepsin treatment time, i.e. a balance between chromosome preservation and its digestion must be found. Beginners: start with target samples that are not limited in availability, thus, enough backup material is on-hand in case of loss of metaphases due to too strong and/or too long pepsination.
11. Denaturation time of only 2–3 min is suggested for the maintenance of available metaphase chromosomes; similarly here (*see* **Note 9**), a balance between chromosome preservation and its denaturation must be found.
12. For directly labeled fluorescence probes avoid direct light exposure; it can lead to fluorochrome fading and weakening the FISH signals. Additional steps that may help reduce fading include wrapping probe-set solution tubes in foil and dimming the light in the laboratory work area.

Acknowledgments

Supported in parts by IZKF Jena, Evangelische Studienwerk e.V. Villigst, Ernst-Abbe-Stiftung, DAAD and Deutsche Fanconi-Anämie-Hilfe e.V.

References

1. Dekaban A (1965) Persisting clone of cells with an abnormal chromosome in a woman previously irradiated. J Nucl Med 10:740–746
2. Magenis RE, Hecht F, Lovrien EW (1970) Heritable fragile site on chromosome 16: probable localization of haptoglobin locus in man. Science 170:85–87
3. Mrasek K, Schoder C, Teichmann AC et al (2010) Global screening and extended nomenclature for 230 aphidicolin-inducible fragile sites, including 61 yet unreported ones. Int J Oncol 36:929–940
4. Durkin SG, Glover TW (2007) Chromosome fragile sites. Annu Rev Genet 41:169–192
5. Ma K, Qiu L, Mrasek K et al (2012) Common fragile sites: genomic hotspots of DNA damage and carcinogenesis. Int J Mol Sci 13:11974–11999
6. Liehr T, Kosayakova N, Schröder J et al (2011) Evidence for correlation of fragile sites and chromosomal breakpoints in carriers of constitutional balanced chromosomal rearrangements. Balk J Med Genet 14:13–16
7. Schoder C, Liehr T, Velleuer E et al (2010) New aspects on chromosomal instability: chromosomal break-points in Fanconi anemia patients co-localize on the molecular level with fragile sites. Int J Oncol 36:307–312
8. Weise A, Liehr T (2009) Pre- and postnatal diagnostics and research on peripheral blood, chorion, amniocytes, and fibroblasts. In: Liehr T (ed) Fluorescence in situ hybridization (FISH) – application guide. Springer, Berlin, pp 113–122

Chapter 16

Application of BAC-Probes to Visualize Copy Number Variants (CNVs)

Anja Weise, Moneeb A.K. Othman, Samarth Bhatt, Sharon Löhmer, and Thomas Liehr

Abstract

Copy number variations (CNVs) are structural variations of the human genome. These alterations result in variant copy numbers of certain stretches of DNA. In other words, some regions may be present in more or less copies than in a reference genome; however, these copy number changes do not have any impact on the phenotype. Also, CNVs may be extremely large and cytogenetically detectable or submicroscopic but still spanning several megabasepairs (Mb). In the recent years, array technology has identified especially the latter ones as so-called copy number variant (CNV) polymorphisms. These CNVs are detected in ~12 % of the human genome sequences and may comprise several hundred kilobasepairs. CNVs contribute significantly to the inter-individual differences in humans, and can range between 0.5 and 1.5 Mb amongst different genomes, well within the level of detection using cytogenetics techniques. Thus, they can be visualized by FISH using bacterial artificial chromosomes (BACs) as probes. Here we describe a method that enables discrimination of individual homologous chromosomes at the single cell level based on CNVs in the genome, called parental origin determination fluorescence in situ hybridization (POD-FISH). Possible fields of applications of this single cell-directed approach are in analyses of the parental origin of single chromosomes in inherited and acquired chromosomal aberrations.

Key words Parental origin determination fluorescence in situ hybridization (POD-FISH), Copy number variant (CNV), Homologous chromosomes, Bacterial artificial chromosome (BAC), Single cell level

1 Introduction

Decades ago, chromosome analyses revealed the existence of cytogenetically visible copy number variations in the human genome [1]. The submicroscopic copy number variants (CNVs) are a more recent finding and are nowadays understood as gain or loss of chromosomal material. CNVs can either be the cause of a disease or just one of the many possible dormant genetic variants in man [2]. CNVs of sizes between 1 kilobasepair (kb) and several megabasepairs (Mb) have been reported. They are distributed

Kumaran Narayanan (ed.), *Bacterial Artificial Chromosomes*, Methods in Molecular Biology, vol. 1227, DOI 10.1007/978-1-4939-1652-8_16, © Springer Science+Business Media New York 2015

along the entire human genome. CNV regions may show variable copy numbers when compared to a reference genome and may include deletions and duplications or multi-copies of genomic loci. In banding cytogenetics these variants normally remain in the submicroscopic, nonvisible level. However, CNVs with sizes of several Mb are within the resolution of fluorescence in situ hybridization (FISH) and can be detected.

Array-comparative genomic hybridization (aCGH) has enabled the detection of hundreds of CNVs in the last decade [3]. It turned out that CNVs may encompass as much as ~12 % of the human genome [4]. It is also known that every individual carries thousands of CNVs that are different from other individuals [5]. This can lead to more or fewer copies of dosage-sensitive genes included in the CNVs, and may then result in phenotypic variability, complex behavioral traits, disease susceptibility or predisposition such as HIV, adipositas, and psoriasis [6]. However, most CNVs are considered to be benign and are usually inherited from one of the parents [7, 8]. When appearing de novo, such CNVs/ genomic imbalances are more likely to be pathologic [9].

Inter-individual differences can be traced (1) on a cytogenetic visible level, (2) on the level of submicroscopic CNVs [10], and/or (3) by considering the DNA sequence (like single nucleotide polymorphisms (SNPs), mini- and micro-satellites [8], small insertion or deletion polymorphisms (INDELS, <1 kb), etc. [10]). All of those different-sized DNA variants may be used to identify individual chromosomes, i.e. in case one chromosome carries polymorphic variant A (e.g. as a SNP) and the homologous chromosome carries variant B these two chromosomes may eventually be tracked as being inherited from the father or the mother. Yet, it was only possible to distinguish individual chromosomes based on chromosomal heteromorphisms or microsatellites.

Interestingly, CNVs being sites of deletion or gain of euchromatic material in normal individuals make up >60 % of the reported structural variations collected in the database of genomic variants (http://projects.tcag.ca/variation/) [11]. Thus, CNVs, besides microsatellites and SNPs, can also be used for distinguishing individual chromosomes. For this task, bacterial artificial chromosomes (BACs) are utilized in FISH to span the genomic region that corresponds to the CNVs.

Consequently, we recently reported the so-called "parental origin determination" (POD)-FISH-approach [11]. This approach allows a FISH-based analysis of submicroscopic CNVs. Thus, similar results as obtained by microsatellite analysis can be achieved using the POD-FISH approach. However, the major advantage of POD-FISH (compared to microsatellite analysis) is that it is single cell-directed. Thus, also chromosomal aberrations present in mosaic may be tracked back on a parental origin. Furthermore, in each POD-FISH studied metaphase individual chromosomes may be identified as being derived from father or mother.

Like microsatellite analysis POD-FISH may result in informative and in noninformative results. POD-FISH-results are informative if there is a detectable difference of FISH-signal intensities due to differences in the detected CNVs on the individual chromosomes. In order to achieve highly rates of informative results CNVs known to be prone to deletions were selected. POD-FISH can be used to discriminate between individual homologous chromosomes on a single cell level [11]. This approach is extremely helpful for parent of origin studies of individual chromosomes particularly in trisomic, monosomic or rearranged clinical cases or even in tumor samples [11–13].

2 Materials

2.1 Chromosome Preparation from Peripheral Blood

1. Carnoy's fixative: methanol/glacial acetic acid 3:1, freshly prepared, at 4 °C.
2. Cell culture medium: RPMI 1640 medium with glutamine, 20 % fetal calf serum, 1 U/ml penicillin, 1 mg/ml streptomycin, 0.1 mg/ml phytohemagglutinin.
3. Colcemide.
4. Hypotonic solution: 0.075 M KCl, freshly prepared.

2.2 Slide and Probe Pre-treatment

1. 24 × 50 mm and 24 × 24 mm coverslips.
2. 24 × 50 mm microscopic slides, pre-cleaned.
3. 1 M $MgCl_2$-solution.
4. 3 M Sodium-acetate solution.
5. 100 % Ethanol.
6. 1 % Formalin-buffer: 5 ml formaldehyde solution, 4.5 ml 1× PBS, 500 μl $MgCl_2$-solution.
7. 1× PBS: PBS Dulbecco w/o Ca, w/o Mg.
8. Pepsin stock solution: 1.0 g pepsin, 50 ml distilled water, aliquot and store at −20 °C.
9. Pepsin-solution: 95 ml distilled water (37 °C) in a Coplin jar, 5 ml of 0.2 N HCl, 0.5 ml pepsin stock solution (make fresh when required).
10. Denaturation buffer: 70 ml formamide, 10 ml 20× SSC, 30 ml distilled water (adjust at pH 7–7.5).
11. Dextransulfate-solution: 2 g dextransulfate, 2 ml 20× SSC, add distilled water to obtain a final volume of 10 ml.
12. Hybridization buffer: 10 ml dextransulfate-solution, 10 ml formamide.

13. POD-FISH-probe set:
    (a) Selection (*see* **Note 1**) and combination of suitable BAC-probes (*see* **Note 2**).
    (b) Labeling of BAC-probes: labeling can be done either by PCR-methods as described [11], or more easily using Nick Translation (performed according to the manufacturer's protocol, Roche); the latter is recommended.
14. C_0t-1 DNA (Roche), 1 μg/μl.

2.3 FISH-Procedure and Washing

1. Rubber cement, Fixogum.
2. 4× SSC/0.2 % Tween solution: 100 ml 20× SSC, 400 ml ddH_2O, 250 μl Tween 20, adjusted to pH 7–7.5.
3. 1× SSC-solution, prepared from 20×SSC.

2.4 Microscopic Evaluation

1. Fluorescence microscope (e.g. Zeiss) with appropriate number of fluorescence filters equipped with a CCD camera. Evaluation assistant software is recommended, e.g. AxioVision Quanti FISH (Zeiss).

3 Methods

POD-FISH is normally performed on metaphase spreads; however, signal differences may also be detected and evaluated on interphase nuclei (*see* **Note 3**). Here we describe how to prepare metaphase spreads (and interphase nuclei) from peripheral blood [14], pre-treat slides with spread metaphases (and interphase nuclei), the POD-FISH probe set before hybridization, and how to perform the FISH-procedure itself and evaluate the results.

3.1 Chromosome Preparation from Peripheral Blood

1. Add 1 ml of heparinized blood to 9 ml of cell culture medium, mix carefully, and incubate for 72 h at 37 °C/5 % CO_2 (*see* **Note 4**).
2. 30 min before harvesting the cells, add 1 mg of colcemide, mix gently, and incubate at 37 °C/5 % CO_2 (*see* **Note 4**).
3. Transfer fluid into a 15 ml tube (*see* **Note 5**).
4. Centrifuge the solution at room temperature for 8 min at 1,000×*g*, and discard the supernatant by sucking it off carefully with a glass pipette (*see* **Note 6**).
5. For hypotonic treatment, the pellet is resuspended in 1 ml hypotonic solution (37 °C) and incubated at 37 °C for 20 min.
6. Slowly add 0.6 ml of Carnoy's fixative (4 °C), mix the solution carefully and repeat **step 4**.
7. Resuspend the pellet in 10 ml of fixative (4 °C), incubate at 4 °C for 20 min and repeat **step 4**.

8. Resuspend the pellet in 5 ml of fixative (4 °C) and repeat **step 4**.
9. Repeat **step 8** twice.
10. Depending on the density of the suspension, the pellet is finally resuspended in 0.3–1 ml of fixative (*see* **Note 7**).
11. Drop 1–2 drops of the suspension onto a clean and humid slide using a glass pipette and let the slide dry at room temperature.
12. After incubation overnight at room temperature, the slides can be subjected FISH, stored dust-free at room temperature for several weeks, or frozen at −20 °C for several months.

3.2 Probe Preparation for 24 × 24 Coverslip

1. Add 200 ng of each labeled BAC-DNAs (*see* **Note 8**) to 1 μg of C_0t-1-DNA per BAC by adding 5 μl of 3 M sodium-acetate and 90 μl of 100 % ethanol and mix.
2. Incubate for 12–24 h at −20 °C or for 30–60 min at −80 °C.
3. Centrifuge for 20 min at 4 °C and 15,000 × *g*.
4. Remove supernatant and dry pellet in a vacuum centrifuge for 10 min.
5. Dissolve in 10 μl of hybridization buffer; probe-set solution is ready for use or may be stored at −20 °C for at least 1 year now (*see* **Note 9**).

3.3 Slide and Probe Pretreatment

1. Incubate slides with metaphase spreads in 100 ml 1× PBS at room temperature for 2 min on a shaking platform.
2. Put slides for 5–10 min in pepsin-solution at 37 °C in a Coplin jar (*see* **Note 10**).
3. Repeat **step 1** twice.
4. Postfix metaphase spreads on the slide surface by replacing 1× PBS with 100 ml of formalin-buffer for 10 min (room temperature, with gentle agitation).
5. Repeat **step 1** twice.
6. Dehydrate slides in an ethanol series (70, 90, 100 %, for 1 min each) and air-dry.
7. Add 100 μl of denaturation buffer to each slide and cover with a 24 × 50 mm coverslip.
8. Incubate slides on a hot plate for 2–3 min at 75 °C (*see* **Note 11**).
9. Remove the coverslip immediately and place slides in a Coplin jar filled with 70 % ethanol (4 °C) to conserve target DNA as single strands.
10. Dehydrate slide in ethanol (70, 90, 100 %, 4 °C, 1 min each) and air-dry.
11. Denature the probe set from Subheading 3.2, **step 5** in a thermocycler at 75 °C for 5 min, pre-anneal at 37 °C for 30 min and cool down to 4 °C.

3.4 FISH-Procedure, Washing, and Detection

1. Add probe-solution onto denatured slide from Subheading 3.3, **step 11**, cover with an appropriate coverslip and seal with rubber cement (*see* **Notes 9** and **12**).
2. Incubate slides overnight at 37 °C in a humid chamber.
3. Take the slides out of 37 °C humid chamber, remove rubber cement with forceps, and the coverslips by letting them swim off in 4× SSC/0.2 % Tween (room temperature, 100 ml Coplin jar). During FISH washing steps prevent the slide surface drying out, as otherwise background problems may arise.
4. Wash the slides 1 × 5 min in 1× SSC-solution (62–65 °C) with gentle agitation followed by 3 × 5 min in 4× SSC/0.2 % Tween (room temperature, with gentle agitation).
5. Dehydrate slide in ethanol (70, 90, 100 %, 4 °C, 3 min each) and air-dry.
6. Counter-stain the slides with DAPI-solution (100 ml in a Coplin jar, room temperature) for 8 min.
7. Wash slides three times in water for a few seconds and air-dry.
8. Add 15 μl of antifade, cover with coverslip and observe the results under a fluorescence microscope.

3.5 Evaluation

Image capturing is performed with a digital camera device and suitable software. The analysis can be done (1) directly by eyes using a fluorescence microscope, (2) by analyzing after image acquisition with corresponding FISH-evaluation software (Fig. 1), and (3) by measuring signal intensity and area with a software that has been shown to be suitable for measuring FISH intensity signal before, like the SCION [11] or Axiovision QuantiFISH. At least 10–20 metaphase spreads or 100 interphase nuclei need to be evaluated (*see* **Note 13**).

4 Notes

1. Genome-wide CNV-specific BAC-probes were published in 2008 [11]. As nowadays much more CNVs are identified and also special chromosomes or chromosomal regions are of interest for individual studies we refrain from suggesting here any specific probes. Therefore, we strongly encourage selecting corresponding regions of interest to you. Chromosome-specific sets may be based on those described previously [11]. For individual probe selection one can refer to the database of genomic variants (http://projects.tcag.ca/variation/). In order to identify the suitable BAC clones from the CNV region of interest, the selected region should not be smaller than 150 kb and the CNV should have been reported in over 20 % of the investigated individuals. The latter assists in obtaining a more informative hybridization

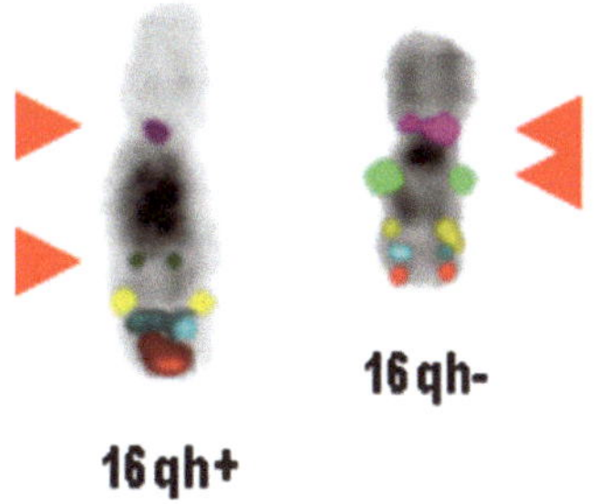

Fig. 1 POD-FISH applied on chromosomes 16 with enlarged and diminished band 16q11.2. POD-FISH set for chromosome 16. Single CNV BACs are labeled in different fluorochromes and hybridized together. The probe set shows signal intensity differences for the *green* and the *pink* labeled BACs indicated by *arrows* on both homologues chromosomes. The signal intensities were the same along all 20 metaphases evaluated for the chromosome 16 with enlarged band 16q11.2 (16qh+) and diminished band 16q11.2 (16qh−). The BACs depicted in *yellow*, *blue* and *red* were noninformative, i.e. did not show any signal intensity differences. Please note: the cytogenetic visible CNV 16qh+is not targeted by the BACs used in POD-FISH

pattern along the chromosome. To order and purchase the selected BACs, there are several sources, e.g. BACPAC Resources Center, Children's Hospital Oakland Research Institute (CHORI), Oakland, CA, USA (http://bacpac.chori.org/home.htm) and Roswell Park Cancer Institute (http://www.roswellpark.org/). As BACs are normally sold as plasmids in *Escherichia coli* with a specific antibiotic resistance, usually an overnight culture, followed by plasmid isolation (e.g. with Qiagen "QIAprep Spin Miniprep Kit" performed according to the manufacturer's protocol), amplification and labeling should give a useful locus-specific FISH probe.

2. CNV-specific BACs can, but must not be combined in a chromosome-specific manner. According to the question studied different chromosomes may be targeted simultaneously too. The number of probes to be combined in a single hybridization step depends on the number of available fluorescence microscope filters. An example of four-color POD-FISH approach is shown in Fig. 1. When working with more BAC clones at the same time it is more convenient to split the probe sets into two or more successive hybridizations [11].
3. While doing the interphase evaluation, it is necessary to perform it with respect to corresponding cut-off values. Best suited are the cut-offs obtained for nonpolymorphic probes in interphase cytogenetics in the same lab. Interphase FISH is very sensitive to methodical variations and also the signal intensities are also sometimes different. Usage times of microscopic lamps,

as well as different focus levels caused by the flattened character of previously spheric interphase nuclei may have an unexpected impact here.

4. Must be performed under sterile conditions to avoid contamination.
5. Sterile conditions no longer need to be observed from now on.
6. Between 1 and 1.5 ml of supernatant is left in the tube to avoid loss of material.
7. Remove as much of the suspension as necessary, i.e. after resuspension the suspension should be slightly gray to white.
8. For labeling of BACs either use a Nick-translation commercial kit (e.g. Roche) or use DOP-PCR (degenerate oligonucleotide primed-polymerase chain reaction) [15]; therefore you need following primer-mix: 5′ CCG ACT CGA GNN NNN NAT GTG G 3′, usual PCR-mix, and dNTPs, which are complemented with a dUTP carrying the fluorochrome or other hapten of your choice. As PCR-program use (1) 95 °C for 3 min, (2) 94 °C for 1 min, (3) 56 °C for 1 min, (4) 72 °C for 2 min—repeat steps 2 to 4 20 times.
9. Coverslips larger than 24 × 24 mm can also be used. The ratio of "probe:solution" has to be adapted according to the coverslip size. In case the amount of hybridization buffer applied on the coverslip is not enough, it may damage the metaphases on the slide.
10. This pepsin pretreatment is performed in the laboratory of the authors as a standard in every FISH approach. It involves pepsin treatment followed by post-fixation with formalin-buffer in order to reduce the background. Each laboratory has to adapt an optimum pepsin treatment time, i.e. a balance between chromosome preservation and its digestion must be found. Beginners: start with target samples that are not limited in availability, thus, enough backup material is on-hand in case of loss of metaphases due to too strong and/or too long pepsination.
11. Denaturation time of only 2–3 min is suggested for the maintenance of available metaphase chromosomes; similarly here (*see* **Note 9**) a balance between chromosome preservation and its denaturation must be found.
12. For directly labeled fluorescence probes avoid direct light exposure; it can lead to fluorochrome fading and weaken the FISH signals. Wrapping probe-set solution tubes in foil may be done. Also dimming the light in the laboratory by switching off superfluous light sources may help.
13. Note that not all applied BAC clones from CNV regions show a distinctive pattern in one individual. In our own POD-FISH

experiments, the probability of finding differences was increased by employing only CNV-BACs with a frequency of over 20 % in the population frequency. Here, results of the POD-FISH approach have similarities to the expected results of a microsatellite [11]. In addition, signal differences not directly visible by eye can be measured by software approaches like SCION or Axiovision QuantiFISH. Once a distinctive signal pattern in one individual is spotted, the parents can be tested with the same POD-FISH set in order to get hints on chromosome segregation. This can also be used to help differentiate between different cell lines, e.g. in follow up studies after bone marrow transplantations, or in cases of maternal contamination in prenatal diagnostics [11].

Acknowledgments

Supported in parts by Stefan-Morsch-Stiftung and Carl Zeiss MicroImaging GmbH, Germany.

References

1. Tjio JH, Puck TT, Robinson A (1960) The human chromosomal satellites in normal persons and in two patients with Marfan's syndrome. Proc Natl Acad Sci U S A 46: 532–539
2. Weise A, Mrasek K, Klein E et al (2012) Microdeletion and microduplication syndromes. J Histochem Cytochem 60:346–358
3. Feuk L, Marshall CR, Wintle RF et al (2006) Structural variants: changing the landscape of chromosomes and design of disease studies. Hum Mol Genet 15:57–66
4. Redon R, Ishikawa S, Fitch KR et al (2006) Global variation in copy number in the human genome. Nature 444:444–454
5. Conrad DF, Pinto D, Redon R et al (2010) Origins and functional impact of copy number variation in the human genome. Nature 464: 704–712
6. Canales CP, Walz K (2011) Copy number variation and susceptibility to complex traits. EMBO Mol Med 3:1–4
7. Lee C (2005) Vive la difference! Nat Genet 37:660–661
8. Lee C, Iafrate AJ, Brothman AR (2007) Copy number variations and clinical cytogenetic diagnosis of constitutional disorders. Nat Genet 39:S48–S54
9. Tyson C, Harvard C, Locker R et al (2005) Submicroscopic deletions and duplications in individuals with intellectual disability detected by array-CGH. Am J Med Genet A 139: 173–185
10. Schlötterer C (2004) The evolution of molecular markers-just a matter of fashion? Nat Rev Genet 5:63–69
11. Weise A, Gross M, Mrasek K (2008) Parental origin-determination fluorescence in situ hybridization distinguishes homologous human chromosomes on a single-cell level. Int J Mol Med 21:189–200
12. Polityko AD, Khurs OM, Kulpanovich AI et al (2009) Paternally derived der(7)t(Y;7)(p11.1 approximately 11.2;p22.3)dn in a mosaic case with Turner syndrome. Eur J Med Genet 52: 207–210
13. Mkrtchyan H, Gross M, Hinreiner S et al (2010) The human genome puzzle—the role of copy number variation in somatic mosaicism. Curr Genomics 11:426–431
14. Weise A, Liehr T (2009) Pre- and postnatal diagnostics and research on peripheral blood, chorion, amniocytes, and fibroblasts. In: Liehr T (ed) Fluorescence in situ hybridization (FISH)—application guide. Springer, Berlin, pp 113–122
15. Telenius H, Carter NP, Bebb CE et al (1992) Degenerate oligonucleotide primed PCR: general amplification of target DNA by a single degenerate primer. Genomics 13:718–725

Chapter 17

Site-specific Integration of Bacterial Artificial Chromosomes into Human Cells

Bradley McColl, Sara Howden, and Jim Vadolas

Abstract

Gene therapy of inherited diseases requires long-term maintenance of the corrective transgene. Stable integration of the introduced DNA molecule into one of the host cell chromosomes is the simplest strategy for achieving this. However, genotoxicity resulting from random insertion of the transgene raises serious safety concerns that must be addressed if gene therapy is to enter the clinical mainstream. The following method makes use of the Rep integrase of adeno-associated virus to insert a transgene into the human AAVS1 site, a known "safe harbor" region within the human genome. This approach has the potential for application to novel gene therapy strategies for improved safety. In addition, with this method it is also possible to create cell lines carrying BAC transgenes in the AAVS1 site.

Key words Adeno-associated virus, Site-specific integration, Rep integrase, Gene therapy, Genotoxicity

1 Introduction

The development of bacterial artificial chromosomes (BACs) permitted the cloning of large (~300 kb) fragments of genomic DNA, advantageous for genomic sequencing or the study of distal genetic regulatory elements [1, 2]. In addition, BAC libraries provide a "ready-made" collection of genomic sequences with the potential to be used as corrective transgenes for gene therapy of inherited disorders [3, 4]. Delivery of a BAC clone containing the coding region and native regulatory elements of the gene of interest avoids the necessity for creation of a transgenic expression construct, and ensures that the introduced gene is likely to be regulated in a physiologically appropriate manner.

Long-term maintenance of the transgene requires that the introduced DNA sequence be replicated and passed to both daughter cells with each cell division. The simplest method for achieving this is via integration of the transgene into the host cell genome. Viral vectors capable of genomic integration, such as those derived from the gammaretroviruses and lentiviruses are the current method

Kumaran Narayanan (ed.), *Bacterial Artificial Chromosomes*, Methods in Molecular Biology, vol. 1227, DOI 10.1007/978-1-4939-1652-8_17, © Springer Science+Business Media New York 2015

of choice for delivery and genomic integration of a transgene. However, the untargeted nature of the integration event resulting from infection with these viruses carries a high risk of genotoxic effects [5, 6]. The serious nature of this risk was demonstrated by the development of leukemias in severe combined immunodeficiency patients who took part in a clinical trial of a retroviral gene therapy construct [7, 8]. Subsequent study revealed the neoplasias to be associated with activation of proto-oncogenes as a result of transgene insertion.

Consequently, viral vectors have been modified so as to reduce the potential for activation of nearby genes as a result of transgene insertion, although as yet no effective method has been found to address the random aspect of the insertion event.

The alternative approach, development of methods for site-specific integration of transgenes has the potential to avoid altogether the complications of random transgene insertion. The use of TALENs and zinc finger nucleases for targeted DNA integration has been demonstrated in numerous laboratory studies, although so far the targeting efficiency of these approaches is not high [9]. In the context of gene therapy, this approach may employ a direct replacement strategy, whereby the transgene is targeted for insertion within the locus of the mutated gene. Alternatively, the inserted DNA may be targeted into so-called "safe harbors," regions of the genome where introduction of foreign DNA can occur without genotoxic effects [10].

Adeno-associated virus (AAV) provides an example of naturally occurring nonpathological site-specific insertion of foreign DNA into the human genome. Upon infection, the action of the virally encoded Rep integrase results in insertion of the AAV genome at a site on chromosome 19 known as AAVS1 [11]. No human pathology is associated with AAV insertion, and the AAVS1 site is now considered to be a useful target for safe transgene insertion [10]. The precise molecular mechanism by which the Rep integrase functions are yet to be fully understood, however it is known to act by binding to the inverted terminal repeat (ITR) sequences on the viral genome and to the Rep binding site (RBS) of the AAVS1 site. Nicking downstream of the Rep binding element leads to insertion of viral DNA into AAVS1 [11].

A fragment of the AAV genome encompassing the viral P5 promoter has been shown to contain an additional Rep binding element (RBE) (outside the ITRs) and is capable of mediating targeted insertion of foreign DNA into AAVS1 (Fig. 1) [12]. By inserting this region known as the P5 integration efficiency element (P5IEE) into the vector sequences of a human BAC clone, we were able to achieve high efficiency Rep-mediated integration of a 200 kb BAC into the AAVS1 site [13] (*see* **Note 1**).

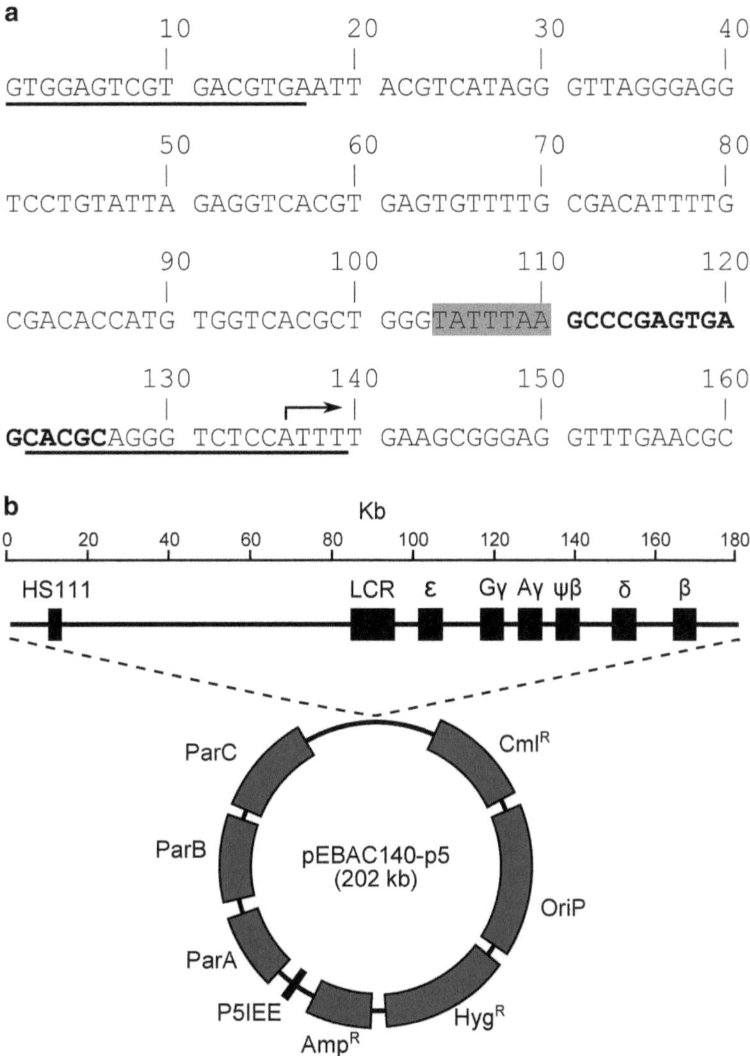

Fig. 1 (**a**) Sequence of the P5IEE region of AAV2. *Underlined* regions denote locations of PCR primers used to amplify the P5 region for insertion into BAC vectors. The RBE sequence is shown in *bold* [12]. Locations of the Rep gene TATA box (*shaded*) and transcription start site (*arrow*) are also shown. (**b**) Schematic diagram of BAC pEBAC140-p5 containing the intact multigenic β-globin locus and the vector backbone featuring the P5IEE. Recombineering was used to insert the P5IEE into the vector backbone [13]

Rep activity was achieved within target cells by delivery of Rep mRNA in *trans* along with the BAC clone. This approach ensures transient expression of the integrase to levels sufficient for integration of the BAC, but avoids the potential for genotoxic effects that may result from long-term expression of the Rep integrase.

2 Materials

For RNA preparation, solutions are prepared using ultrapure (18 MΩ/cm) water, treated with diethyl pyrocarbonate (DEPC) to inactivate RNases. Use autoclaved glass and plasticware for preparation of all reagents.

2.1 BAC DNA Purification

1. Luria-Bertani medium: Bacto-tryptone 10 g/l, Bacto-yeast extract 5 g/l, NaCl 10 g/l. Dissolve in ddH_2O, sterilize by autoclaving. Store at room temperature.
2. Chloramphenicol: 20 μg/ml (stock: 20 mg/ml in 100 % ethanol). Store at −20 °C.
3. Solution 1: 50 mM Tris-Cl, pH 8.0; 10 mM EDTA, 100 μg/ml RNAse A.
4. Solution 2: 200 mM NaOH, 1 % sodium dodecyl sulfate.
5. Solution 3: 3.0 M potassium acetate, pH 5.5. Dissolve in ddH_2O, adjust pH with glacial acetic acid. Store at 4 °C.
6. TE buffer: 10 mM Tris-Cl, 1 mM EDTA pH 8.0. Dissolve in MilliQ H_2O, adjust pH with HCl and sterilize by autoclaving. For CsCl gradient preparation, dissolve CsCl in TE immediately prior to the ultracentrifugation step.
7. Ethidium bromide solution: 10 μg/ml (stock: 10 mg/ml).
8. Quick-Seal Ultra-Clear tubes (Beckman Coulter).
9. Slide-A-Lyzer 7 K MWCO Dialysis Cassettes (Thermo Fisher Scientific).
10. Low Range PFGE Marker (New England Biolabs).

2.2 In Vitro transcription

1. Primers are given for both Rep isoforms (*see* **Note 2**). Underline denotes SP6 promoter.

   Rep78ivtF:CATACG<u>ATTTAGGTGACACTATAG</u>AATACA GGC TCGACAATTGACTAGTG

   Rep78ivtR:TTTTTTTTTTTTTTTTTTTTTTTTTTTTTT CTC GAGATTACCCTGTTATC

   Rep68ivtF:CATACG<u>ATTTAGGTGACACTATAG</u>AATACA GTG AGCACGCAGGGTCTC

   Rep68ivtR:TTTTTTTTTTTTTTTTTTTTTTTTTTTTTC CTTGTC GAGTCCGTTGAAGR

2. Standard PCR reagents (thermostable DNA polymerase, buffer, $MgCl_2$, deoxynucleotide triphosphates). For fidelity of amplification, a proofreading DNA polymerase such as Platinum Pfx (Invitrogen) is recommended.
3. mMessage mMachine in vitro transcription and capping kit (Invitrogen) (*see* **Note 3**).

2.3 Co-transfection of Target Cells with BAC DNA and Rep mRNA

1. Purified Rep mRNA and BAC DNA containing the P5IEE (*see* **Note 4**).
2. K562 human erythroid cell line and media (DMEM + 10 % fetal calf serum).
3. Opti MEM serum free media (Invitrogen).
4. 4 mm electroporation cuvettes and electroporation apparatus (BioRAD GenePulser).
5. Tissue culture plasticware and humidified 37 °C, 5 % CO_2 incubator.

2.4 Fluorescence In Situ Hybridization

1. Colcemid.
2. Nick translation kit for generation of fluorescent probes, spectrum red-dUTP and spectrum green-dUTP (Abbott Molecular).
3. SSC (20×): 3 M NaCl, 300 mM trisodium citrate ($Na_3C_6H_5O_7$). Adjust pH to 7.0 with HCl.
4. Dextran sulfate: Dissolve in MilliQ H_2O.
5. Formamide.
6. Cytospin centrifuge (Shandon Inc.).
7. Probe hybridization buffer: (50 % formamide, 10 % dextran sulfate, 2× SSC).

2.5 Immunohistochemical Detection of Rep protein

1. Phosphate buffered saline (PBS): 137 mM NaCl, 2.7 mM KCl, 10 mM Na_2HPO_4, 2.0 mM KH_2PO_4.
2. 0.1 % Triton X-100/PBS: Dilute Triton X-100 to 0.1 % (vol/vol) in PBS. Prepare fresh before use.
3. 4 % (w/vol) paraformaldehyde. Dissolve in PBS.
4. Tris-buffered normal saline (TBS): 50 mM Tris–Cl pH 7.6, 150 mM NaCl pH 7.6. Adjust pH with HCl.
5. Blocking reagent (Roche Applied Science): For use dissolve at 1 % (w/vol) in TBS.
6. Anti-Rep antibody (American Research Products Inc.): Dilute 1:50 in TBS prior to use.
7. Anti-mouse IgG-rhodamine (Jackson ImmunoResearch Laboratories Inc.): Dilute 1:100 in TBS prior to use.
8. Vectashield/(DAPI 500 ng/μl) (Vector Laboratories).

3 Methods

3.1 Large-Scale Purification of BAC DNA by Cesium Chloride Gradient Centrifugation

1. Centrifuge 250 ml of overnight bacterial culture grown in Luria-Bertani medium supplemented with 12.5 μg/ml chloramphenicol (or other appropriate selective antibiotic) at 5,000 × *g* for 5 min at 4 °C (*see* **Note 5**).

2. Resuspend bacterial pellet in 110 ml Solution 1 followed by the addition of 110 ml S2. Mix gently by inversion until lysis is visible and incubate on ice for 10 min.
3. Add 110 ml of Solution 3 and mix gently by inversion. Incubate for 30 min on ice.
4. Centrifuge at 30,000 × *g* for 30 min at 4 °C. Decant supernatant and filter through sterilized gauze. The pellet may be discarded.
5. Add isopropanol (0.7 volumes) to the supernatant and mix gently by inversion. Precipitate DNA by centrifugation at 20,000 × *g* for 30 min.
6. Wash the DNA pellet in 70 % ethanol and centrifuge at 20,000 × *g* for 5 min. Decant the supernatant and air dry the pellet for 5–10 min.
7. Gently dissolve the DNA pellet in 10 ml TE buffer pH 8.0 containing 1.1 g/ml CsCl. Add ethidium bromide to a final concentration of 250 μg/ml. Spin solution at 7,500 × *g* for 5 min at room temperature. After centrifugation a floating layer will form above a clear red solution of DNA/CsCl/ethidium bromide. Remove the clear, red solution for further purification and discard the floating layer.
8. Transfer solution into 13.5 ml Quick-Seal tubes. Fill tubes completely by addition of TE pH 8.0 containing 1.1 g/ml CsCl and balance tubes to within 0.05 gm. Centrifuge tubes for 18 h at 285,000 × *g* (55,000 rpm, e.g., using a Beckman Coulter 80Ti rotor) at 14 °C.
9. Visualize the BAC DNA band with UV light. The upper band is composed of bacterial chromosomal DNA and the lower band is composed of the supercoiled BAC DNA. Carefully extract the BAC DNA using a 20-gauge needle and 1 ml syringe.
10. Extract the ethidium bromide from the BAC DNA solution using equal volume of saturated NaCl/TE pH 8.0/isopropanol. Repeat the extraction four times.
11. Dialyse BAC DNA overnight in 1 liter of TE pH 8.0 at 4 °C to remove CsCl. Repeat this step two times over a 24–48 h period.
12. Evaluate the quality of DNA by digesting 0.5–1 μg of BAC DNA and run on pulse field gel electrophoresis (PFGE) together with undigested BAC DNA and Low Range PFGE Marker. The anticipated yield of BAC DNA from this procedure is expected to be in the range of 100–200 μg. Yield of BAC DNA may vary due to DNA sequence.
13. Store purified BAC DNA at 4 °C. Freeze thawing of purified BAC DNA is not recommended. Purified BAC DNA is stable for several months at 4 °C.

3.2 Synthesis of Rep mRNA

In vitro transcription of mRNA can be achieved using as a template a linearized plasmid vector containing the Rep cDNA under the control of an SP6 (or similar) promoter, followed by a poly A tract (e.g., pSP64, Promega). Alternatively, a PCR product may be used as template, with the necessary promoter sequences and poly-A tail incorporated into oligonucleotide primers used for amplification. This approach has the advantage of avoiding the necessity to subclone the gene to be transcribed into pSP64. However if using a plasmid template, the vector should be linearized by restriction digestion and the appropriate fragment purified by isolation from agarose gel. PCR-generated templates should be purified by DNA affinity chromatography.

3.2.1 In Vitro Transcription

The in vitro transcription reaction is described as per the mMessage mMachine kit protocol.

1. Mix supplied with in vitro transcription reaction components:

2× NTP/CAP	10 μl
10× reaction buffer	2 μl
Linear template DNA	0.1–1 μg
Enzyme mix	2 μl
Nuclease-free water	To 20 μl total

   Mix gently, incubate at 37 °C for 1–2 h (extended incubation may improve yield). Following incubation, template DNA may be removed via DNase treatment, however this step is optional.

2. Precipitate RNA with lithium chloride:
   To the in vitro transcription reaction, add 30 μl of nuclease-free water and 30 μl of the supplied LiCl precipitation solution. Mix and chill for >30 min at −20 °C. Centrifuge at 4 °C for 15 min at 13,000 ×*g*, then wash pellet with 70 % ethanol and centrifuge as before. Carefully decant supernatant and allow pellet to air dry. Resuspend in nuclease-free water. RNA can now be quantified by spectrophotometry and quality checked by denaturing agarose gel electrophoresis (Fig. 2a).

3.3 Co-transfection of Purified BAC DNA and Rep mRNA

The method presented below is suitable for co-transfection by electroporation of the human erythroid cell line K562 with Rep mRNA and P5IEE BAC (Fig. 2b, c) (Reproduced from (2007) with permission from (John Wiley & Sons, Ltd)). For other cell lines, optimal transfection techniques must be determined empirically (*see* **Note 6**).

1. Wash K562 cells twice with serum-free OptiMEM and resuspend at a concentration of 10^7 cells/ml.
2. Mix 500 μl of cell suspension with 8 μg of BAC DNA and 20 μg of Rep mRNA in 0.4 cm electroporation cuvette.

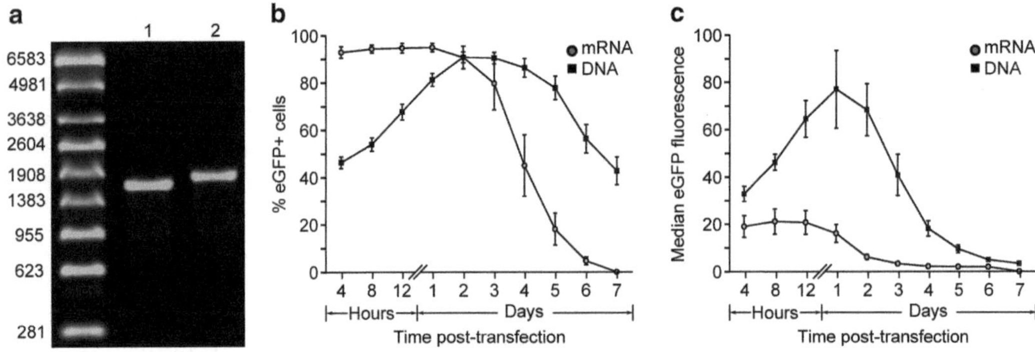

Fig. 2 (**a**) Denaturing agarose gel electrophoresis of Rep in vitro transcription products. *Lane 1*, Rep68; *2*, Rep78. Sizes in base pairs shown at *left*. (**b**, **c**) Flow cytometry analysis of transient eGFP expression driven by synthetic mRNA. K562 cells were electroporated with either 20 μg of an EGFP expression plasmid (pEGFPN2) or 20 μg of 5′-m7GpppG-capped and polyadenylated in vitro transcribed eGFP mRNA. The percentage of fluorescent cells and median fluorescence intensity was measured at multiple time points as shown. Note the earlier peak of expression and more rapid decline that results from mRNA transfection rather than DNA

3. Electroporate at 220 V, 25 μF using a Bio-Rad gene pulser.
4. Plate cells in DMEM/10 % FCS, incubate at 37 °C/5 % CO_2 for 48 h.
5. Seed cells into 96-well plates at 100 cells/well, in medium supplemented with the appropriate antibiotic to select for the BAC, in order to establish stably transfected cell lines. Maintain cells in medium supplemented with antibiotic over a 2- to 4-week period.

3.4 Immunohistochemical Detection of Rep Expression

1. Following transfection with Rep mRNA, centrifuge 1×10^5 cells onto slides using a Cytospin centrifuge (Shandon Inc.). Air-dry slides and dehydrate cells with methanol for 5 min.
2. Permeabilize cells in 0.1 % Triton X-100/PBS for 5 min, then wash 3× in PBS.
3. Fix cells with 4 % paraformaldehyde in PBS at room temperature for 30 min.
4. Wash cells 3× in PBS and block with 1 % Blocking Reagent at room temperature for 30 min.
5. Add anti-Rep antibody onto cells and incubate at room temperature for 1 h.
6. Wash slides 3× in PBS then incubate with anti-mouse IgG-rhodamine secondary antibody at room temperature for 1 h.
7. Wash cells in PBS, mount with Vectashield/DAPI and visualize by fluorescence microscopy.

3.5 Fluorescence In Situ Hybridization Analysis of Sites of Integration

Following selection of cells with the appropriate antibiotic, the site of BAC integration within the genome can be determined by fluorescence in situ hybridization (FISH). Following co-transfection with Rep mRNA and BAC DNA, metaphase spreads of antibiotic-resistant cells are hybridized to fluorescently labeled DNA probes complementary to the AAVS1 region and to the integrated BAC. Fluorescence microscopy is then used to identify cell lines in which the BAC and AAVS1 sequence co-localize (*see* **Note 7**).

3.5.1 Preparation of Metaphase Spreads

1. Add colcemid (final concentration 100 ng/ml) to cultures of cells to be analyzed and incubate for 1 h at 37 °C.
2. Harvest arrested cells by centrifugation and resuspend gently in 0.075 M KCl. Incubate for 10 min at 37 °C.
3. Fix cells in 3:1 methanol:acetic acid solution for at least 12 h at –20 °C.
4. Replace fixative with fresh methanol:acetic acid solution before dropping fixed cells gently onto clean glass slides. Allow slides to air-dry and inspect for quality of preparations under phase-contrast microscopy before proceeding.

3.5.2 Probe Preparation and Hybridization

Probes are prepared by the nick-translation method using the kit from Abbott Molecular to incorporate fluorophore-conjugated dNTPs (Spectrum Red-dUTP and Spectrum Green-dUTP) (*see* **Note 8**). Mix the following:

0.2 mM Spectrum Red or Spectrum Green dUTP	2.5 μl
100 μM dTTP	5 μl
dNTP mix	10 μl
10× nick translation buffer	5 μl
Nick translation enzyme	10 μl
Template DNA	1 μg
Nuclease-free water	To 50 μl total

Mix tube gently and centrifuge briefly. Incubate 8–16 h at 15 °C. Stop the reaction by heating to 70 °C for 10 min and store on ice/4 °C until required.

Following labeling of AAVS1 and BAC-specific probes, metaphase spreads are hybridized to the fluorescently tagged probes and the locations of hybridizing sequences within the karyotype are identified by fluorescence microscopy.

1. Precipitate labeled probes with ethanol and resuspend in hybridization buffer at 30 ng/μl. Denature probes by heating to 72 °C for 8 min.

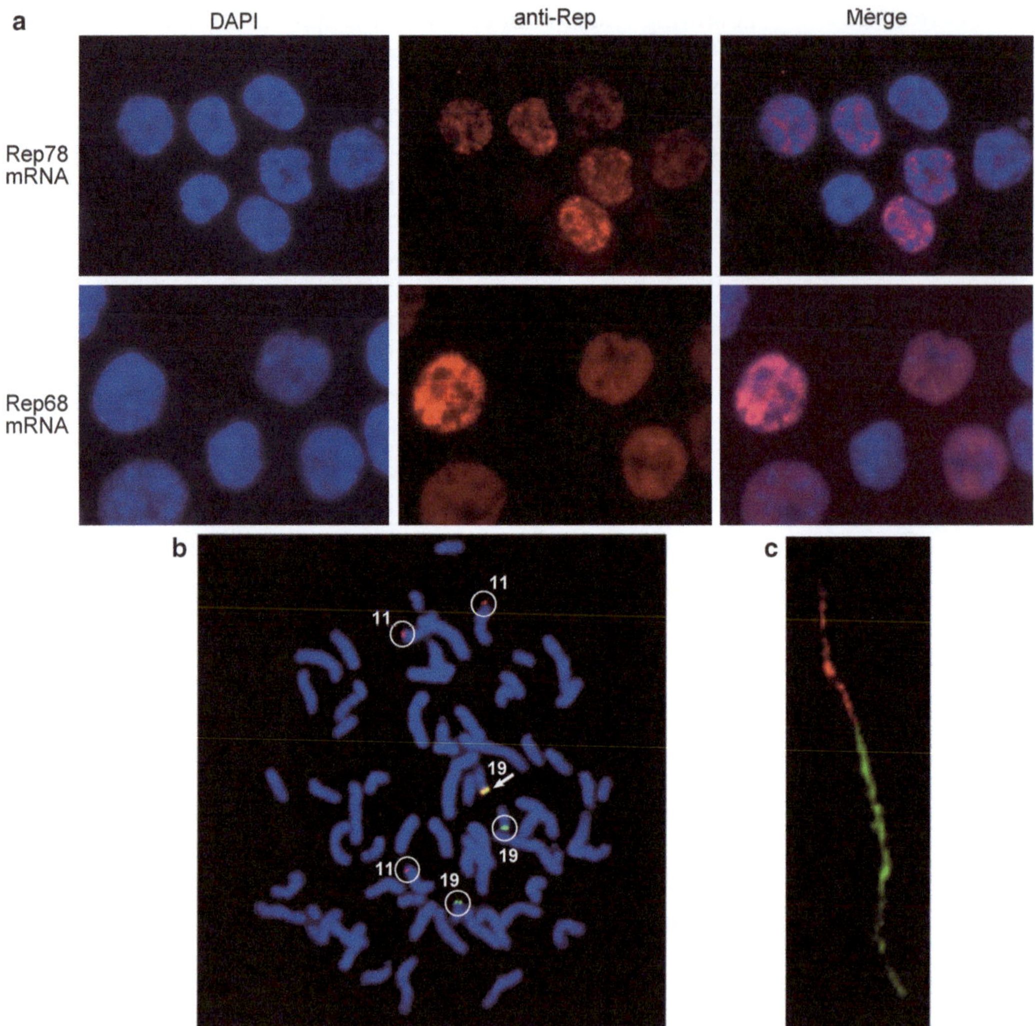

Fig. 3 (**a**) Immunohistochemistry of K562 cells, showing Rep protein expression following transfection with synthetic mRNA encoding Rep78 (*upper*) or Rep68 (*lower*). Rep protein expression was detected using an antibody specific to both Rep68 and Rep78 proteins, followed by staining with an anti-mouse IgG secondary antibody conjugated to rhodamine. Slides were mounted in Vectashield containing DAPI and examined using an Olympus BX60 epifluorescence microscope. (**b**) FISH analysis of metaphase spreads from K562 cells to determine site of β-globin BAC integration. *Circles* indicate locations of probe hybridization on chromosome 11 and 19. AAVS1 genomic region is labeled with *green* probe, β-globin BAC sequence with *red*. Note that K562 cells are trisomic for chromosome 11 (location of the native β-globin gene) and chromosome 19 (location of AAVS1). Arrow indicates site of BAC integration. (**c**) Fiber FISH analysis of K562 genomic DNA showing adjacent AAVS1 sequence (*green*) and β-globin locus (*red*), indicating site-specific integration of the BAC

2. Denature metaphase spreads in 70 % formamide/30 % 2× SSC at 70 °C for 2 min, then dehydrate through an ethanol series (5 min each in 70 %, 90 % and 100 % ethanol).

3. Add 2 μl of denatured probe to denatured metaphase spreads beneath a 10 mm coverslip. Seal coverslips with rubber cement and allow hybridization reaction to proceed overnight at 37 °C.

4. Gently remove coverslips and wash metaphase spreads in 1× SSC at 70 °C for 10 min. Mount fresh coverslips (including antifade reagent) and examine by fluorescence microscopy (Fig. 3) (Reproduced from (2008) with permission from (Nature Publishing Group)).

4 Notes

1. This method is dependent upon the presence of the RBE within the AAVS1 site on chromosome 19. This feature seems to be a relatively recent product of evolution, being restricted to primates. Therefore cells and cell lines derived from mice and other nonprimate species are not applicable to this approach, although it may be possible that within the genomes of other species sequences exist with sufficient homology to the RBE to function as substrates for Rep-mediated recombination.
2. Rep is expressed as both 68 and 78 kDa isoforms, both of which are capable of catalyzing recombination at the AAVS1 site. Oligonucleotide primers are given for amplification of both isoforms and incorporate SP6 promoter and polyA tail into the PCR product for subsequent transcription. For amplification of Rep68 and 78 sequences, we used plasmids pHIV68 [14] (provided by Jude Samulski, University of North Carolina, Chapel Hill, NC, USA) and pREP10 (provided by Horace Drew, CSIRO, North Ryde, NSW, Australia). The primer sequences provided are specific to these vectors; researchers wishing to use other Rep vectors will need to design oligonucleotides specific for those. Vectors containing the AAV Rep78 coding sequence are available from Cell Biolabs Inc. (San Diego, CA, USA). Primers can also be designed to generate a template for in vitro transcription of eGFP for control experiments. In our hands, both Rep68 and Rep78 isoforms have been used to achieve site-specific integration at the AAVS1 site. A recent paper has shown that Rep78 alone is sufficient to target transgenes to AAVS1 [15].
3. Other commercial in vitro transcription kits are available, or the necessary reagents may be prepared individually. For simplicity, the mMessage mMachine kit is recommended. The mMessage mMachine kit generates transcripts incorporating a 7-methyl-guanine 5′ cap analogous to native mRNA for improved stability.
4. The P5IEE may be incorporated into the BAC DNA by either traditional restriction enzyme-based cloning, or by recombineering should there be no suitable restriction enzyme sites (Fig. 1). Note also that many BAC vectors contain loxP recombination sites and may therefore be modified by cre-mediated recombination. Regardless of the method used, researchers must bear in mind the fact that Rep-mediated integration

results in a break in the DNA sequence at the RBE, followed by joining to chromosome 19 DNA. Therefore the P5IEE should be inserted within the BAC vector sequence, to avoid reorganization of sequences relating to the gene of interest upon insertion into the host cell genome. BAC DNA should be prepared via the cesium chloride method to obtain highly purified sample suitable for transfection.

5. This method is suitable for large-scale purification of plasmids >10 kb in size. Although slightly more technically complex than common anion-exchange methods of DNA purification, in our experience cesium chloride purification provides better yields (100–200 μg). For BAC preparation, the method is modified by use of greater volumes of solutions 1, 2 and 3 than are the norm for plasmid preparation. To prevent shearing of BAC DNA, use wide bore tips when resuspending or transferring BAC DNA in solution.
6. A control experiment should be carried out using the transcript of a readily detectable reporter gene such as the enhanced green fluorescent protein (eGFP). In this way researchers can readily establish the optimal conditions for transfection of cells with mRNA and subsequent expression. Rep expression can be detected by immunohistochemistry (Fig. 3) (Reproduced from (2008) with permission from (Nature Publishing Group)).
7. Fluorescence *in situ* hybridization is used to characterize the K562 cell lines generated by this technique. By FISH analysis, it is possible to determine the chromosomal co-localization of the AAVS1 and BAC transgene sequences. A more advanced technique, fiber FISH permits visualization of adjacent AAVS1 and BAC transgene sequences within a single extended DNA strand. In our experience, analysis of metaphase spreads from K562 cells co-transfected with a 20 kb BAC and Rep mRNA targeted the AAVS1 region at 32 % efficiency, while a 200 kb BAC and Rep mRNA targeted the AAVS1 site at 20 % efficiency [16, 13]. Furthermore, Southern blotting demonstrated that the majority of insertions at the AAVS1 site consisted of the entire BAC, without rearrangement or deletion.
8. For preparation of the AAVS1 probe, our laboratory uses a BAC clone (RP11-463M24) from a human genomic library [17] incorporating the appropriate region of chromosome 19.

Acknowledgments

This work was supported by the National Health and Medical Research Council; the Murdoch Children's Research Institute; the Victorian Government's Operational Infrastructure Support Program; and Thalassaemia Australia and Thalassaemia Society of New South Wales.

References

1. Shizuya H, Birren B, Kim UJ, Mancino V, Slepak T, Tachiiri Y, Simon M (1992) Cloning and stable maintenance of 300-kilobase-pair fragments of human DNA in Escherichia coli using an F-factor-based vector. Proc Natl Acad Sci U S A 89:8794–8797
2. Shizuya H, Kouros-Mehr H (2001) The development and applications of the bacterial artificial chromosome cloning system. Keio J Med 50:26–30
3. Wade-Martins R, Smith ER, Tyminski E, Chiocca EA, Saeki Y (2001) An infectious transfer and expression system for genomic DNA loci in human and mouse cells. Nat Biotechnol 19:1067–1070
4. Vadolas J, Wardan H, Bosmans M, Zaibak F, Jamsai D, Voullaire L, Williamson R, Ioannou PA (2005) Transgene copy number-dependent rescue of murine beta-globin knockout mice carrying a 183 kb human beta-globin BAC genomic fragment. Biochim Biophys Acta 1728:150–162
5. Trobridge GD (2011) Genotoxicity of retroviral hematopoietic stem cell gene therapy. Expert Opin Biol Ther 11:581–593
6. Wu C, Dunbar CE (2011) Stem cell gene therapy: the risks of insertional mutagenesis and approaches to minimize genotoxicity. Front Med 5:356–371
7. Hacein-Bey-Abina S, Von Kalle C, Schmidt M, McCormack MP, Wulffraat N, Leboulch P, Lim A, Osborne CS, Pawliuk R, Morillon E et al (2003) LMO2-associated clonal T cell proliferation in two patients after gene therapy for SCID-X1. Science 302:415–419
8. Hacein-Bey-Abina S, Garrigue A, Wang GP, Soulier J, Lim A, Morillon E, Clappier E, Caccavelli L, Delabesse E, Beldjord K et al (2008) Insertional oncogenesis in 4 patients after retrovirus-mediated gene therapy of SCID-X1. J Clin Invest 118:3132–3142
9. Wirt SE, Porteus MH (2012) Development of nuclease-mediated site-specific genome modification. Curr Opin Immunol 24:609–616
10. Sadelain M, Papapetrou EP, Bushman FD (2011) Safe harbours for the integration of new DNA in the human genome. Nat Rev Cancer 12:51–58
11. Linden RM, Ward P, Giraud C, Winocour E, Berns KI (1996) Site-specific integration by adeno-associated virus. Proc Natl Acad Sci U S A 93:11288–11294
12. Philpott NJ, Gomos J, Berns KI, Falck-Pedersen E (2002) A p5 integration efficiency element mediates Rep-dependent integration into AAVS1 at chromosome 19. Proc Natl Acad Sci U S A 99:12381–12385
13. Howden SE, Voullaire L, Wardan H, Williamson R, Vadolas J (2008) Site-specific, Rep-mediated integration of the intact beta-globin locus in the human erythroleukaemic cell line K562. Gene Ther 15:1372–1383
14. Amiss TJ, McCarty DM, Skulimowski A, Samulski RJ (2003) Identification and characterization of an adeno-associated virus integration site in CV-1 cells from the African green monkey. J Virol 77:1904–1915
15. Van Rensburg R, Beyer I, Yao X-Y, Wang H, Denisenko O, Li Z-Y, Russell DW, Miller DG, Gregory P, Holmes M et al (2013) Chromatin structure of two genomic sites for targeted transgene integration in induced pluripotent stem cells and hematopoietic stem cells. Gene Ther 20:201–214
16. Howden SE, Voullaire L, Vadolas J (2008) The transient expression of mRNA coding for Rep protein from AAV facilitates targeted plasmid integration. J Gene Med 10:42–50
17. Osoegawa K, Mammoser AG, Wu C, Frengen E, Zeng C, Catanese JJ, de Jong PJ (2001) A bacterial artificial chromosome library for sequencing the complete human genome. Genome Res 11: 483–496

Chapter 18

Generation of BAC Reporter Cell Lines for Drug Discovery

Betty R. Kao, Bradley McColl, and Jim Vadolas

Abstract

Bacterial artificial chromosome (BAC) reporter cell lines are generated through stable transfection of a BAC reporter construct wherein the gene of interest is tagged with a reporter gene such as eGFP. The large capacity of BACs (up to 350 kb of genomic sequence) enables the inclusion of all regulatory elements that ensure appropriate regulation of the gene of interest. Furthermore, the reporter gene allows the expression of the gene of interest to be readily detected by flow cytometry. Cell lines can also be easily cultured for extended periods with minimal cost. These features of BAC reporter cell lines make them highly amenable for use in high-throughput screening of large drug libraries for compounds that induce the expression of the gene of interest.

This chapter describes a method for generation of BAC reporter cell lines that are suitable as cellular assay systems in high-throughput screening. Briefly, this method involves (A) generation of cell clones stably transfected with a BAC reporter construct, (B) selection of "candidate" cell clones based on the responsiveness to known inducers, (C) confirmation of the integrity of the BAC reporter construct integrated within the candidate clones, and (D) assessment of the developmental regulation of the BAC reporter construct. As an example, we describe the generation of a BAC reporter cell line containing the human β-globin locus modified to express γ-globin as eGFP for use as a cellular reporter assay for screening of drugs that can reactivate expression of developmentally silenced γ-globin for the treatment of β-hemoglobin disorders.

Key words Bacterial artificial chromosomes, Stable transfection, Drug discovery, High-throughput screening, Fluorescent reporter cell line, Cellular assay system

1 Introduction

A bacterial artificial chromosome (BAC) reporter cell line is created by transfection of a cell line with a BAC in which the gene of interest is fused to or substituted by a reporter gene such as eGFP. The reporter gene enables easy detection of expression of the gene of interest using flow cytometry. The use of a BAC ensures appropriate regulation of the gene of interest given that BACs can carry genomic sequences up to 350 kb, enabling the inclusion of most if not all long-range regulatory elements [1]. Furthermore,

Kumaran Narayanan (ed.), *Bacterial Artificial Chromosomes*, Methods in Molecular Biology, vol. 1227, DOI 10.1007/978-1-4939-1652-8_18, © Springer Science+Business Media New York 2015

genes expressed from BACs are protected from positional effects influenced by the site of chromosomal integration and are stably expressed for prolonged periods [1]. A BAC reporter cell line is therefore a useful tool for investigating expression of a gene.

Modern drug discovery routinely uses high-throughput screening of drug libraries to test for their ability to modify expression of the gene of interest, and thereby rapidly identify candidate drugs for further investigation. A BAC reporter cell line is an ideal cellular assay system for high-throughput screening given that the gene of interest is appropriately regulated, readily detected by flow cytometry, and because a cell line can be maintained in culture for prolonged periods with minimal labor and financial expense [1–4].

This chapter will describe our method for generating BAC reporter cell lines suitable for use in high-throughput screening for drug discovery. Murine and human erythroleukemia cell lines harboring BACs containing the human β-globin locus (Fig. 1) are used to illustrate the generation of cellular model systems for drug discovery. For example, murine erythroleukemic (MEL) cells harboring the eGFP-modified β-globin locus (Construct 1; Fig. 1b) are an ideal cellular assay system for evaluating drugs that increase expression of the developmentally silenced γ-globin gene, for the treatment of β-hemoglobin disorders [2, 3]. MEL cells possess an adult globin expression profile, displaying very low expression of human γ-globin and high expression of human β-globin [5]. Therefore, the low expression of γ-globin makes MEL cells an appropriate cellular assay system in which to screen for drugs that increase expression of γ-globin.

Generation of a BAC reporter cell line involves preparation of the BAC reporter construct, transfection of the BAC reporter construct into an appropriate cell line by either electroporation or

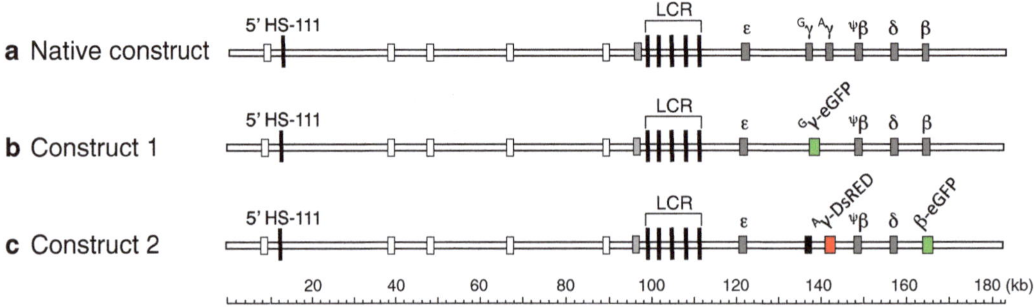

Fig. 1 BAC constructs modified to incorporate one or more fluorescent reporters. The native construct (**a**) containing 183 kb of the β-globin locus was modified to produce fluorescent reporter constructs 1 and 2. In Construct 1 (**b**), eGFP replaces the γ-globin genes. In Construct 2 (**c**), dsRED replaces γ-globin and eGFP replaces β-globin

chemical methods, selection of stably transfected cells, and lastly, generation of single cell clones through limiting cell dilution. Upon transfection with Construct 1 (Fig. 1b), transcription from the γ-globin promoter will be readily detected as eGFP by flow cytometry, and high-throughput screening of drug libraries can be carried out in 96-well or 386-well format to identify drugs that increase expression of γ-globin/eGFP [2, 3]. This approach can easily be adapted for any drug discovery experiment, simply requiring an appropriate BAC reporter construct and cell line for the gene of interest.

The generation of stably transfected BAC reporter cell lines does not guarantee that they will be suitable for use in high-throughput screening. The responsiveness of each clone to inducers of the gene of interest must be assessed using known inducers in order to evaluate their usefulness. For example, K562 cells, a embryonic/fetal human erythroleukemia cell line derived from a patient with chronic myeloid leukemia, can be induced to express γ-globin using hemin [6, 7]. Only clones that respond as expected to known inducers of the gene of interest, and with adequate dynamic range, should be considered for use in screening.

Further characterization of the clones should be carried out prior to using the cell line in a screen, including Southern blot and PCR analysis to confirm the integrity of the BAC reporter construct, as well as assessment of the developmental regulation of the BAC reporter construct [2]. The latter is an important consideration particularly if the BAC reporter construct is regulated in a developmental stage-specific manner. Therefore, the BAC reporter constructs containing the human β-globin locus, such as Constructs 1 and 2 (Fig. 1b, c), must be assessed in both an adult and fetal erythroid environment to ensure the reporter genes are appropriately regulated in a developmental stage-specific manner [2].

We describe methods for generating BAC reporter cell lines suitable for high-throughput screening of drug libraries aimed to identify small molecules that can reactivate developmentally silenced γ-globin as a therapy for β-hemoglobin disorders.

2 Materials

2.1 Purification of BAC DNA by Cesium Chloride Density Gradient Centrifugation

1. BAC constructs may be obtained from the BACPAC Resources Center (BPRC) at the Children's Hospital Oakland Research Institute, Oakland, California. This is a not-for-profit resource center with collections of genomic DNA libraries including BAC and PAC clones.
2. Luria-Bertani (LB) medium: Bacto-tryptone 10 g/L, Bacto-yeast extract 5 g/L, NaCl 10 g/L. Dissolve in dH_2O and sterilize by autoclaving. Store at room temperature.

3. Chloramphenicol: 20 μg/mL (stock: 20 mg/mL in 100 % ethanol). Store at –20 °C.
4. Solution 1: 50 mM Tris-Cl, pH 8.0, 10 mM EDTA, 100 μg/mL RNAse A.
5. Solution 2: 200 mM NaOH, 1 % sodium dodecyl sulfate.
6. Solution 3: 3 M potassium acetate, pH 5.5. Dissolve in dH_2O, adjust pH with glacial acetic acid. Store at 4 °C.
7. TE buffer: 10 mM Tris–HCl pH 8.0, 1 mM EDTA pH 7.5. Dissolve in MilliQ H_2O, adjust pH with 1 M HCl and sterilize by autoclaving. For CsCl gradient preparation, dissolve CsCl in TE pH 8.0 immediately prior to the ultracentrifugation step. Store at room temperature.
8. Ethidium bromide solution: 10 μg/mL (stock: 10 mg/mL).
9. Quick-Seal Ultra-Clear tubes (Beckman Coulter).
10. Tube Sealer (Beckman Coulter).
11. Slide-A-Lyzer 7K MWCO Dialysis Cassettes (Thermo Fisher Scientific).
12. Low Range PFGE Marker (New England Biolabs).
13. Avanti® J series centrifuge with JLA 10.500 and JLA 16.250 rotors (Beckman Coulter).
14. 3 M sodium acetate, pH 5.2.
15. 100 % ethanol.

2.2 Linearization of BAC DNA for Transfection

1. Appropriate restriction enzyme and restriction enzyme buffer.
2. NanoDrop.
3. Heat block at 37 °C.
4. SeaKem Gold agarose.
5. CHEF-DRII pulsed-field gel electrophoresis apparatus (BIO-RAD).
6. 5× Tris-borate-EDTA (TBE) buffer: 445 mM Tris base, 445 mM boric acid, 10 mM EDTA (store at room temperature).
7. Ethidium bromide solution: 0.5 μg/mL (stock: 10 mg/mL).
8. UV transilluminator.

2.3 Transfection of BAC DNA into Cell Lines by Electroporation

1. 10 μg of linearized BAC construct purified using cesium chloride density gradient centrifugation.
2. Appropriate cell line, for example, K562 cells or MEL cells.
3. Opti-MEM serum-free media.
4. Dulbecco's Modified Eagle Medium (DMEM) supplemented with 10 % Fetal Calf Serum (FCS).

5. 4 mm electroporation cuvettes and electroporation apparatus (BIO-RAD GenePulser).
6. T25 flask.
7. Humidified 37 °C, 5 % CO_2 incubator.

2.4 Transfection of BAC DNA into Cell Lines by Lipofectamine

1. At least 0.8 μg of linearized BAC construct purified using cesium chloride density gradient centrifugation.
2. Appropriate cell line, for example, MEL cells.
3. Opti-MEM serum-free media.
4. DMEM supplemented with 10 % FCS.
5. Lipofectamine 2000 (Invitrogen).
6. 24-well plate.
7. Humidified 37 °C, 5 % CO_2 incubator.

2.5 Selection of Stably Transfected Cells

1. Selection of stably transfected cells can be achieved using an appropriate antibiotic at an optimal concentration (*see* **Note 1**). For example, selection of MEL cells stably transfected with Construct 1 (Fig. 1b) or Construct 2 (Fig. 1c) requires hygromycin. Hygromycin (Sigma) 50 mg/mL stock solution: prepare in PBS and store at 4 °C. Use at 500 μg/mL in DMEM supplemented with 10 % FCS.
2. Appropriate growth media. For MEL cells and K562 cells, this is DMEM supplemented with 10 % FCS.
3. MoFlo cell sorter (Beckman Coulter).

2.6 Generation of Single Cell Clones

1. 8-channel pipette with capacity up to 200 μL.
2. Growth media (with or without antibiotics).
3. Three 96-well flat-bottom plates.
4. Cell suspension at 2×10^4 cells/mL.
5. Humidified 37 °C, 5 % CO_2 incubator.

2.7 Selection of Candidate Clones Based on Responsiveness to Known Inducers

1. Known inducer(s) of the gene of interest. For example, hemin is used as an inducer of γ-globin in K562 cells [7].
   (a) Hemin (Sigma) 5 mM stock solution: dissolve 32.5 mg hemin in 0.5 mL 1 M NaOH for 30 min, then mix with 0.5 mL of 0.5 M Tris base followed by 10 mL of 10 % bovine serum albumin. Add 0.5 mL of 1 M HCl to neutralize the mixture. Filter through a 0.45 μm filter and store at 4 °C for no more than 1 week before use.
2. A method to measure induction of the gene of interest.
   (a) If a fluorescent reporter is available, use flow cytometric analysis (for example, using a BD LSRII flow cytometer)

to assess induction by measuring both the intensity of fluorescence (median peak fluorescence) as well as percentage of fluorescent cells.

(b) If a fluorescent reporter is not available, quantitative PCR can be used to measure induction of the gene of interest.

2.8 Southern Blotting to Confirm the Integrity of BAC DNA Integrated Within Candidate Clones

2.8.1 Preparation of Genomic Agarose Plugs for Southern Blotting

1. Candidate clones.
2. Ice cold phosphate buffered saline (PBS).
3. InCert agarose (GMC BioProducts).
4. Chef Mapper and 50-well disposable plug molds (BIO-RAD).
5. Proteinase K lysis solution: 100 mM EDTA at pH 8.0, 0.2 % sodium deoxycholate and 1 % sodium lauryl sarcosine with 0.4 mg/mL of proteinase K.
6. 0.1 mM phenylmethylsulfonyl fluoride solution prepared with TE buffer pH 8.0.
7. Water bath at 50 °C.
8. TE buffer: 10 mM Tris-Cl, 1 mM EDTA pH 8.0. Dissolve in MilliQ H_2O, adjust pH with HCl and sterilize by autoclaving.

2.8.2 Restriction Digest and Pulsed-Field Gel Electrophoresis for Southern Blotting

1. Appropriate restriction enzyme and restriction enzyme buffer.
2. Heat block at 37 °C.
3. SeaKem Gold agarose.
4. CHEF-DRII pulsed-field gel electrophoresis apparatus (BIO-RAD).
5. 5× Tris-borate-EDTA (TBE) buffer: 445 mM Tris base, 445 mM boric acid, 10 mM EDTA (store at room temperature).
6. Ethidium bromide solution: 0.5 μg/mL (stock: 10 mg/mL).
7. UV transilluminator.

2.8.3 ^{32}P-Labeling of DNA Probe for Southern Blotting

1. Radprime DNA labeling system kit (Life Technologies).
2. 25 ng DNA generated by PCR or restriction digest.
3. TE buffer: 10 mM Tris-Cl, 1 mM EDTA pH 8.0. Dissolve in MilliQ H_2O, adjust pH with HCl and sterilize by autoclaving.
4. G-50 spin column (GE Healthcare).
5. Heat block at 37 °C.
6. Heat block at 95 °C.

2.8.4 Southern Blot

1. 250 mM HCl.
2. Denaturation buffer: 1.5 M NaCl, 0.5 M NaOH, pH 13.

3. Neutralization buffer: 0.5 M Tris, 1.5 M NaCl, pH 7.5.
4. Hybond N+ nitrocellulose membrane (Amersham).
5. Blotting paper.
6. A weight to put on top of the capillary transfer stack, for example, a tray holding books.
7. 20× SSC solution: 3 M NaCl, 0.3 M sodium citrate.
8. UVC 500 UV Crosslinker 115 VAC (GE Healthcare).
9. Glass hybridization tube.
10. Church buffer: 0.5 M, Na_2HPO_4 pH 7.2, 1 mM EDTA, 7 % (w/v) SDS, 1 % bovine serum albumin.
11. Rolling hybridization oven at 65 °C.
12. ^{32}P-labeled DNA probe at concentration 3×10^6 cpm/mL in Church buffer.
13. Wash buffer 1: 2× SSC, 0.1 % SDS.
14. Wash buffer 2: 1× SSC, 0.1 % SDS.
15. Wash buffer 3: 0.5× SSC, 0.1 % SDS.
16. Plastic wrap/cling film (for example, Glad wrap or Saran wrap).
17. Biomax film (Amersham) and hyper cassette.
18. Automatic film processor.

2.9 Extraction of Genomic DNA from Candidate Clones and Use of PCR Analysis to Confirm Integrity of the BAC DNA Integrated Within Candidate Clones

1. Phosphate buffered saline (PBS).
2. Digestion buffer: 100 mM NaCl, 10 mM Tris–Cl, pH 8.0, 25 mM EDTA, pH 8.0, 0.5 % sodium dodecyl sulphate (SDS). Store at room temperature. Add 0.1 mg/mL proteinase K prior to use.
3. Benchtop microcentrifuge.
4. Phenol chloroform pH 8.0.
5. 100 % ethanol.
6. TE buffer: 10 mM Tris-Cl, 1 mM EDTA pH 8.0. Dissolve in MilliQ H_2O, adjust pH with 1 M HCl and sterilize by autoclaving.
7. Standard PCR reagents: thermostable DNA polymerase, buffer, $MgCl_2$, deoxynucleotide triphosphates.
8. Primers spanning various regions of the genomic region carried by the BAC, including BAC ends and specific internal features. For example, to confirm the integrity of Construct 2 (Fig. 1c), use primers that correspond to the 5′-end, HS111, LTR-5′, LTR-3′ and 5′HS5 regions within the β-globin locus (Table 2).

3 Methods

3.1 Generation of a BAC Reporter Cell Line

3.1.1 Purification of BAC DNA by Cesium Chloride Density Gradient Centrifugation

1. Centrifuge 250 mL of overnight bacterial culture grown in LB medium supplemented with 12.5 μg/mL chloramphenicol (or other appropriate selective antibiotic) at 5,500 × *g* for 15 min at 4 °C using the JLA 10.500 rotor (*see* **Note 2**).
2. Resuspend bacterial pellet in 110 mL of Solution 1.
3. Add 110 mL of Solution 2 and mix gently by inversion until lysis is visible. Incubate on ice for 10 min.
4. Add 110 mL of Solution 3 and mix gently by inversion. Incubate on ice for 30 min.
5. Centrifuge at 20,000 × *g* for 30 min at 4 °C using the JLA 16.250 rotor. Decant supernatant and filter through sterilized gauze. The pellet may be discarded.
6. Add isopropanol (0.7 volumes) to the supernatant and mix gently by inversion. Precipitate DNA by centrifugation at 20,000 × *g* for 30 min at 4 °C using the JLA 16.250 rotor.
7. Wash the DNA pellet in 70 % ethanol and centrifuge at 20,000 × *g* for 5 min using the JLA 16.250 rotor. Decant the supernatant and air dry the pellet for 5–10 min.
8. Gently dissolve the DNA pellet in 10 mL TE buffer pH 8.0 containing 1.1 g/mL CsCl and transfer into a 15 mL Falcon tube. Add ethidium bromide to a final concentration of 250 μg/mL.
9. Transfer into 13.5 mL Quick-Seal tubes. Fill tubes completely by addition of TE pH 8.0 containing 1.1 g/mL CsCl and balance tubes to within 0.05 g. Heat seal the tubes using a Tube Sealer (Beckman Coulter) (*see* **Note 3**).
10. Centrifuge tubes for 18 h at 285,000 × *g* (55,000 rpm when using a Beckman Coulter 80Ti rotor) at 14 °C.
11. Visualize the BAC DNA band in the Quick-Seal tube directly with UV light. The upper band is composed of bacterial chromosomal DNA and the lower band is composed of the supercoiled BAC DNA. Carefully extract the BAC DNA by using a 23-gauge needle and 5 mL syringe to pierce through the tube. Transfer into a 15 mL Falcon tube.
12. Extract the ethidium bromide from the BAC DNA solution using an equal volume of saturated NaCl/TE pH 8.0/isopropanol. The ethidium bromide will form a separate top layer. Centrifuge for 3 min at 300 × *g* in a benchtop centrifuge and remove the top red layer. Repeat the extraction four times or until the top layer is no longer red.
13. Transfer the BAC DNA solution into the Dialysis Cassette (Thermo Scientific) and dialyze overnight in 1 L of TE pH 8.0

at 4 °C to remove CsCl. Repeat this twice over a 24- to 48-h period.

14. Transfer the dialyzed BAC DNA solution into a 15 mL Falcon tube and ethanol precipitate the DNA. Add 1/10 volume 3 M sodium acetate and 2.5 volumes of 100 % ethanol. Centrifuge at 15,000 × *g* for 30 min at 4 °C using the JLA 16.250 rotor.
15. Decant supernatant and wash pellet with 10 mL 70 % ethanol. Air dry the pellet for 5–10 min.
16. Resuspend the pellet in 500 μL TE pH 8.0 and store at 4 °C (*see* **Note 4**).

3.1.2 Preparation of BAC DNA for Transfection

1. Quantitate the BAC DNA using a NanoDrop (*see* **Note 5**).
2. Linearize 10 μg of the BAC DNA using a restriction enzyme that cuts at an appropriate position within the BAC. The cut should not be within the genomic fragment of the BAC or be within an antibiotic resistance gene that may be required for selection later (*see* **Note 6**). This can be achieved by either (a) or (b).
   (a) Complete digest with a restriction enzyme that only cuts once within the BAC DNA (*see* **Note 7**).
   (b) Partial digest with a restriction enzyme that cuts more than once within the BAC DNA (*see* **Note 8**). Conduct a restriction enzyme titration to determine the amount of enzyme required to cut the BAC DNA only once (Fig. 2).
   - Set up a 1:2 serial dilution with the enzyme from 1 unit through to 1/32 units.

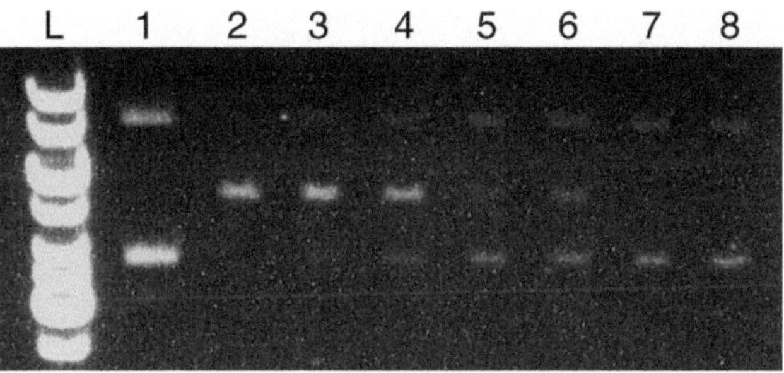

Fig. 2 Enzyme titration to determine the optimal amount of restriction enzyme required for a partial digest. L: 1 kb ladder, 1: uncut pcDNA3 plasmid, 2–8: plasmid cut with decreasing units of NotI enzyme, 2: 10 units, 3: 1 unit, 4: 1/2 units, 5: 1/4 units, 6: 1/8 units, 7: 1/16 units, 8: 1/32 units. 1/32 units to 10 units of NotI were used to cut 500 ng pcDNA3 which contains two NotI cut sites. Ten units of NotI produced a complete digestion with only one band, while the uncut control had two bands. The optimal amount of enzyme to use in this example is 1/8 units because this is the minimum amount of enzyme required to generate bands that result from a complete digestion, but uncut bands remain visible

- Set up seven 50 μL reactions with 10 μg of the BAC DNA, the appropriate restriction enzyme buffer and the dilution series of enzyme units. Include one reaction with an excess amount of restriction enzyme, for example, 10 units.
- Incubate at 37 °C for 1 h.

3. Heat-inactivate the restriction enzyme as per the manufacturer's recommendations.
4. Confirm the BAC DNA is linearized by pulse-field gel electrophoresis (PFGE).
   (a) Run the digest(s) on a 1 % SeaKem Gold agarose gel using CHEF-DRII pulsed-field gel electrophoresis apparatus set at 180 V, angle 120°, ramping 1-20 s. Include an uncut sample and appropriate size marker. Run the gel in 0.5× TBE buffer at 14 °C for 16 h.
   (b) Stain the gel in an ethidium bromide bath (0.5× TBE containing 0.5 μg/mL ethidium bromide) for 30 min.
   (c) De-stain the gel in dH_2O for 30 min to remove unbound ethidium bromide.
   (d) Visualize the gel by UV transillumination. A single, linearized band should be visible if the method in **step 2a** was used. Multiple bands should be visible across the restriction enzyme titration if the method in **step 2b** was used (Fig. 2).
5. Store the linearized BAC DNA at 4 °C. This DNA can be used for transfection into cell lines without further purification.

3.1.3 Transfection of BAC DNA into K562 or MEL Cells Using Electroporation

The transfection conditions described here are suitable for K562 cells or MEL cells as specified. Transfection conditions for other cell lines must be determined empirically.

1. Pellet cells (300 × *g* for 5 min) and wash twice in serum-free Opti-MEM.
2. Resuspend at 1×10^7 cells/mL for K562 cells or 2×10^8 cells/mL for MEL cells.
3. Add 10 μg of linearized BAC construct to 500 μL of cell suspension.
4. Electroporate in a 4 mm cuvette using a Bio-Rad gene pulser with the following parameters:
   (a) K562 cells
   - Resistance: ∞ Ω
   - Capacitance 950 μF
   - Charge 226 V

(b) MEL cells
   - Resistance: ∞ Ω
   - Capacitance 1,000 μF
   - Charge 280 V

5. Transfer cells to a T25 flask and culture in 9.5 mL DMEM supplemented with 10 % FCS.
6. Incubate in a humidified 37 °C, 5 % CO_2 incubator for 72 h before adding selective antibiotics or using fluorescence-activated cell sorting to sort for cells expressing a fluorescent reporter.

3.1.4 Alternative Method Using Lipofectamine for Transfection of BAC DNA into the MEL Cells

The transfection conditions described here are optimized for MEL cells. Transfection conditions for other cell lines must be determined empirically.

1. Pellet cells ($300 \times g$ for 5 min) and resuspend at 8×10^5 cells/mL in DMEM supplemented with 10 % FCS.
2. Aliquot 500 μL of cell suspension per well of a 24-well plate.
3. Prepare the DNA–Lipofectamine complex:
   (a) Dilute 0.8–1.2 μg linearized BAC construct in 50 μL Opti-MEM (*see* **Note 9**).
   (b) Dilute 5 μL Lipofectamine2000 in 45 μL optimum and incubate at room temperature for 5 min.
   (c) Combine the diluted BAC from **step 3a** with the diluted Lipofectamine (*see* **Note 10**) from **step 2b** (total volume 100 μL) and incubate at room temperature for 20 min.
4. Add the DNA–Lipofectamine complex from **step 2c** to cells and mix gently by rocking the plate back and forth.
5. Incubate in a humidified 37 °C, 5 % CO_2 incubator for 72 h before adding selective antibiotics or fluorescence activated cell sorting.

3.1.5 Selection of Stably Transfected Cells

Antibiotics can be added to the growth media 72 h post-transfection. Antibiotic selection should be maintained for at least 1 month to select for stably transfected cells (*see* **Note 11**). If a fluorescent reporter is available, selection can also be carried out using fluorescence-activated cell sorting (FACS) (*see* **Note 12**). Ideally both antibiotic selection and FACS should be carried out to ensure both the antibiotic resistance gene and fluorescent reporter genes are intact and expressed. The method described here is suitable for MEL cells transfected with Construct 1 (Fig. 1b), which contains both a hygromycin resistance gene and a fluorescent reporter (eGFP).

1. At 72 h post-transfection, add hygromycin at a final concentration of 500 μg/mL to growth media (DMEM supplemented with 10 % FCS).
2. Culture the cells in growth media containing hygromycin for at least 1 month. Replace the growth media containing hygromycin every 3–5 days as appropriate for the rate of cell proliferation.
3. When cells begin to proliferate rapidly (approximately 1 month post-transfection), use the MoFlo cell sorter to isolate at least 1×10^3 eGFP-positive cells.
4. Seed the eGFP-positive cells in one well of a 12-well plate with 2 mL of growth media containing hygromycin. Allow the cells to regrow for 2–3 days prior to serial dilution to generate single cell clones.

3.1.6 Limiting Cell Dilution to Generate Single Cell Clones

1. Use an 8-channel pipette to dispense 100 μL growth media to each well of a 96-well flat-bottom plate (*see* **Note 13**).
2. Remove media from well A1 (Fig. 3) and replace with 200 μL of cell suspension at 2×10^4 cells/mL.
3. Set up a 1:2 serial dilution down column 1 by transferring 100 μL from well A1 to well B1. Gently mix the cell suspension in B1 before transferring 100 μL to C1. Repeat this through to H1. Discard 100 μL from H1.
4. Set up a 1:2 serial dilution along columns 1–12 using an 8-channel pipette to gently mix the cell suspension (avoid

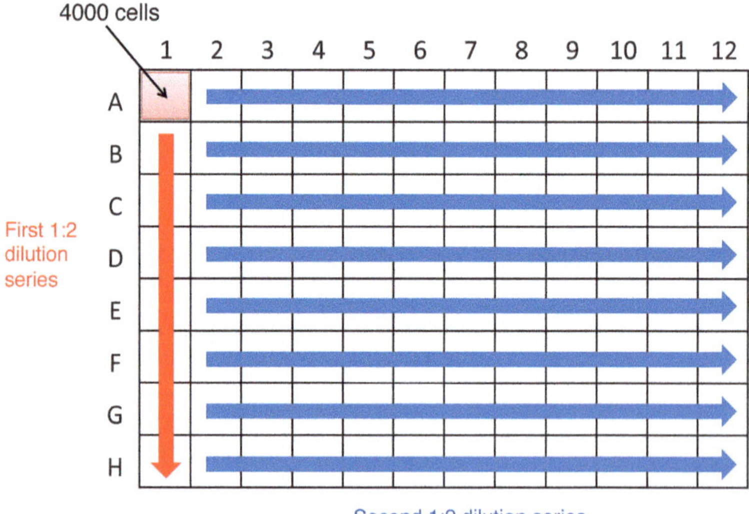

Fig. 3 Schematic diagram of the limiting cell dilution used to generate single cell clones. Well A1 contains 4,000 cells, which undergo two 1:2 dilution series; the first 1:2 dilution series down column 1 as shown in *red*, and the second 1:2 dilution series across columns 2–12 as shown in *blue*

creating bubbles) in column 1 before transferring 100 μL to column 2. Continue mixing and transferring 100 μL through to column 12. Discard 100 μL from column 12.

5. Incubate the plate for 6 h to allow cells to settle before scoring which wells contain a single cell.
6. Maintain single cells in the 96-well plate until confluent before transferring to a larger vessel. Replenish growth media as required during this period.

3.2 Select Candidate Clones Based on Responsiveness to Known Inducers

After generating at least 20 clones using limiting cell dilution, select at least two candidate clones that would be suitable for use in a drug discovery screen. Candidate clones can be selected based on responsiveness of the gene of interest to known inducers. For example, hemin is a known inducer of γ-globin and can be used to select candidate clones in K562 cells transfected with Construct 1 (Fig. 1b) [4]. The method described here is suitable for selecting candidate clones in K562 cells transfected with Construct 1 (Fig. 1b).

1. Pellet and resuspend clones at a density of 6×10^5 cells/mL in growth media (DMEM supplemented with 10 % FCS).
2. Pipette 500 μL of each clone into six wells of a 24-well plate.
3. Add 100 μM hemin (H100) to three wells of each clone, leaving the remaining three wells as negative controls (H0).
4. Replace growth media as required for 5 days post-hemin treatment.
5. At 5 days post-hemin treatment, measure median peak fluorescence (MPF) and percentage of eGFP expressing cells using flow cytometry.
6. Analyze the data by comparing induction of eGFP/γ-globin between the H0 and H100 cells. Refer to Table 1.
7. Select candidate clones based on high responsiveness (increased eGFP expression) to treatment with hemin (*see* **Note 14**).

3.3 Confirm Integrity of BAC DNA Integrated Within Candidate Clones by Southern Blot

3.3.1 Preparation of Genomic Agarose Plugs for Southern Blotting

1. Pellet cells (300 × *g* for 5 min) and wash twice in ice-cold PBS.
2. Resuspend cells at 4×10^7 cells/mL.
3. Mix 500 μL cells with molten 1.2 % InCert agarose and aliquot into 90 μL plug molds. Chill plugs on ice for 1 h.
4. Transfer plugs into 50 mL Falcon tubes containing 20 mL of proteinase K lysis solution and incubate at 50 °C for 24 h.
5. Replace with fresh proteinase K lysis solution and incubate at 50 °C for another 24 h.
6. Inactivate the proteinase K by washing plugs twice for 2 h with 20 mL of 2 mM phenylmethylsulfonyl fluoride solution prepared with TE buffer.
7. Rinse plugs with dH_2O and store at 4 °C in TE buffer.

Table 1
Analysis of eGFP induction in K562 clones stably transfected with Construct 1 (Fig. 1b) [4]

	EGFP (MPF)		MPF
Clone	H0	H100	% Increase
Stable clones			
1	23	137	496
2	312	1,080	246
3	88	245	178
4	154	610	296
5	49	250	410
6	151	333	120
7	57	213	274
8	232	730	215
9	27	186	589
10	28	404	1,343
11	49	234	378
12	242	2,054	749
13	76	275	262
14	126	922	632
15	92	431	369

The responsiveness of 15 clones was tested using hemin as a known inducer of γ-globin (eGFP). Median peak fluorescence (MPF) was compared between no hemin treatment (H0) and 100 μM hemin treatment (H100). The increase in MPF with hemin treatment is also expressed as a percentage value

3.3.2 Restriction Digest and Pulse Field Gel Electrophoresis for Southern Blotting

1. Place plugs in 1.5 mL microfuge tubes and incubate with up to 300 μL of the appropriate restriction enzyme buffer for 5 h.
2. Add up to 100 units of restriction enzyme for each plug and incubate at 37 °C for up to 12 h.
3. Run the digest as described in **step 4** in Subheading 3.1.2.

3.3.3 Synthesis of ^{32}P-Labeled DNA Probe for Southern Blotting

1. Generate probe(s) to target specific region(s) of the BAC integrated within the candidate clones by PCR or restriction digest (*see* **Note 15**).
2. Incubate 25 ng of the DNA to be labeled in 5–20 μL of TE buffer at 95 °C for 5 min and immediately set on ice.
3. Keep the tube on ice while adding: 20 μL of RadPrime buffer, 1 μL each of dATP, dGTP and dTTP (500 μM) and 5 μL of [α-^{32}P]dCTP (3,000 Ci/mmol).

4. Adjust reaction volume to 49 μL with dH_2O.
5. Add 1 μL Klenow fragment (40 U/μL) and mix thoroughly.
6. Incubate at 37 °C for 30–60 min.
7. Remove unincorporated dNTPs by passing the reaction through a Sephadex G-50 spin column.
8. Incubate at 95 °C for 5 min then immediately ice.

3.3.4 Southern Blot to Detect Specific Regions of the BAC Integrated Within Candidate Clones

1. Wash the gel twice in 250 mM HCl for 15 min at room temperature, then rinse in dH_2O.
2. Incubate the gel in denaturation buffer for 45 min at room temperature then rinse in dH_2O.
3. Incubate the gel in neutralization buffer for 45 min at room temperature.
4. Transfer the DNA fragments from the gel to Hybond N+ nitrocellulose membrane (*see* **Note 16**) using capillary transfer in 20× SSC solution:
   (a) Cover the gel with the Hybond N+ nitrocellulose membrane. Be sure to remove any bubbles between the gel and the membrane using wet gloves.
   (b) Cover with several layers of blotting paper soaked in 20× SSC solution.
   (c) Cover with a stack of dry blotting paper and place a tray holding books to provide weight on top of the stack.
   (d) Let the transfer proceed overnight at room temperature.
5. Carefully remove the membrane and place in the UV Crosslinker. Cross-link the DNA fragments to the membrane by UV irradiation according to the UV Crosslinker manufacturer's recommendations.
6. Place the membrane in a glass hybridization tube with 25 mL Church buffer prewarmed to 65 °C and incubate in a rolling hybridization oven at 65 °C for 1 h.
7. Add the ^{32}P-labeled probe at a concentration of 3×10^6 cpm/mL in Church buffer and incubate overnight at 65 °C.
8. Wash the membrane twice with 100 mL wash buffer 1 at 65 °C for 15 min.
9. Wash the membrane twice with 100 mL buffer 2 at 65 °C for 15 min.
10. Wash the membrane twice with 100 mL buffer 3 at 65 °C for 15 min.
11. Wrap the membrane in plastic wrap to prevent it from drying out and expose to Biomax film in a hyper cassette.
12. Develop the Biomax film using an automatic film processor.

3.4 Confirming the Integrity of the BAC DNA Integrated Within Candidate Clones by PCR Analysis

The method described is appropriate for confirming the integrity of BAC DNA in candidate clones of cells stably transfected with a BAC containing the β-globin locus. This can be achieved by extracting genomic DNA from candidate clones and PCR screening with primers corresponding to the 5′-end, HS111, LTR-5′, LTR-3′ and 5′HS5 regions of the β-globin BAC [2].

1. Pellet 10×10^6 cells from each candidate clone and wash twice in PBS. Transfer cells to an eppendorf tube.
2. Resuspend cells in 500 μL digestion buffer and incubate at 50 °C overnight.
3. Add 500 μL phenol chloroform pH 8.0 and allow to mix gently on a rotator for 30 min.
4. Centrifuge for 5 min at maximum speed (>15,000 × *g*).
5. Carefully transfer the supernatant to another eppendorf tube with a wide bore pipette.
6. Repeat Steps 3–5.
7. Add 1 mL of 100 % ethanol. Invert to mix and incubate on ice for 15 min.
8. Centrifuge for 5 min at maximum speed (>15,000 × *g*) at 4 °C.
9. Remove the supernatant and wash the pellet with 300 μL 70 % ethanol. Air dry the pellet for 5–10 min.
10. Add 1 mL TE buffer and let the pellet resuspend overnight at 4 °C.
11. Dilute the DNA up to 1 in 500 for use in PCR reactions.
12. Set up PCR reactions as per the manufacturer's recommendations using primers corresponding to the 5′-end, HS111, LTR-5′, LTR-3′ and 5′HS5 regions of the β-globin BAC. Set the annealing temperature to 52 °C. Primer sequences are shown in Table 2.
13. Run the PCR products on an agarose gel. Stain the gel with ethidium bromide and visualize using UV transillumination. PCR products should be present for the 5′-end, HS111, LTR-5′, LTR-3′ and 5′HS5 regions. The expected sizes of the PCR products are shown in Table 2. Presence of all of these PCR products is required to confirm integrity of the β-globin locus within the BAC integrated within candidate clones [2].

3.5 Assess Developmental Regulation of the BAC

The BAC itself should be assessed to determine whether it is appropriately regulated in a developmental stage-specific manner. This can be achieved by assessing expression of the BAC in cell lines with different developmental expression profiles. The method described is suitable for assessing developmental regulation of Construct 2 (Fig. 1c) [2]. This is achieved by analyzing the expression of Construct 2 (Fig. 1c) after transfection into two cell lines

Table 2
Primers used in PCR analysis of cell lines containing the β-globin locus [2]

Primer	Sequence (5′–3′)	Product (bp)
5′-end_F	GACCAGGATAGAAAAAGATGA	239
5′-end_R	TATCCAGAAGGCAACTTTATT	
HS111_F	GGAACCTCTATCCAGAATTGA	379
HS111_R	CCTATCTCTCTCCATCAATTT	
LTR-5′_F	CAGGGTGTGGGATAATGCTA	778
LTR-5′_R	GTGTTGATTGGTGCATTCACAA	
LTR-3′_F	GGGATTGTAAATACACCAATT	1,120
LTR-3′_R	GCTGTTGGAAATTGTTGAGAAG	
5′HS5_F	GTTCGTCATAATATGGGTTTT	763
5′HS_R	TGTGGGGAAAGAAATTAAATTA	

These primers, corresponding to pertinent features of the β-globin locus, were used for confirming the integrity of the β-globin BAC integrated within candidate clones. Presence of all PCR products confirms the integrity of the β-globin BAC integrated within candidate clones

(K562 cells and MEL cells) that possess different expression profiles (*see* **Note 17**).

1. Transfect Construct 2 (Fig. 1c) into K562 cells and MEL cells using the method described in Subheadings 3.1.2–3.1.5.
2. At 1 month post-transduction, use flow cytometry to measure expression of both fluorescent reporters (dsRED and eGFP) in both cell lines. Measure both median peak fluorescence (MPF) and percentage of fluorescent reporter expressing cells.
3. Analyze the expression of the fluorescent reporters in both cell lines to determine whether the BAC is appropriately regulated in a developmental stage-specific manner. In K562 cells, expression of dsRED should be high, while expression of eGFP should be low (Fig. 4). In MEL cells, expression of dsRED should be low while expression of eGFP should be high (Fig. 4).

4 Notes

1. An optimal concentration is defined as the concentration at which all untransfected cells are dead after 4 days of antibiotic selection. This should be empirically determined for each cell line by a titration of antibiotic concentrations on untransfected cells.
2. This method is suitable for large-scale purification of plasmids greater than 10 kb in size. Cesium chloride purification is more

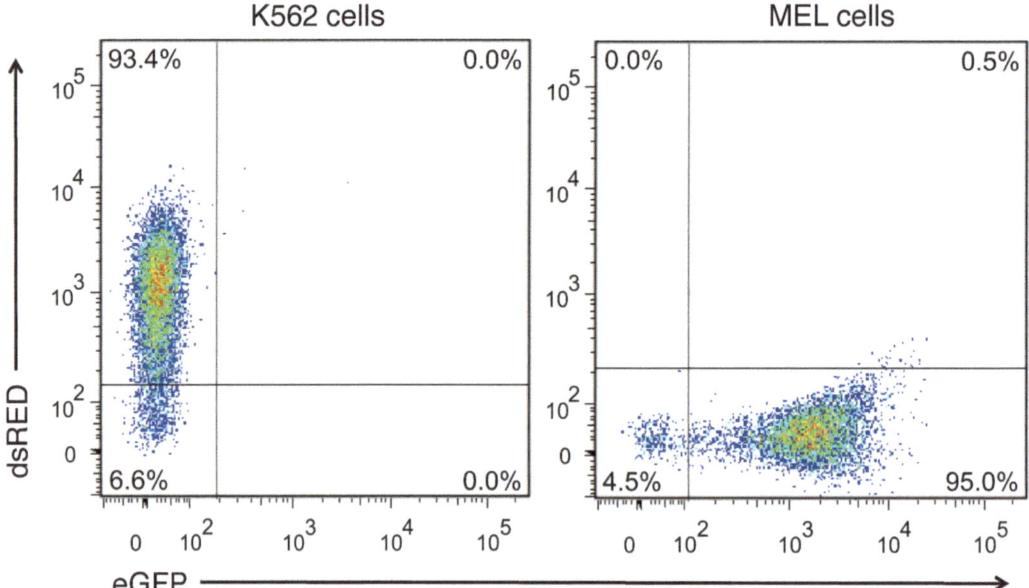

Fig. 4 Flow cytometric analysis of stage-specific regulation of dsRED and eGFP reporters in Construct 2 (Fig. 1c) [2]. Developmental expression of fluorescent reporters in Construct 2 (Fig. 1c) was confirmed by transfection into two cell lines with different developmental globin expression profiles. K562 cells have an embryonic globin expression profile with high expression of γ-globin (dsRED) and no β-globin (eGFP), while MEL cells have an adult globin expression profile with low expression of γ-globin (dsRED) and high expression of β-globin (eGFP)

technically complex than common anion-exchange methods of DNA purification, but provides better yields (100–200 μg) in our experience. For BAC preparation, the method was modified by use of greater volumes of solutions 1, 2 and 3 than for plasmid preparation. To prevent shearing of BAC DNA, use wide bore tips when resuspending or transferring BAC DNA in solution.

3. Ensure the tube is completely filled before sealing. There must not be any air bubbles in the tube when sealed. Balance tubes by addition of excess CsCl/TE solution, or with mineral oil if a solution of lower density is required to achieve the appropriate mass. Mineral oil must be used for this purpose, as it will not mix with the CsCl/TE buffer, leaving the essential CsCl concentration unaltered.

4. Freeze thawing of purified cesium chloride BAC DNA is not recommended because it can cause breaks in the DNA strand. The purified BAC DNA is stable for several months at 4 °C.

5. Heat the BAC DNA to 65 °C for 10 min prior to pipetting with a wide bore tip. Heating may help to relax supercoiled DNA and thus prevent shearing of BAC DNA. Always handle BAC DNA gently and do not pipette excessively.

6. An ideal position to cut would be the boundary of the BAC vector backbone and the genomic fragment.
7. A complete digest using a restriction enzyme that cuts the BAC only once without disturbing the genomic fragment or any antibiotic resistance genes that may be required later is the ideal option. If this is not feasible, a partial digest using a restriction enzyme that cuts more than once is an alternative solution.
8. A partial digest using a restriction enzyme that cuts more than once is an alternative solution when a restriction enzyme that cuts the BAC only once is not available. For example, NotI has two cut sites within the β-globin BAC, both at positions that do not disrupt the genomic fragment or hygromycin resistance gene that will be required later. Therefore, linearization of the β-globin BAC DNA can be achieved through a partial digest using the minimal amount of NotI enzyme required to cut the BAC only once. Refer to Fig. 2 for an example of an enzyme titration to determine the optimal amount for a partial digest.
9. Multiple concentrations of BAC, for example 0.8, 1.0 and 1.2 μg, can be tested in parallel in a single experiment to determine the optimal amount of BAC to use to achieve the highest transfection efficiency. Transfection efficiency can be determined 72 h post-transfection using flow cytometry if a fluorescent reporter is available, or through an antibiotic kill curve if a fluorescent reporter is not available.
10. Combine diluted Lipofectamine with DNA within 30 min as longer incubation times may result in decreased activity.
11. It may take up to 1 month simply for cells to recover and start expanding given that transfection efficiency of large BAC constructs is very low (<1 %).
12. Fluorescence-activated cell sorting (FACS) allows physical separation of a cell with a particular physical or chemical property from a heterogeneous population [8]. In the scenario described in this chapter, it is separation of eGFP-positive cells from a heterogeneous population of eGFP-positive and eGFP-negative cells. eGFP fluorescence in MEL cells transfected with Construct 1 (Fig. 1b) is detectable at 24 h post-transfection, however the median peak fluorescence (MPF) is very low. MPF is higher at 48 h post-transfection, therefore it is better to wait at least 48 h post-transfection before using FACS to select for stably transfected cells that express eGFP.
13. At least three 96-well plates should be set up to generate a minimum of 20 clones. Note that flat-bottom plates should be used instead of round-bottom plates as it is easier to count cells in flat-bottom plates.

14. Candidate clones should behave as expected; display uniform reporter gene expression in their un-induced state and display adequate dynamic range in their induced state. In the example given (refer to Table 1), all clones responded with increased eGFP median peak fluorescence (MPF), ranging from a modest 120 % increase in clone 6–1,343 % increase in clone 10 (Table 1). Clones 12 and 14 were selected as candidate clones based on their high responsiveness to hemin.
15. Generate probes that target pertinent features of your BAC. For example, probes suitable for Construct 2 (Fig. 1c) include the eGFP gene and the dsRED gene.
16. Gloves must always be used to handle the Hybond N+ nitrocellulose membrane because oils on the skin will affect the transfer. The membrane must never be allowed to dry out at any point as this fixes the probe to the membrane and produces unwanted background signal.
17. K562 cells possess an embryonic globin expression profile with expression of γ-globin but absence of β-globin [6, 7, 9]. When transfected with Construct 2 (Fig. 1c), they should express high levels of dsRED and low levels of eGFP. MEL cells possess an adult globin expression profile with low expression of embryonic globins and high expression of adult globins. When transfected with Construct 2 (Fig. 1c), they should express low levels of dsRED and high levels of eGFP (Fig. 4).

Acknowledgments

This work was supported by the National Health and Medical Research Council; the Murdoch Children's Research Institute; the Victorian Government's Operational Infrastructure Support Program; and Thalassaemia Australia and Thalassaemia Society of New South Wales.

References

1. Vadolas J, Wardan H, Orford M, Voullaire L, Zaibak F, Williamson R, Ioannou PA (2002) Development of sensitive fluorescent assays for embryonic and fetal hemoglobin inducers using the human beta -globin locus in erythropoietic cells. Blood 100:4209–4216
2. Chan KS, Xu J, Wardan H, McColl B, Orkin S, Vadolas J (2012) Generation of a genomic reporter assay system for analysis of gamma- and beta-globin gene regulation. FASEB J 26:1736–1744
3. Roosjen M, McColl B, Kao B, Gearing LJ, Blewitt ME, Vadolas J (2013) Transcriptional regulators Myb and BCL11A interplay with DNA methyltransferase 1 in developmental silencing of embryonic and fetal beta-like globin genes. FASEB J 28(4):1610–1620
4. Vadolas J, Wardan H, Orford M, Williamson R, Ioannou PA (2004) Cellular genomic reporter assays for screening and evaluation of inducers of fetal hemoglobin. Hum Mol Genet 13: 223–233

5. Friend C, Scher W, Holland JG, Sato T (1971) Hemoglobin synthesis in murine virus-induced leukemic cells in vitro: stimulation of erythroid differentiation by dimethyl sulfoxide. Proc Natl Acad Sci U S A 68:378–382
6. Lozzio CB, Lozzio BB (1975) Human chronic myelogenous leukemia cell-line with positive Philadelphia chromosome. Blood 45:321–334
7. Rutherford TR, Clegg JB, Weatherall DJ (1979) K562 human leukaemic cells synthesise embryonic haemoglobin in response to haemin. Nature 280:164–165
8. Davies D (2012) Cell separations by flow cytometry. Methods Mol Biol 878:185–199
9. Benz EJ Jr, Murnane MJ, Tonkonow BL, Berman BW, Mazur EM, Cavallesco C, Jenko T, Snyder EL, Forget BG, Hoffman R (1980) Embryonic-fetal erythroid characteristics of a human leukemic cell line. Proc Natl Acad Sci U S A 77:3509–3513

Index

Kumaran Narayanan (ed.), *Bacterial Artificial Chromosomes*, Methods in Molecular Biology, vol. 1227, DOI 10.1007/978-1-4939-1652-8, © Springer Science+Business Media New York 2015

Y

Z

MIX
Papier aus verantwortungsvollen Quellen
Paper from responsible sources
FSC® C105338

If you have any concerns about our products,
you can contact us on
ProductSafety@springernature.com

In case Publisher is established outside the EU,
the EU authorized representative is:
Springer Nature Customer Service Center GmbH
Europaplatz 3, 69115 Heidelberg, Germany

Printed by Libri Plureos GmbH
in Hamburg, Germany